Lecture Notes in Computer Science 14958

Founding Editors

Gerhard Goos
Juris Hartmanis

Editorial Board Members

Elisa Bertino, *Purdue University, West Lafayette, IN, USA*
Wen Gao, *Peking University, Beijing, China*
Bernhard Steffen , *TU Dortmund University, Dortmund, Germany*
Moti Yung , *Columbia University, New York, NY, USA*

The series Lecture Notes in Computer Science (LNCS), including its subseries Lecture Notes in Artificial Intelligence (LNAI) and Lecture Notes in Bioinformatics (LNBI), has established itself as a medium for the publication of new developments in computer science and information technology research, teaching, and education.

LNCS enjoys close cooperation with the computer science R & D community, the series counts many renowned academics among its volume editors and paper authors, and collaborates with prestigious societies. Its mission is to serve this international community by providing an invaluable service, mainly focused on the publication of conference and workshop proceedings and postproceedings. LNCS commenced publication in 1973.

Lorraine Goeuriot · Philippe Mulhem ·
Georges Quénot · Didier Schwab ·
Giorgio Maria Di Nunzio · Laure Soulier ·
Petra Galuščáková ·
Alba García Seco de Herrera ·
Guglielmo Faggioli · Nicola Ferro
Editors

Experimental IR Meets Multilinguality, Multimodality, and Interaction

15th International Conference of the CLEF Association, CLEF 2024
Grenoble, France, September 9–12, 2024
Proceedings, Part I

Editors
Lorraine Goeuriot
Université Grenoble Alpes, CNRS
Grenoble, France

Philippe Mulhem
Université Grenoble Alpes, CNRS
Grenoble, France

Georges Quénot
Université Grenoble Alpes, CNRS
Grenoble, France

Didier Schwab
Univ. Grenoble Alpes, CNRS
Grenoble, France

Giorgio Maria Di Nunzio
University of Padova
Padua, Italy

Laure Soulier
Sorbonne University
Paris, France

Petra Galuščáková
University of Stavanger
Stavanger, Norway

Alba García Seco de Herrera
University of Essex
Colchester, UK

Guglielmo Faggioli
University of Padova
Padua, Italy

Nicola Ferro
University of Padova
Padua, Italy

ISSN 0302-9743   ISSN 1611-3349 (electronic)
Lecture Notes in Computer Science
ISBN 978-3-031-71735-2   ISBN 978-3-031-71736-9 (eBook)
https://doi.org/10.1007/978-3-031-71736-9

© The Editor(s) (if applicable) and The Author(s), under exclusive license
to Springer Nature Switzerland AG 2024, corrected publication 2024

This work is subject to copyright. All rights are solely and exclusively licensed by the Publisher, whether the whole or part of the material is concerned, specifically the rights of translation, reprinting, reuse of illustrations, recitation, broadcasting, reproduction on microfilms or in any other physical way, and transmission or information storage and retrieval, electronic adaptation, computer software, or by similar or dissimilar methodology now known or hereafter developed.
The use of general descriptive names, registered names, trademarks, service marks, etc. in this publication does not imply, even in the absence of a specific statement, that such names are exempt from the relevant protective laws and regulations and therefore free for general use.
The publisher, the authors and the editors are safe to assume that the advice and information in this book are believed to be true and accurate at the date of publication. Neither the publisher nor the authors or the editors give a warranty, expressed or implied, with respect to the material contained herein or for any errors or omissions that may have been made. The publisher remains neutral with regard to jurisdictional claims in published maps and institutional affiliations.

This Springer imprint is published by the registered company Springer Nature Switzerland AG
The registered company address is: Gewerbestrasse 11, 6330 Cham, Switzerland

If disposing of this product, please recycle the paper.

Preface

Since 2000, the *Conference and Labs of the Evaluation Forum* (CLEF) has played a leading role in stimulating research and innovation in the domain of multimodal and multilingual information access. Initially founded as the *Cross-Language Evaluation Forum* and running in conjunction with the *European Conference on Digital Libraries* (ECDL/TPDL), CLEF became a standalone event in 2010 combining a peer-reviewed conference with a multi-track evaluation forum. The combination of the scientific program and the track-based evaluations at the CLEF conference creates a unique platform to explore information access from different perspectives, in any modality and language.

The CLEF conference has a clear focus on experimental information retrieval (IR) as seen in evaluation forums (like the CLEF Labs, TREC, NTCIR, FIRE, MediaEval, RomIP, TAC) with special attention to the challenges of multimodality, multilinguality, and interactive search, ranging from unstructured to semi-structured and structured data. The CLEF conference invites submissions on new insights demonstrated by the use of innovative IR evaluation tasks or in the analysis of IR test collections and evaluation measures, as well as on concrete proposals to push the boundaries of the Cranfield/TREC/CLEF paradigm.

CLEF 2024[1] was organized by the University of Grenoble Alpes, Grenoble, France, from 9 to 12 September 2024. CLEF 2024 was the 15th year of the CLEF Conference and the 25th year of the CLEF initiative as a forum for IR Evaluation, so it marked an important anniversary for CLEF. The conference format remained the same as in past years and consisted of keynotes, contributed papers, lab sessions, and poster sessions, including reports from other benchmarking initiatives from around the world. All sessions were organized in presence but also allowing for remote participation for those who were not able to attend physically. The CLEF 25th anniversary paper, a kind of *una tantum* paper to celebrate the event, was reviewed by one of the Program Chairs. As usual, the lab overview papers were reviewed by the Lab Chairs.

CLEF 2024 continued the initiative introduced in the 2019 edition, during which the *European Conference on Information Retrieval (ECIR)* and CLEF joined forces: ECIR 2023[2] hosted a special session dedicated to CLEF Labs where lab organizers presented the major outcomes of their Labs and their plans for ongoing activities, followed by a poster session to favour discussion during the conference. This was reflected in the ECIR 2024 proceedings, where CLEF Lab activities and results were reported as short papers. The goal was not only to engage the ECIR community in CLEF activities but also to disseminate the research results achieved during CLEF evaluation cycles as papers submitted to ECIR.

The following scholars were invited to give a keynote talk at CLEF 2024: Paula Carvalho (INESC-ID, Lisboa, Portugal) and Aurélie Névéol (Université Paris-Saclay, LISN, CNRS, France).

---

[1] https://clef2024.clef-initiative.eu/.
[2] https://ecir2024.org/.

CLEF 2024 received a total of 25 scientific submissions, of which a total of 11 papers (7 long, 3 short & 1 position) were accepted. Each submission was reviewed in double-blind fashion by at least two program committee members, and the program chairs oversaw the reviewing and follow-up discussions. Several papers were a product of international collaboration. This year, researchers addressed the following important challenges in the community: factual reporting and political bias; sexism, discrimination, and misinformation; information retrieval and recommendation; information retrieval for decision making; document sanitization for information release and retrieval; evaluation dataset for knowledge acquisition; evaluation with gen-IR; medical entity linking; and classification with large language models.

Like in previous editions, since 2015, CLEF 2024 continued inviting CLEF lab organizers to nominate a "best of the labs" paper, among those submitted in the CLEF 2023 labs, that was reviewed as a full paper submission to the CLEF 2024 conference, according to the same review criteria and PC. 6 full papers were accepted for this "best of the labs" section.

The conference integrated a series of workshops presenting the results of lab-based comparative evaluations. A total of 23 lab proposals were received and evaluated in peer review based on their innovation potential and the quality of the resources created. The 14 selected labs represented scientific challenges based on new datasets and real-world problems in multimodal and multilingual information access. These datasets provide unique opportunities for scientists to explore collections, to develop solutions for these problems, to receive feedback on the performance of their solutions, and to discuss the challenges with peers at the workshops. In addition to these workshops, the labs reported results of their year-long activities in overview talks and lab sessions. Overview papers describing each of the labs are provided in this volume. The full details for each lab are contained in a separate publication, the Working Notes[3].

The 14 labs running as part of CLEF 2024 comprised mainly labs that continued from previous editions at CLEF (BioASQ, CheckThat!, eRisk, EXIST, iDPP, ImageCLEF, JOKER, LifeCLEF, LongEval, PAN, SimpleText, and Touché) and new pilot/workshop activities (ELOQUENT and qCLEF). In the following we give a few details for each of the labs organized at CLEF 2024 (presented in alphabetical order):

BioASQ: Large-scale biomedical semantic indexing and question answering[4] aimed to push the research frontier towards systems that use the diverse and voluminous information available online to respond directly to the information needs of biomedical scientists. It offered the following tasks. *Task 1 - b: Biomedical Semantic Question Answering*: benchmark datasets of biomedical questions, in English, along with gold standard (reference) answers constructed by a team of biomedical experts. The participants had to respond with relevant articles, and snippets from designated resources, as well as exact and "ideal" answers. *Task 2 - Synergy: Question Answering for developing problems*: biomedical experts posed unanswered questions for developing problems, such as COVID-19, received the responses provided by the participating systems, and provided feedback, together with updated questions in an iterative procedure that aimed

---

[3] Faggioli, G., Ferro, N., Galuščáková, P., and García Seco de Herrera, A. editors (2024). *CLEF 2024 Working Notes*. CEUR Workshop Proceedings (CEUR-WS.org), ISSN 1613-0073.

[4] http://www.bioasq.org/workshop2024

to facilitate the incremental understanding of developing problems in biomedicine and public health. *Task 3 - MultiCardioNER: Multiple clinical entity detection in multilingual medical content*: focused on the automatic detection and normalization of mentions of four clinical entity types, namely diseases, symptoms, procedures, and medications, in cardiology clinical case documents in Spanish, English, Italian, and Dutch. *BioNNE: Nested NER in Russian and English*: dealt with nested named-entity recognition (NER) in PubMed abstracts in Russian and English. The train/dev datasets included annotated mentions of disorders, anatomical structures, chemicals, diagnostic procedures, and biological functions. Participants were encouraged to apply cross-language (Russian to English) and cross-domain techniques.

CheckThat! Lab on Checkworthiness, Subjectivity, Persuasion, Roles, Authorities and Adversarial Robustness[5] provided a diverse collection of challenges to the research community interested in developing technology to support and understand the journalistic verification process. The tasks went from core verification tasks such as assessing the check-worthiness of a text to understanding the strategies used to influence the audience and identifying the stance of relevant characters on questionable affairs. It offered the following tasks. *Task 1 - Check-worthiness estimation*: asked to assess whether a statement, sourced from either a tweet or a political debate, warrants fact-checking. *Task 2 - Subjectivity*: given a sentence from a news article, it asks to determine whether it is subjective or objective. *Task 3 - Persuasion Techniques*: given a news article and a list of 23 persuasion techniques organized into a 2-tier taxonomy, including logical fallacies and emotional manipulation techniques that might be used to support flawed argumentation, it asked to identify the spans of texts in which each technique occurs. *Task 4 - Detecting hero, villain, and victim from memes*: asked to determine the roles of entities within memes, categorizing them as "hero", "villain", "victim", or "other" through a multi-class classification approach that considers the systematic modelling of multimodal semiotics. *Task 5 - Authority Evidence for Rumor Verification*: given a rumor expressed in a tweet and a set of authorities for that rumor, it asked to retrieve up to 5 evidence tweets from the authorities' timelines, and determine whether the rumor is supported, refuted, or unverifiable according to the evidence. *Task 6 - Robustness of Credibility Assessment with Adversarial Examples*: the task was realised in five domains: style-based news bias assessment (HN), propaganda detection (PR), fact checking (FC), rumour detection (RD), and COVID-19 misinformation detection (C19). For each domain, the participants were provided with three victim models, trained for the corresponding binary classification task, as well as a collection of 400 text fragments. Their aim was to prepare adversarial examples which preserve the meaning of the original examples, but were labelled differently by the classifiers.

ELOQUENT shared tasks for evaluation of generative language model quality[6] provided a set of tasks for evaluating the quality of generative language models. It offered the following tasks. *Task 1 - Topical competence*: tested and verified a model's understanding of an application domain and specific topic of interest. *Task 2 - Veracity and hallucination*: tested how the truthfulness or veracity of automatically generated text can be assessed. *Task 3 - Robustness*: tested the capability of a model to handle input

---

[5] http://checkthat.gitlab.io/.

[6] https://eloquent-lab.github.io/.

variation – e.g., dialectal, sociolectal, and cross-cultural – as represented by a set of equivalent but non-identical varieties of input prompts. *Task 4 - Voight Kampff*: explored whether automatically generated text can be distinguished from human-authored text. This task was organized in collaboration with the PAN lab at CLEF.

eRisk: Early Risk Prediction on the Internet[7] explored the evaluation methodology, effectiveness metrics, and practical applications (particularly those related to health and safety) of early risk detection on the Internet. It offered the following tasks. *Task 1 - Search for symptoms of depression*: consisted of ranking sentences from a collection of user writings according to their relevance to a depression symptom. The participants had to provide rankings for the 21 symptoms of depression from the BDI Questionnaire. *Task 2 - Early Detection of Signs of Anorexia*: consisted in performing a task on early risk detection of anorexia. The challenge consisted of sequentially processing pieces of evidence to detect early traces of anorexia as soon as possible. *Task 3 - Measuring the severity of the signs of Eating Disorders*: consisted of estimating the level of features associated with a diagnosis of eating disorders from a thread of user submissions. For each user, the participants were given a history of postings and the participants had to fill in a standard eating disorder questionnaire.

EXIST: sEXism Identification in Social neTworks[8] aimed to capture and categorize sexism, from explicit misogyny to other subtle behaviours, in social networks. Participants were asked to classify tweets in English and Spanish according to the type of sexism they enclose and the intention of the persons that wrote the tweets. It offered the following tasks. *Task 1 - Sexism Identification in Tweets*: was a binary classification. The systems had to decide whether or not a given tweet contains sexist expressions or behaviours (i.e., it is sexist itself, describes a sexist situation, or criticizes a sexist behaviour). *Task 2 - Source Intention in Tweets*: aimed to categorize the message according to the intention of the author, which provides insights in the role played by social networks on the emission and dissemination of sexist messages. *Task 3 - Sexism Categorization in Tweets*: many facets of a woman's life may be the focus of sexist attitudes including domestic and parenting roles, career opportunities, sexual image, and life expectations, to name a few. Automatically detecting which of these facets of women are being more frequently attacked in social networks will facilitate the development of policies to fight against sexism. *Task 4 - Sexism Identification in Memes*: was a binary classification task consisting of deciding whether or not a given meme is sexist. *Task 5 - Source Intention in Memes*: aimed to categorize the meme according to the intention of the author, which provides insights in the role played by social networks in the emission and dissemination of sexist messages.

iDPP: Intelligent Disease Progression Prediction[9] Amyotrophic Lateral Sclerosis (ALS) and Multiple Sclerosis (MS) are chronic diseases characterized by progressive or alternate impairment of neurological functions (motor, sensory, visual, cognitive). Patients have to manage alternating periods in hospital with care at home, experiencing a constant uncertainty regarding the timing of the disease acute phases and facing a considerable psychological and economic burden that also involves their caregivers.

---

[7] https://erisk.irlab.org/.
[8] http://nlp.uned.es/exist2024/.
[9] https://brainteaser.health/open-evaluation-challenges/idpp-2024/.

Clinicians, on the other hand, need tools able to support them in all the phases of the patient treatment, to suggest personalized therapeutic decisions, and to indicate urgently needed interventions. It offered the following tasks. *Task 1 – Predicting ALSFRS-R score from sensor data (ALS)*: focused on predicting the ALSFRS-R score (ALS Functional Rating Scale - Revised), assigned by medical doctors roughly every three months, from the sensor data collected via the app. The ALSFRS-R score is a somehow "subjective" evaluation performed by a medical doctor and this task will help in answering a currently open question in the research community, i.e., whether it could be derived from objective factors. *Task 2 – Predicting patient self-assessment score from sensor (ALS)*: focused on predicting the self-assessment score assigned by patients from the sensor data collected via the app. If the self-assessment performed by patients, more frequently than the assessment performed by medical doctors every three months or so, can be reliably predicted by sensor and app data, we can imagine a proactive application which, monitoring the sensor data, alerts the patient if an assessment is needed. *Task 3 – Predicting relapses from EDDS sub-scores and environmental data (MS))*: focused on predicting a relapse using environmental data and EDSS (Expanded Disability Status Scale) sub-scores. This task will allow us to assess whether exposure to different pollutants is a useful variable in predicting a relapse.

ImageCLEF: Multimedia Retrieval[10] aimed at evaluating the technologies for annotation, indexing, classification, and retrieval of multimodal data. Its main objective resided in providing access to large collections of multimodal data for multiple usage scenarios and domains. Considering the experience of the last four successful editions, ImageCLEF 2024 continued to address a diversity of applications, namely medical, social media, and Internet, and recommending, giving to the participants the opportunity to deal with interdisciplinary approaches and domains. It offered the following tasks. *Task 1 - ImageCLEFmedical*: continued the tradition of bringing together several initiatives for medical applications fostering cross-exchanges, namely: (i) caption task with medical concept detection and caption prediction, (ii) GAN task on synthetic medical images generated with GANs, (iii) MEDVQA-GI task for medical images generation based on text input, and (iv) Mediqa task with a new use-case on multimodal dermatology response generation. *Task 2 - Image Retrieval/Generation for Arguments*: given a set of arguments, asked to return for each argument several images that help to convey the argument's premise, that is, suitable images to depict what is described in the argument. *Task 3 - ImageCLEFrecommending*: focused on content recommendation for cultural heritage content. Despite current advances in content-based recommendation systems, there is limited understanding of how well these perform and how relevant they are for the final end-users. This task aimed to fill this gap by benchmarking different recommendation systems and methods. *Task 4 - ImageCLEFtoPicto*: aimed to provide a translation in pictograms from a natural language, either from (i) text or (ii) speech understandable by the users, in this case, people with language impairments, as pictogram generation is an emerging and significant domain in natural language processing, with multiple potential applications, enabling communication with individuals who have disabilities, aiding in medical settings for individuals who do not speak the language of a country, and also enhancing user understanding in the service industry..

---

[10] https://www.imageclef.org/2024.

JOKER: Automatic Humour Analysis[11] aimed to foster research on automated processing of verbal humour, including tasks such as retrieval, classification, interpretation, generation, and translation. It offered the following tasks. *Task 1 - Humour-aware information retrieval*: aimed at retrieving short humorous texts from a document collection. *Task 2 - Humour classification according to genre and technique*: aimed at classifying short texts of humour among the different classes such as Irony, Sarcasm, Exaggeration, Incongruity, Absurdity, etc. *Task 3 - Pun translation*: aimed to translate English punning jokes into French preserving wordplay form and wordplay meaning.

LifeCLEF: species identification and prediction[12] was dedicated to the large-scale evaluation of biodiversity identification and prediction methods based on artificial intelligence. It offered the following tasks. *Task 1 - BirdCLEF*: bird species recognition in audio soundscapes. *Task 2 - FungiCLEF*: fungi recognition from images and metadata. *Task 3 - GeoLifeCLEF*: remote sensing-based prediction of species. *Task 4 - PlantCLEF*: global-scale plant identification from images. *Task 5 - SnakeCLEF*: snake species identification in medically important scenarios.

LongEval: Longitudinal Evaluation of Model Performance[13] focused on evaluating the temporal persistence of information retrieval systems and text classifiers. The goal was to develop temporal information retrieval systems and longitudinal text classifiers that survive through dynamic temporal text changes, introducing time as a new dimension for ranking models' performance. It offered the following tasks. *Task 1 - LongEval-Retrieval*: aimed to propose a temporal information retrieval system which can handle changes over time. The proposed retrieval system should demonstrate temporal persistence on Web documents. This task had 2 sub-tasks focusing on short-term and long-term persistence. *Task 2 - LongEval-Classification* aimed to propose a temporal persistence classifier which can mitigate performance drop over short and long periods of time compared to a test set from the same time frame as training. This task had 2 sub-tasks focusing on short-term and long-term persistence.

PAN: Digital Text Forensics and Stylometry[14] aimed to advance the state of the art and provide for an objective evaluation on newly developed benchmark datasets in those areas. It offered the following tasks. *Task 1 - Multi-Author Writing Style Analysis*: given an English document, asked to determine at which paragraphs the author changes. Examples varied in difficulty from easy to hard depending on the topical homogeneity of the paragraphs. *Task 2 - Multilingual Text Detoxification*: given a toxic piece of text, asked to re-write it in a non-toxic way while saving the main content as much as possible. Texts were provided in 7 languages. *Task 3 - Oppositional Thinking Analysis*: given an English or Spanish online message, asked to determine whether it is a conspiracy theory or critical thinking. In former case, find the core elements of the conspiracy narrative. *Task 4 - Generative AI Authorship Verification*: given a document, asked to determine whether the author is a human or a language model. In collaboration with the ELOQUENT lab.

---

[11] http://joker-project.com/.
[12] http://www.lifeclef.org/.
[13] https://clef-longeval.github.io/.
[14] http://pan.webis.de/.

qCLEF: QuantumCLEF[15] Quantum Computing (QC) is a rapidly growing field, involving an increasing number of researchers and practitioners from different backgrounds who develop new methods that leverage quantum computers to perform faster computations. QuantumCLEF provided an evaluation infrastructure to design and develop QC algorithms and, in particular, for Quantum Annealing (QA) algorithms, for Information Retrieval and Recommender Systems. It offered the following tasks. *Task 1 - Feature Selection*: focused on applying quantum annealers to find the most relevant subset of features to train a learning model, e.g., for ranking. This problem is very impactful, since many IR and RS systems involve the optimization of learning models, and reducing the dimensionality of the input data can improve their performance. *Task 2 - Clustering*: focused on using quantum annealing to cluster different documents in the form of embeddings to ease the browsing process of large collections. Clustering can be helpful for organizing large collections, helping users to explore a collection and providing similar search results to a given query. Furthermore, it can be helpful to divide users according to their interests or build user models with the cluster centroids speeding up the runtime of the system or its effectiveness for users with limited data. Clustering is however a very complex task in the case of QA since it is possible to perform clustering only considering a limited number of items and clusters due to the architecture of quantum annealers. A baseline using K-medoids clustering with cosine distance was used as an overall alternative.

SimpleText: Improving Access to Scientific Texts for Everyone[16] addressed technical and evaluation challenges associated with making scientific information accessible to a wide audience, students, and experts. Appropriate reusable data and benchmarks were provided for scientific text summarization and simplification. *Task 1 - Retrieving passages to include in a simplified summary*: given a popular science article targeted to a general audience, aimed at retrieving passages which can help to understand this article, from a large corpus of academic abstracts and bibliographic metadata. Relevant passages should relate to any of the topics in the source article. *Task 2 - Identifying and explaining difficult concepts*: aimed to decide which concepts in scientific abstracts require explanation and contextualization in order to help a reader understand the scientific text. *Task 3 - Simplify Scientific Text*: aimed to provide a simplified version of sentences extracted from scientific abstracts. Participants were provided with popular science articles and queries and matching abstracts of scientific papers, split into individual sentences. *Task 4 - Tracking the State-of-the-Art in Scholarly Publications*: aimed to develop systems which, given the full text of an AI paper, are capable of recognizing whether an incoming AI paper indeed reports model scores on benchmark datasets, and if so, to extract all pertinent (Task, Dataset, Metric, Score) tuples presented within the paper.

Touché: Argumentation Systems[17] aimed foster the development of technologies that support people in decision-making and opinion-forming and to improve our understanding of these processes. It offered the following tasks. *Task 1 - Human Value Detection*: given a text, for each sentence, asked to detect which human values the sentence refers to and whether this reference (partially) attains or (partially) constrains the value.

---

[15] https://qclef.dei.unipd.it/.

[16] http://simpletext-project.com/.

[17] https://touche.webis.de/.

Task 2 - Ideology and Power Identification in Parliamentary Debates: given a parliamentary speech in one of several languages, asked to identify the ideology of the speaker's party and identify whether the speaker's party is currently governing or in opposition.

Task 3 - Image Retrieval for Arguments: given an argument, asked to retrieve or generate images that help to convey the argument's premise.

The success of CLEF 2024 would not have been possible without the huge effort of several people and organizations, including the CLEF Association[18], the Program Committee, the Lab Organizing Committee, the reviewers, and the many students and volunteers who contributed.

We thank the Friends of SIGIR program for covering the registration fees for a number of student delegates.

July 2024

Lorraine Goeuriot
Philippe Mulhem
Georges Quénot
Didier Schwab
Giorgio Maria Di Nunzio
Laure Soulier
Petra Galuščáková
Alba García Seco de Herrera
Guglielmo Faggioli
Nicola Ferro

---

[18] https://www.clef-initiative.eu/#association.

Organization

General Chairs

Lorraine Goeuriot	Université Grenoble Alpes, France
Philippe Mulhem	Université Grenoble Alpes, France
Georges Quénot	Université Grenoble Alpes, France
Didier Schwab	Université Grenoble Alpes, France

Program Chairs

Giorgio Maria Di Nunzio	University of Padua, Italy
Laure Soulier	Sorbonne Université, France

Lab Chairs

Petra Galuščáková	University of Stavanger, Norway
Alba García Seco de Herrera	University of Essex, UK

Lab Mentorship Chair

Liana Ermakova	Université de Bretagne Occidentale, France
Florina Piroi	TU Wien, Austria

Proceedings Chairs

Guglielmo Faggioli	University of Padua, Italy
Nicola Ferro	University of Padua, Italy

CLEF Steering Committee

Steering Committee Chair

Nicola Ferro   University of Padua, Italy

Deputy Steering Committee Chair for the Conference

Paolo Rosso   Universitat Politècnica de València, Spain

Deputy Steering Committee Chair for the Evaluation Labs

Martin Braschler   Zurich University of Applied Sciences, Switzerland

Members

Avi Arampatzis   Democritus University of Thrace, Greece
Alberto Barrón-Cedeño   University of Bologna, Italy
Khalid Choukri   Evaluations and Language resources Distribution Agency, France
Fabio Crestani   Università della Svizzera italiana, Switzerland
Carsten Eickhoff   University of Tübingen, Germany
Norbert Fuhr   University of Duisburg-Essen, Germany
Anastasia Giachanou   Utrecht University, The Netherlands
Lorraine Goeuriot   Université Grenoble Alpes, France
Julio Gonzalo   National Distance Education University (UNED), Spain
Donna Harman   National Institute for Standards and Technology, USA
Bogdan Ionescu   University "Politehnica" of Bucharest, Romania
Evangelos Kanoulas   University of Amsterdam, The Netherlands
Birger Larsen   University of Aalborg, Denmark
David E. Losada   Universidade de Santiago de Compostela, Spain
Mihai Lupu   Vienna University of Technology, Austria
Maria Maistro   University of Copenhagen, Denmark
Josiane Mothe   IRIT, Université de Toulouse, France
Henning Müller   University of Applied Sciences Western Switzerland (HES-SO), Switzerland

Jian-Yun Nie	Université de Montréal, Canada
Gabriella Pasi	University of Milano-Bicocca, Italy
Eric SanJuan	University of Avignon, France
Giuseppe Santucci	Sapienza University of Rome, Italy
Laure Soulier	Pierre and Marie Curie University (Paris 6), France
Theodora Tsikrika	Information Technologies Institute, Centre for Research and Technology Hellas (CERTH), Greece
Christa Womser-Hacker	University of Hildesheim, Germany

Past Members

Paul Clough	University of Sheffield, UK
Djoerd Hiemstra	Radboud University, The Netherlands
Jaana Kekäläinen	University of Tampere, Finland
Séamus Lawless	Trinity College Dublin, Ireland
Carol Peters	ISTI, National Council of Research (CNR), Italy
Emanuele Pianta	Centre for the Evaluation of Language and Communication Technologies, Italy
Maarten de Rijke	University of Amsterdam, The Netherlands
Jacques Savoy	University of Neuchâtel, Switzerland
Alan Smeaton	Dublin City University, Ireland

Supporters and Sponsors

SIGIR
Special Interest Group
on Information Retrieval

Contents – Part I

CLEF 25th Anniversary

What Happened in CLEF... For Another While? 3
Nicola Ferro

Conference Papers

Sexism Identification on TikTok: A Multimodal AI Approach with Text, Audio, and Video 61
Iván Arcos and Paolo Rosso

Knowledge Acquisition Passage Retrieval: Corpus, Ranking Models, and Evaluation Resources 74
Artemis Capari, Hosein Azarbonyad, Georgios Tsatsaronis, Zubair Afzal, Judson Dunham, and Jaap Kamps

Assessing Document Sanitization for Controlled Information Release and Retrieval in Data Marketplaces 88
Luca Cassani, Giovanni Livraga, and Marco Viviani

The Impact of Web Search Result Quality on Decision-Making 100
Jan Heinrich Merker, Lena Merker, and Alexander Bondarenko

Improving Laypeople Familiarity with Medical Terms by Informal Medical Entity Linking 113
Annisa Maulida Ningtyas, Alaa El-Ebshihy, Florina Piroi, and Allan Hanbury

Mapping the Media Landscape: Predicting Factual Reporting and Political Bias Through Web Interactions 127
Dairazalia Sánchez-Cortés, Sergio Burdisso, Esaú Villatoro-Tello, and Petr Motlicek

Under-Sampling Strategies for Better Transformer-Based Classifications Models 139
Marcin Sawiński, Krzysztof Węcel, and Ewelina Księżniak

Classification of Social Media Hateful Screenshots Inciting Violence and Discrimination 152
Davide Buscaldi, Paolo Rosso, Berta Chulvi, and Ting Wang

SessionPrint: Accelerating kNN via Locality-Sensitive Hashing
for Session-Based News Recommendation 159
   Mozhgan Karimi

Who Will Evaluate the Evaluators? Exploring the Gen-IR User Simulation
Space 166
   *Johannes Kiesel, Marcel Gohsen, Nailia Mirzakhmedova,
Matthias Hagen, and Benno Stein*

De-noising Document Classification Benchmarks via Prompt-Based Rank
Pruning: A Case Study 172
   Matti Wiegmann, Benno Stein, and Martin Potthast

Best of CLEF 2023 Labs

Best of Touché 2023 Task 4: Testing Data Augmentation and Label
Propagation for Multilingual Multi-target Stance Detection 181
   Jorge Avila, Álvaro Rodrigo, and Roberto Centeno

Leveraging LLM-Generated Data for Detecting Depression Symptoms
on Social Media 193
   Ana-Maria Bucur

From Sentence Embeddings to Large Language Models to Detect
and Understand Wordplay 205
   Ryan Rony Dsilva

Replicability Measures for Longitudinal Information Retrieval Evaluation 215
   Jüri Keller, Timo Breuer, and Philipp Schaer

SimpleText Best of Labs in CLEF-2023: Scientific Text Simplification
Using Multi-prompt Minimum Bayes Risk Decoding 227
   Andrianos Michail, Pascal Severin Andermatt, and Tobias Fankhauser

Large Language Model Cascades and Persona-Based In-Context Learning
for Multilingual Sexism Detection 254
   Lin Tian, Nannan Huang, and Xiuzhen Zhang

Correction to: Under-Sampling Strategies for Better Transformer-Based
Classifications Models C1
   Marcin Sawiński, Krzysztof Węcel, and Ewelina Księżniak

Author Index 267

Contents – Part II

Lab Overviews

Overview of BioASQ 2024: The Twelfth BioASQ Challenge
on Large-Scale Biomedical Semantic Indexing and Question Answering 3
*Anastasios Nentidis, Georgios Katsimpras, Anastasia Krithara,
Salvador Lima-López, Eulàlia Farré-Maduell, Martin Krallinger,
Natalia Loukachevitch, Vera Davydova, Elena Tutubalina,
and Georgios Paliouras*

Overview of the CLEF-2024 CheckThat! Lab: Check-Worthiness,
Subjectivity, Persuasion, Roles, Authorities, and Adversarial Robustness 28
*Alberto Barrón-Cedeño, Firoj Alam, Julia Maria Struß,
Preslav Nakov, Tanmoy Chakraborty, Tamer Elsayed, Piotr Przybyła,
Tommaso Caselli, Giovanni Da San Martino, Fatima Haouari,
Maram Hasanain, Chengkai Li, Jakub Piskorski, Federico Ruggeri,
Xingyi Song, and Reem Suwaileh*

Overview of ELOQUENT 2024—Shared Tasks for Evaluating Generative
Language Model Quality 53
*Jussi Karlgren, Luise Dürlich, Evangelia Gogoulou, Liane Guillou,
Joakim Nivre, Magnus Sahlgren, Aarne Talman, and Shorouq Zahra*

Overview of eRisk 2024: Early Risk Prediction on the Internet 73
*Javier Parapar, Patricia Martín-Rodilla, David E. Losada,
and Fabio Crestani*

Overview of EXIST 2024 — Learning with Disagreement for Sexism
Identification and Characterization in Tweets and Memes 93
*Laura Plaza, Jorge Carrillo-de-Albornoz, Víctor Ruiz, Alba Maeso,
Berta Chulvi, Paolo Rosso, Enrique Amigó, Julio Gonzalo,
Roser Morante, and Damiano Spina*

Intelligent Disease Progression Prediction: Overview of iDPP@CLEF 2024 ... 118
Giovanni Birolo, Pietro Bosoni, Guglielmo Faggioli, Helena Aidos, Roberto Bergamaschi, Paola Cavalla, Adriano Chiò, Arianna Dagliati, Mamede de Carvalho, Giorgio Maria Di Nunzio, Piero Fariselli, Jose Manuel García Dominguez, Marta Gromicho, Alessandro Guazzo, Enrico Longato, Sara C. Madeira, Umberto Manera, Stefano Marchesin, Laura Menotti, Gianmaria Silvello, Eleonora Tavazzi, Erica Tavazzi, Isotta Trescato, Martina Vettoretti, Barbara Di Camillo, and Nicola Ferro

Overview of the ImageCLEF 2024: Multimedia Retrieval in Medical Applications 140
Bogdan Ionescu, Henning Müller, Ana-Maria Drăgulinescu, Johannes Rückert, Asma Ben Abacha, Alba García Seco de Herrera, Louise Bloch, Raphael Brüngel, Ahmad Idrissi-Yaghir, Henning Schäfer, Cynthia Sabrina Schmidt, Tabea M. G. Pakull, Hendrik Damm, Benjamin Bracke, Christoph M. Friedrich, Alexandra-Georgiana Andrei, Yuri Prokopchuk, Dzmitry Karpenka, Ahmedkhan Radzhabov, Vassili Kovalev, Cécile Macaire, Didier Schwab, Benjamin Lecouteux, Emmanuelle Esperança-Rodier, Wen-Wai Yim, Yujuan Fu, Zhaoyi Sun, Meliha Yetisgen, Fei Xia, Steven A. Hicks, Michael A. Riegler, Vajira Thambawita, Andrea Storås, Pål Halvorsen, Maximilian Heinrich, Johannes Kiesel, Martin Potthast, and Benno Stein

Overview of the CLEF 2024 JOKER Track: Automatic Humour Analysis 165
Liana Ermakova, Anne-Gwenn Bosser, Tristan Miller, Victor Manuel Palma Preciado, Grigori Sidorov, and Adam Jatowt

Overview of LifeCLEF 2024: Challenges on Species Distribution Prediction and Identification 183
Alexis Joly, Lukáš Picek, Stefan Kahl, Hervé Goëau, Vincent Espitalier, Christophe Botella, Diego Marcos, Joaquim Estopinan, Cesar Leblanc, Théo Larcher, Milan Šulc, Marek Hrúz, Maximilien Servajean, Hervé Glotin, Robert Planqué, Willem-Pier Vellinga, Holger Klinck, Tom Denton, Ivan Eggel, Pierre Bonnet, and Henning Müller

Overview of the CLEF 2024 LongEval Lab on Longitudinal Evaluation of Model Performance 208
Rabab Alkhalifa, Hsuvas Borkakoty, Romain Deveaud, Alaa El-Ebshihy, Luis Espinosa-Anke, Tobias Fink, Petra Galuščáková, Gabriela Gonzalez-Saez, Lorraine Goeuriot, David Iommi, Maria Liakata, Harish Tayyar Madabushi, Pablo Medina-Alias, Philippe Mulhem, Florina Piroi, Martin Popel, and Arkaitz Zubiaga

Overview of PAN 2024: Multi-author Writing Style Analysis, Multilingual Text Detoxification, Oppositional Thinking Analysis, and Generative AI Authorship Verification Condensed Lab Overview 231
   Abinew Ali Ayele, Nikolay Babakov, Janek Bevendorff, Xavier Bonet Casals, Berta Chulvi, Daryna Dementieva, Ashaf Elnagar, Dayne Freitag, Maik Fröbe, Damir Korenčić, Maximilian Mayerl, Daniil Moskovskiy, Animesh Mukherjee, Alexander Panchenko, Martin Potthast, Francisco Rangel, Naquee Rizwan, Paolo Rosso, Florian Schneider, Alisa Smirnova, Efstathios Stamatatos, Elisei Stakovskii, Benno Stein, Mariona Taulé, Dmitry Ustalov, Xintong Wang, Matti Wiegmann, Seid Muhie Yimam, and Eva Zangerle

Overview of QuantumCLEF 2024: The Quantum Computing Challenge for Information Retrieval and Recommender Systems at CLEF 260
   Andrea Pasin, Maurizio Ferrari Dacrema, Paolo Cremonesi, and Nicola Ferro

Overview of the CLEF 2024 SimpleText Track: Improving Access to Scientific Texts for Everyone 283
   Liana Ermakova, Eric SanJuan, Stéphane Huet, Hosein Azarbonyad, Giorgio Maria Di Nunzio, Federica Vezzani, Jennifer D'Souza, and Jaap Kamps

Overview of Touché 2024: Argumentation Systems 308
   Johannes Kiesel, Çağrı Çöltekin, Maximilian Heinrich, Maik Fröbe, Milad Alshomary, Bertrand De Longueville, Tomaž Erjavec, Nicolas Handke, Matyáš Kopp, Nikola Ljubešić, Katja Meden, Nailia Mirzhakhmedova, Vaidas Morkevičius, Theresa Reitis-Münstermann, Mario Scharfbillig, Nicolas Stefanovitch, Henning Wachsmuth, Martin Potthast, and Benno Stein

Author Index 333

CLEF 25th Anniversary

What Happened in CLEF... For Another While?

Nicola Ferro[✉]

Department of Information Engineering, University of Padua, Padua, Italy
nicola.ferro@unipd.it

Abstract. 2024 marks the 25th birthday for CLEF, an evaluation campaign activity which has applied the Cranfield evaluation paradigm to the testing of multilingual and multimodal information access systems in Europe. This paper provides a summary of the motivations which led to the establishment of CLEF, a description of how it has evolved over the years, and its major achievements.

1 Introduction

Performance measuring is a key to scientific progress. This is particularly true for research concerning complex systems, whether natural or human-built. Multilingual and multimedia information systems are particularly complex: they need to satisfy diverse user needs and support challenging tasks. Their development calls for proper evaluation methodologies to ensure that they meet the expected user requirements and provide the desired effectiveness.

Large-scale worldwide experimental evaluations provide fundamental contributions to the advancement of state-of-the-art techniques through the establishment of common evaluation procedures, the organisation of regular and systematic evaluation cycles, the comparison and benchmarking of proposed approaches, and the spreading of knowledge.

The *Conference and Labs of the Evaluation Forum (CLEF)*[1] is a large-scale *Information Retrieval (IR)* evaluation initiative organised in Europe but involving researchers world-wide. CLEF shares the stage and coordinates with the other major evaluation initiatives in the field, namely: the *Text REtrieval Conference (TREC)*[2] [265], the first large-scale evaluation activity in the field of IR, which began in 1992; the *NII Testbeds and Community for Information access Research (NTCIR)*[3] [542], which promotes research in information access technologies with a special focus on East Asian languages and English; and the *Forum for Information Retrieval Evaluation (FIRE)*[4], whose aim is to encourage research in Indian languages by creating a platform similar to CLEF, providing

---

[1] http://www.clef-initiative.eu/.
[2] http://trec.nist.gov/.
[3] http://research.nii.ac.jp/ntcir/.
[4] http://fire.irsi.res.in/.

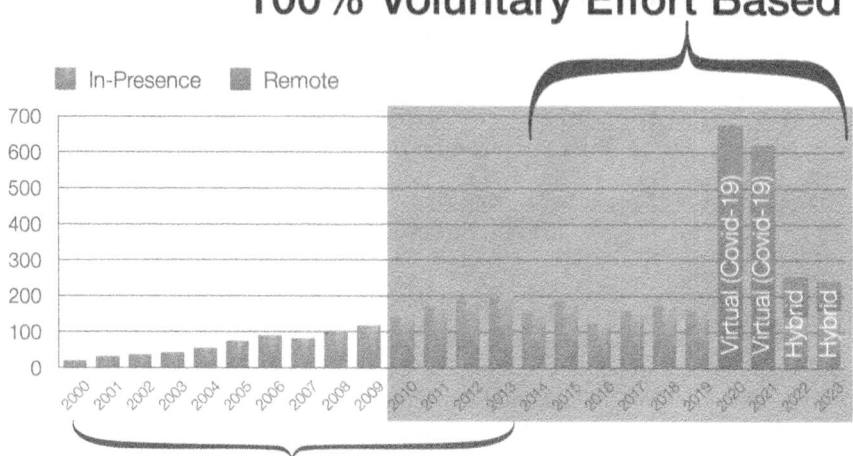

Fig. 1. Participation in CLEF over the years (CLEF "Classic" period un-shaded; CLEF Initiative period shaded).

data and a common forum for comparing models and techniques applied to these languages.

This year marks the 25th birthday of CLEF, which began as an independent activity in 2000. The goal of this report is to provide a short overview of what motivated the setting up of CLEF, what has happened in CLEF during these years, and how CLEF has evolved to keep pace with emerging challenges.

CLEF can be divided into a CLEF "Classic" (2000–2009) period, where CLEF begun and was supported by the European funding, and a "CLEF Initiative" period (2010–2024), where CLEF was profoundly reorganized and became a self-sustained activity not baked by dedicated European funding, thanks to the contribution of the PROMISE network of excellence [94].

Figure 1 shows the attendance to the CLEF event over the years: we can observe a stable and consistent grow in participation, a possible consequence of the capacity of CLEF to renew itself and to attract new communities and expertise in addition to core information retrieval activities. We can also see that in 2020 and 2021 CLEF has gone completely virtual due the Covid-19 pandemic and this caused a spike in attendance. CLEF 2022 and 2023 represent the first editions back in presence after Covid-19 which offered an hybrid modality of participation: we can observed as the in-presence attendance is back to the pre-Covid levels and how it is complemented by a substantial remote attendance as well.

The paper is organized as follows: Sect. 2 describes the beginning and the first period of CLEF, the so-called "CLEF Classic" period; Sect. 3 introduces the second (and current) period of CLEF, known as the "CLEF Initiative" period;

Sects. 4 and 5 give an idea of the spread and extension of CLEF activities by providing a short account of the topics addressed in the conference, tracks and labs over the years together with pointers to papers providing more details; Sect. 6 attempts to provide an assessment of the impact of CLEF in the IR community and beyond; Sect. 7 summarizes the book which was prepared for celebrating the past 20th anniversary of CLEF; finally, Sect. 8 presents the CLEF Association, the no-profit legal entity committed to sustaining and running CLEF.

2 CLEF "Classic": 2000–2009

The *Cross-Language Evaluation Forum* (CLEF) began as a cross-lingual track at TREC in 1997 [562], moving to an independent activity in 2000 [479], since Europe was felt as a more suitable environment than USA for fully empowering multilinguality.

The underlying motivation for CLEF was the "Grand Challenge" formulated at the Association for the Advancement of Artificial Intelligence (AAAI) 1997 Spring Symposium on Cross-Language and Speech Retrieval [274]. The ambitious goal was the development of fully multilingual and multimodal information access systems capable of:

- processing a query in any medium and any language;
- finding relevant information from a multilingual multimedia collection containing documents in any language and form;
- presenting it in the style most likely to be useful to the user.

The main objective of CLEF has thus been to promote research and stimulate development of multilingual and multimodal IR systems for European (and non-European) languages [480], through:

- the creation of an evaluation infrastructure and the organisation of regular evaluation campaigns for system testing;
- the building of a multidisciplinary research community;
- the construction of publicly available test-suites.

CLEF has pursued this objective by attempting to anticipate the emerging needs of the R&D community and to promote the development of multilingual and multimodal systems that fulfil the demands of the AAAI 1997 Grand Challenge. However, while the first three editions of CLEF were dedicated to mono- and multilingual ad-hoc text retrieval, gradually the scope of activity was extended to include other kinds of text retrieval across languages (i.e., not just document retrieval but question answering and geographic IR as well) and on other media (i.e., collections containing images and speech).

During what is jokingly referred to as the "classic" period of CLEF (2000–2009), several important results were achieved: research activities in previously unexplored areas were stimulated, permitting the growth of IR for languages other than English; evaluation methodologies for different types of *Cross Language Information Retrieval (CLIR)* as well as *MultiLingual Information Access*

(*MLIA*) systems, operating in diverse domains, were studied and implemented; a large set of empirical data about multilingual information access from the user perspective was created; quantitative and qualitative evidence with respect to best practices in cross-language system development was collected; reusable test collections for system benchmarking were developed; language resources for a wide range of European languages, some of which had been little studied, were built. CLEF activities have resulted in the creation of a considerable amount of valuable resources, also for under-represented languages, extremely useful for many types of text processing and benchmarking activities in the IR domain. Perhaps, most important, a strong, multidisciplinary, and active research community focussed mainly, but not only, on IR for European languages came into being.

If we had to summarize the major outcome of CLEF in this period with just one sentence, we could safely say that CLEF has made multilingual IR for European languages a reality, with performances as satisfactory as monolingual ones.

3 The CLEF Initiative: 2010 Onwards

3.1 Scope

The second period of CLEF started with a clear and compelling question: after a successful decade studying multilinguality for European languages, what were the main unresolved issues currently facing us? To answer this question, CLEF turned to the CLEF community to identify the most pressing challenges and to list the steps to be taken to meet them.

The discussion led to the definition and establishment of the *CLEF Initiative*, whose main mission is to promote research, innovation, and the development of information access systems with an emphasis on multilingual and *multimodal* information with various levels of structure.

In the CLEF Initiative an increased focus is on the *multimodal* aspect, intended not only as the ability to deal with information coming in multiple media but also in different modalities, e.g. the Web, social media, news streams, specific domains and so on. These different modalities should, ideally, be addressed in an integrated way; rather than building vertical search systems for each domain/modality the interaction between the different modalities, languages, and user tasks needs to be exploited to provide comprehensive and aggregated search systems.

The continuity with the first period of CLEF on multilinguality and this increased attention for multimodality has led to the definition of a set of action lines for the CLEF Initiative:

– multilingual and multimodal system testing, tuning and evaluation;
– investigation of the use of unstructured, semi-structured, highly-structured, and semantically enriched data in information access;
– creation of reusable test collections for benchmarking;

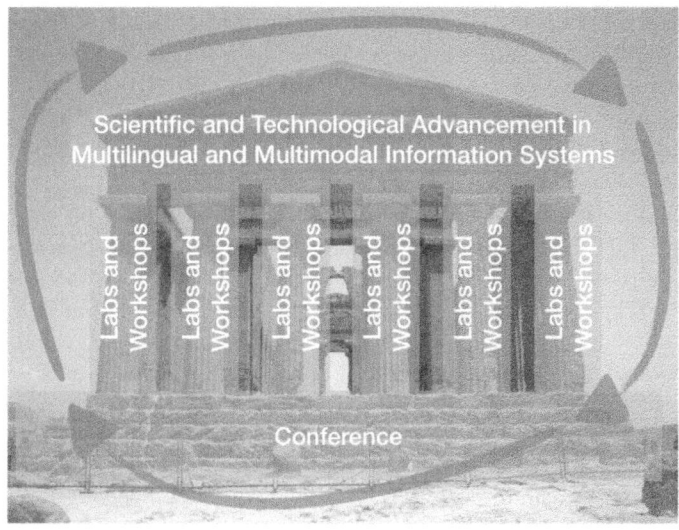

Fig. 2. Structure of the CLEF Initiative.

– exploration of new evaluation methodologies and innovative ways of using experimental data;
– discussion of results, comparison of approaches, exchange of ideas, and transfer of knowledge.

The new challenges and the new organizational structure, described below, have motivated a change of name for CLEF: from the *Cross-Language Evaluation Forum*, of the "classic" period, to *Conference and Labs of the Evaluation Forum*, which now reflects the widened scope.

3.2 Structure

The new challenges for CLEF also called for a renewal of its structure and organization. The annual CLEF meeting is no longer a Workshop, held in conjunction with the European Conference on Digital Libraries (ECDL, now TPDL – Theory and Practice of Digital Libraries), but has become an independent event, held over 3.5–4 days and made up of two interrelated activities: the *Conference* and the *Labs*. As shown in Fig. 2, the *Conference* and the *Labs* are expected to interact together and mutually reinforce each other, bringing new interest and new expertise into CLEF.

More in detail, the *Conference* is a peer-reviewed conference, open to the IR community as a whole and not just to *Lab* participants, and aims at stimulating discussion on innovative evaluation methodologies, fostering a deeper analysis and understanding of experimental results, and promoting multilingual and multimodal information access at large. The *Labs* are the core of the evaluation activities; they are selected on the basis of topical relevance, novelty, potential research impact, the existence of clear real-world use cases, a likely number of participants, and the experience of the organizing consortium. We allow also for a special case of pilot lab activities, called *Workshops*, whose goal is to explore new and "risky" evaluation activities, which are not ready yet for being shipped as full-fledged *Labs* and benefit from an incubation and discussion period to better tune them.

The *Conference* and the *Labs* originate two streams of peer-reviewed publications. The *CLEF Proceedings*[5] are published in the Springer Lecture Notes in Computer Science (LNCS) [10, 36, 37, 55, 64, 104, 117, 138, 208, 210, 218, 239, 240, 304, 311, 419, 479, 482–488, 493, 494, 496] series and contain full and short papers submitted to the *Conference*, condensed overviews of the *Lab* activities, and revised and selected "best of labs" papers from labs in the previous edition of CLEF. The *CLEF Working Notes* are published in the CEUR Workshop Proceedings (CEUR-WS.org)[6] series [18, 51, 87–89, 96, 106–111, 186–188, 209, 211, 434, 435, 489–492, 495, 501] and contain extended lab overviews and detailed papers from the participants in the lab activities. The peer-review process for the *CLEF Proceedings* is ensured by a Programme Committee, which is established for each CLEF edition; the peer-review process for the *CLEF Working Notes* is ensured by dedicated Programme Committees, which are setup separately for each lab of each CLEF edition.

3.3 Organization

In order to favour participation and the introduction of new perspectives, CLEF has introduced a new open-bid process which allows research groups and institutions to bid to host the annual CLEF event and to propose themes. Initially, the bidding process followed a one-year ahead cycle but now, thanks to the interest in and the engagement with CLEF, it follows a three-years ahead cycle, i.e. we are now managing the bids for hosting CLEF 2027.

While in the CLEF "Classic" period the governing body of CLEF was the *Steering Committee*, which was in charge of the overall coordination of CLEF, of selecting the evaluation activities to be carried out in each edition, and of looking ahead for future research directions, the new participatory approach called for a more articulated organization and for a better separation of concerns.

Each edition of CLEF appoints its own *General Chairs*, *Programme Chairs*, and *Lab Chairs*. The General Chairs are responsible for the overall running of the annual CLEF event, i.e. *Conference* and *Labs* meetings, and serve as the chairs of the organizing committee. The Program Chairs are responsible for planning and implementing the technical program of the *Conference*, and therefore their main responsibility is to ensure that the scientific quality of the *Conference* is at the highest possible level. The Labs Chairs are responsible for selecting, planning, and implementing the focused benchmarking activities, and therefore their main

---

[5] https://link.springer.com/conference/clef.
[6] http://ceur-ws.org/.

responsibility is to ensure that the scientific and technical quality of the *Labs* is at the highest possible level.

Finally, as before, the *Steering Committee* is charge of the overall coordination of CLEF: it assists in the appointment of and approves the General Chairs, the Program Chairs and the Labs Chairs for the annual CLEF edition; it devises improvements to the CLEF structure and organization; it manages the bidding process; and, it looks ahead for future research directions to be pursued.

3.4 Beyond CLEF

An aspect of CLEF of which we are particularly proud is the consolidation of a strong community of European researchers in the multidisciplinary context of IR. In occasion of the past 20th anniversary of CLEF in 2019, for the first time, the *European Conference for Information Retrieval (ECIR)* and CLEF have joined forces: ECIR 2019 hosted a session dedicated to CLEF Labs where lab organizers present the major outcomes of their Labs and plans for ongoing activities, followed by a poster session in order to favour discussion during the conference. This is reflected in the ECIR 2019 proceedings, where CLEF Lab activities and results are reported as short papers. Since then, ECIR kept hosting a CLEF session and similar plans are already in place for CLEF 2025 and ECIR 2025.

The goal is not only to engage the ECIR community in CLEF activities, but also to disseminate the research results achieved during CLEF evaluation cycles at ECIR. This collaboration will of course strengthen European IR research even more. However, this European community should not be seen in isolation. CLEF is part of a global community; we have always maintained close links with our peer initiatives in the Americas and Asia. There is a strong bond connecting TREC, NTCIR, CLEF and FIRE, and a continual, mutually beneficial exchange of ideas, experiences and results.

4 The Conference

Figure 3 gives an overview of the topics addressed by the CLEF conference over the years, together with the number of papers published for each topic. Figure 3 clearly shows there is a constant stream of papers in the two core areas of CLEF, namely *evaluation* – broken down into "Experimental Collections", "Evaluation Methods", "Evaluation Measures", and "Evaluation Infrastructures" – and *multilinguality and multimodality* – broken down into "Language Processing, Ranking, and Resources", "Tools, Systems, Applications", and "Multimodality". Moreover, we also have a third focus on less mainstream topics – broken down into "Information Visualization for Evaluation" and "Longitudinal Studies". Figure 3 shows how the *evaluation* and *multilinguality and multimodality* streams attract a good number of papers across all the editions of CLEF. On the other hand, "Information Visualization for Evaluation" attracted a few papers for about ten years (2010–2019) but it has faded away in the last few years.

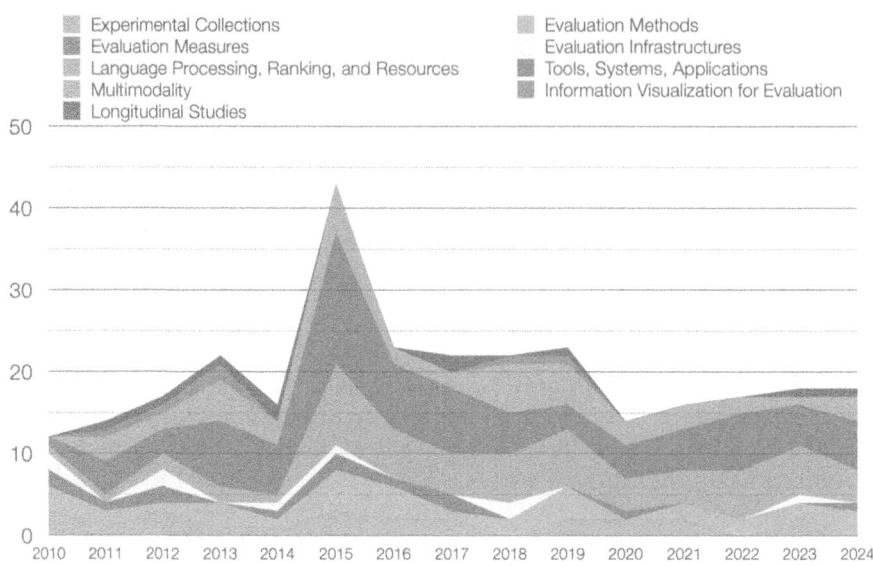

Fig. 3. Topics addressed by the CLEF conference over the years and number of papers published for each topic.

Finally, "Longitudinal Studies" keep appearing at CLEF, even if in a discontinuous way but this is understandable when you consider the additional effort required to conduct a longitudinal study.

Since CLEF 2023 the conference has introduced a new *result-less review process*, inspired by the "Dagstuhl Seminar 23031 on Frontiers of Information Access Experimentation for Research and Education" [60,61], where the reviewing process is split in two parts. Firstly, the papers are reviewed by their methodological contribution, their research questions, and their experimental design – the submitted papers do not contain an experimentation part. Secondly, those papers which pass the first step are then reviewed by their experimentation, analyses, and insights. The purpose of this new review process is: (i) to avoid accepting papers just because performance improvements with respect to some baseline; (ii) to ensure they have a grounded methodology; and, (iii) to verify that the research questions are driven by the methodology and not, post-hoc, by the experimental results.

We briefly summarize below the topics of Fig. 3, covered by the conference over the years, with pointers to the main references:

Experimental Collections explored different issues concerning experimental collections such as: the creation of collections for languages such as Persian, Arabic, Ahmaric; resource-effective creation of pseudo-test collections for specialised tasks; log-based experimental collections; collections for specific domains, e.g. question answering, plagiarism detection, social image tagging; gamification for relevance judgments; early risk detection, such as depression prediction; collections of query features; social media cross-domain corpora; reading comprehension; subjectivity in news articles; spoken Portuguese document corpus; datasets for automatic gene variant interpretation;   [17, 34, 52, 66, 67, 167, 220, 222, 225, 231, 270, 277, 318, 361, 384, 386, 405, 416, 430, 506, 560, 608, 620, 635];

Evaluation Methods studied core problems related to evaluation methodologies and proposed new methods, such as: the reliability of relevance assessments; living labs for product search tasks; evaluation of information extraction and entity profiles; semantic-oriented evaluation of machine translation and summarization; search snippet evaluation and query simulators; news recommendation; teaching; study of long tails in relevance judgments; crowdsourcing; definition of transactional tasks; component-based evaluation; methodologies for authorship verification; accounting for bias; evaluation of user models; creation of ground-truth in text classification and question answering; impact of gold standards on evaluation; longitudinal evaluation of IR systems and prediction of their performance; multidimensional relevance; noise reduction in relevance judgements: benchmarking methodologies for information extraction from documents; evaluation methods for knowledge acquisition; user simulation via generative AI; de-noising benchmarks   [46, 49, 65, 105, 134, 170, 182, 205, 216, 244, 258, 276, 313, 317, 336, 340, 374, 412, 415, 456–458, 523, 533, 554, 557, 559, 569, 577, 619, 627, 629, 640, 644];

Evaluation Measures dealt with the analysis of the features of the evaluation measures and the proposal of new measures such as: formal properties of measures for document filtering; robustness of metrics for patent retrieval; problems with ties in evaluation measures; effort-based measures and measures for speech retrieval; extension of measures to graded relevance; click models; text interestingness and diversity; measures for real-life categorisation and hierarchical clustering; measures for merging multiple assessors; measures for reproducibility and replicability [28, 31, 47, 103, 193, 194, 221, 251, 322, 325, 392, 418, 463];

Evaluation Infrastructures investigated how to design and develop shared infrastructures to support different aspects of IR evaluation such as: automating component-based evaluation; managing and providing access to the experimental outcomes and the related literature; using cloud-base approaches to offer evaluation services in specialised domains; developing proper ontologies to describe the experimental results; and exploiting map-reduce techniques for effective IR evaluation; frameworks for question-answering; tools for replicability and reproducibility; evaluation infrastructures for quantum computing for information access [8, 59, 158, 261, 262, 269, 379, 467];

Ranking, Language Processing, and Resources continued the CLEF interest in multilinguality by dealing with tools, algorithm, and resources for multiple languages such as: lemmatizers, decompounders and normalizers for underrepresented resources using statistical approaches; statistical stemmers; named entity extraction, linking and clustering in cross-lingual settings; exploitation of multiple translation resources; language-independent generation of document snippets; language variety identification; gender identification; text alignment; Web genre identification; sentiment analysis and opinion mining; personality and author profiling; mixed-code script analyzers; text clustering; language and terminology analysis, also for query suggestion; microblog contextualization; early depression detection; personality recognition; stance detection in social media; readability; fact checking, fake news and mis-information; gender and bot identification argument retrieval; pseudo-relevance feedback; medical concept normalization; text analysis for authorship attribution; hate speech detection; query rewriting for health search; irony detection; entity representation using transformers; large language models [1, 35, 48, 77, 86, 99, 101, 116, 121, 132, 133, 135, 137, 139, 140, 150, 159, 166, 183, 192, 213, 215, 224, 229, 255, 263, 273, 312, 352, 355, 360, 364, 365, 378, 380, 385, 410, 411, 414, 442, 455, 470, 502, 524, 527, 528, 543, 544, 547, 549, 552, 556, 567, 570, 586, 595, 596, 602, 611, 625, 632, 645];

Tools, Systems, and Applications covered the design and development of various kinds of algorithms, systems, and applications focused on multi-linguality and specialised domains such as: semantic discovery of resources in cloud-based systems; Arabic question answering; cross-language similarity search using thesauri; automatic annotation of bibliographic references; exploitation of visual context in multimedia translation; sub-topic mining in Web documents and query interpretation; exploiting relevance feedback for building tag-clouds in image search; query expansion for image retrieval; transcript-based video retrieval; *Peer-To-Peer (P2P)* information retrieval; event detection in microblogs; medical information retrieval; citation for scientific publication; news recommendation; image decomposition and captioning; ranking products in e-commerce; conversational search; mathematical retrieval; systematic reviews; data fusion; event detection; stock market prediction; skill extraction for job; tracking news stories in short messages search; studies on exposure of children to search technology; sensitive information classification; dataset recommendation; analysis of crime-related time series; entity linking; gambling disorders detection: prompt engineering; political bias in media; sexism identification; quality of Web search results for decision making; document sanitization; news recommendation [3, 4, 13–16, 19, 22, 42, 57, 58, 69, 70, 79, 80, 85, 115, 118, 131, 136, 142, 149, 151, 153, 162, 165, 172, 184, 185, 207, 214, 223, 226, 232, 233, 243, 250, 254, 259, 266, 287, 314, 326, 330, 334, 339, 344, 346, 350, 353, 362, 363, 371, 372, 375, 383, 403, 407–409, 413, 420, 443, 451, 513, 522, 529, 531, 532, 536, 546, 548, 553, 564, 568, 571, 572, 576, 578, 592, 594, 599, 623, 624, 630, 631, 633, 634, 636, 637, 640, 641, 643];

Multimodality explored multimodality in the sense described in Sect. 3 above, i.e. the aggregation and integration of information in multiple languages, media, and coming from different domains, such as: semantic annotation and question answering in the biomedical domain; selecting success criteria in an academic library catalogue; finding similar content in different scenarios on the Web; interactive information retrieval and formative evaluation for medical professionals; microblog summarization, disambiguation and expansion; multimodal music tagging; multi-faceted IR in multimodal domains; ranking in faceted search; domain adaptation; cross-domain vertical search;

prediction of venues in social media; query expansion for speech retrieval; neural networks for medical image classification; vocal assistant-mediated search; life-logging; voice question answering; image forgery; podcast retrieval; detecting eating disorders using text and images; image captioning; argument classification via images; sexism and hate speech identification via multiple media [2, 12, 23, 30, 38, 44, 68, 98, 100, 102, 114, 120, 122, 123, 125, 144, 145, 161, 225, 256, 257, 260, 267, 306, 315, 323, 324, 332, 333, 345, 354, 366, 370, 381, 391, 400, 441, 452, 453, 540, 563, 566, 580, 583, 584, 593, 642];

Information Visualization for Evaluation opened up a brand new area concerned with exploiting information visualization and visual analytics techniques not only for presenting the results of a search system but also for improving interaction with and exploration of experimental outcomes such as exploiting visual analytics for failure analysis; comparing the relative performances of IR systems; and visualization for sentiment analysis; visualization for patterns; data analytics and visualization for system settings [33, 143, 152, 359, 545, 613];

Longitudinal Studies conducted various kinds of medium and long term analyses such as: the scholarly impact of evaluation initiatives; lessons learned in running evaluation activities and in specific domains; performance trends over the years for multilingual information access; and, component-level analysis across different system configurations; reproducibility of tecnology assisted reviews [160, 163, 164, 195, 196, 204, 406, 444, 603, 606, 628].

5 Tracks and Labs

Figure 4 provides an overview of the tracks and labs offered by CLEF over the years; these are briefly summarized below together with some pointers to relevant literature.

Multilingual Text Retrieval (Ad-hoc, 2000–2009) focused on multilingual information retrieval on news corpora, offering monolingual, bilingual and multilingual tasks, and developed a huge collection in 14 European languages [5, 6, 90–93, 95, 154–156, 200];

Domain Specific Cross-Language IR (DS, 2000–2008) dealt with multilingual information retrieval on structured scientific data from the social sciences domain [91–93, 341–343, 497, 498, 585];

Interactive Cross-Language IR (iCLEF, 2001–2006, 2008–2009) explored different aspects of interactive information retrieval on multilingual and multimedia collections, also using gamification techniques [245–249, 320, 448, 449];

Spoken Document/Speech Retrieval (CLEF SR, 2002–2007) investigated speech retrieval and spoken document retrieval in a monolingual and bilingual setting on automatic speech recognition transcripts [190, 191, 303, 450, 469, 626];

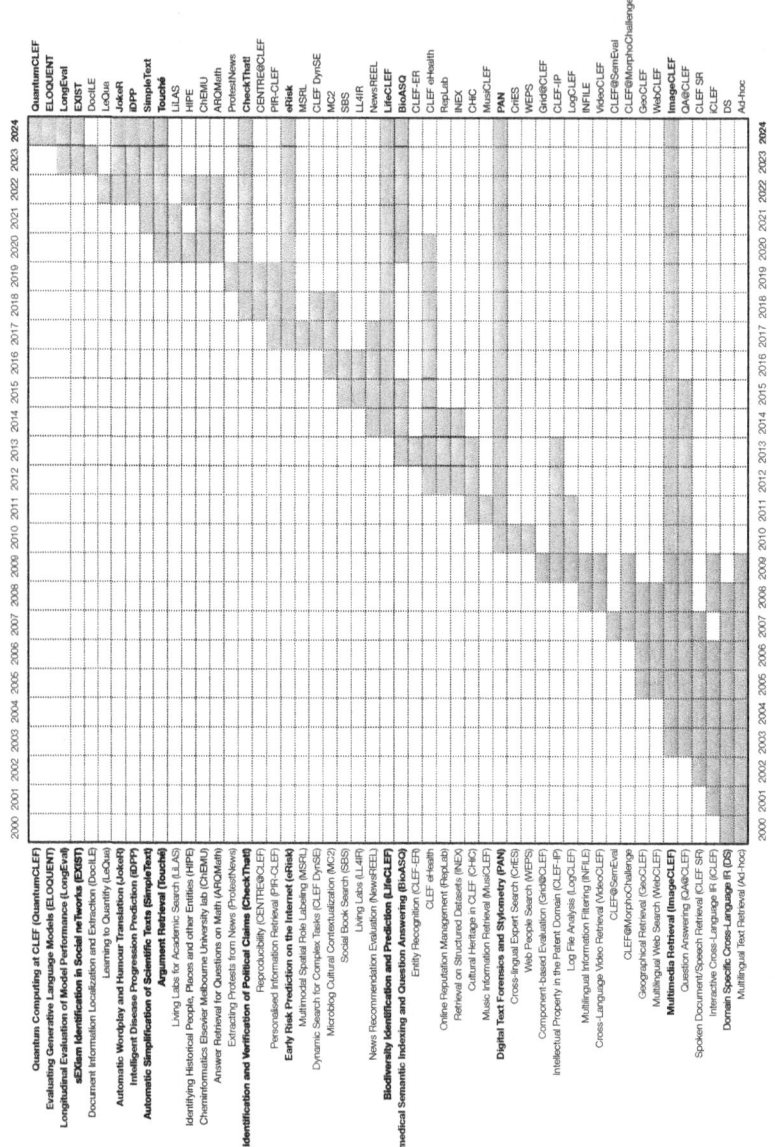

Fig. 4. Labs offered by CLEF over the years. CLEF "Classic" period in green; the CLEF Initiative period in blue; labs active in CLEF 2024 in bold. (Color figure online)

Question Answering (QA@CLEF, 2003–2015) examined several aspects of question answering in a multilingual setting on document collections ranging from news, legal documents, medical documents, linked data [124, 212, 230, 393–395, 417, 471–475, 477, 478, 535, 551, 609, 610, 612];

Multimedia Retrieval (ImageCLEF, 2003–2024) studied the cross-language annotation and retrieval of images to support the advancement of the field of visual media analysis, indexing, classification, and retrieval [41, 112, 113, 126–129, 146–148, 234, 235, 252, 279–286, 307, 373, 404, 422, 423, 425–427, 429, 445–447, 512, 525, 526, 597, 600, 604, 605, 607, 614–616, 639];

Multilingual Web Search (WebCLEF, 2005–2008) addressed multilingual Web search, exploring different faces of navigational queries and known-item search [50, 288, 289, 573];

Geographical Retrieval (GeoCLEF, 2005–2008) evaluated cross-language *Geografic Information Retrieval (GIR)* against search tasks involving both spatial and multilingual aspects [227, 228, 397, 399];

CLEF@SemEval (2007) explored the impact of *Word Sense Disambiguation (WSD)* on multilingual information retrieval [7]; it continued as a sub-task of the Ad Hoc lab in 2008 and 2009;

CLEF@MorphoChallenge (2007–2009) assessed unsupervised morpheme analysis algorithms using information retrieval experiments with the goal of designing statistical machine learning algorithms that discover which morphemes make up words [356–358];

Cross-Language Video Retrieval (VideoCLEF, 2008–2009) aimed at developing and evaluating tasks related to the analysis of and access to multilingual and multimedia content with a special focus on video retrieval [368, 369]; it went on to become the *MediaEval Benchmarking Initiative for Multimedia Evaluation*[7] successful series, dedicated to evaluating new algorithms for multimedia access and retrieval;

Multilingual Information Filtering (INFILE, 2008–2009) experimented with cross-language adaptive filtering systems on news corpora [71, 72];

Log File Analysis (LogCLEF, 2009–2011) investigated the analysis and classification of queries in order to understand search behavior in multilingual contexts and ultimately to improve search systems by offering openly-accessible query logs from search engines and digital libraries [157, 396, 398];

Intellectual Property in the Patent Domain (CLEF-IP, 2009–2013) focused on various aspects of patent search and intellectual property search in a multilingual set using the MAREC collection of patents, gathered from the European Patent Office [504, 507–509, 534];

Component-based Evaluation (Grid@CLEF, 2009) piloted component-based evaluation by allowing participants to exchange the intermediate state of their systems in order to asynchronously compose components coming from different systems and experiment with a larger grid of possibilities [198];

Web People Search (WEPS, 2010) focused on person name ambiguity and person attribute extraction on Web pages and on online reputation management for organizations [26, 43]; the activity continued in the RepLab lab;

---

[7] http://www.multimediaeval.org/.

Cross-lingual Expert Search (CriES, 2010) was run as a brainstorming workshop and addressed the problem of multi-lingual expert search in social media environments [579];

Digital Text Forensics and Stylometry (PAN, 2010–2024) studied plagiarism, authorship attribution, social software misuse, different types of profiling [29, 39, 45, 73–76, 141, 242, 278, 305, 514–517, 519–521, 538, 581, 582];

Music Information Retrieval (MusiCLEF, 2011) was run as a brainstorming workshop to aid the development of novel methodologies for both content-based and contextual-based (e.g. tags, comments, reviews, etc.) access and retrieval of music [454]; this activity has continued as part of MediaEval;

Cultural Heritage in CLEF (CHiC, 2011–2013) promoted systematic and large-scale evaluation of digital libraries and, more in general, cultural heritage information access systems, using the huge Europeana dataset, aggregating information from libraries, museums, and archives [219, 499, 500];

Retrieval on Structured Datasets (INEX, 2012–2014) was a stand-alone initiative pioneering structured and XML retrieval from 2002[8]; it joined forces with CLEF in 2012 to further promote the evaluation of focused retrieval by providing large test collections of structured documents [62, 63, 119, 349, 550, 601, 622];

Online Reputation Management (RepLab, 2012–2014) has been a competitive evaluation exercise for online reputation management systems; the lab focused on the task of monitoring the reputation of entities (companies, organizations, celebrities) on Twitter [24, 25, 27];

CLEF eHealth (2012–2021) focused on *Natural Language Processing (NLP)* and IR for clinical care, such as annotation of entities in a set of narrative clinical reports or retrieval of web pages based on queries generated when reading the clinical reports [236, 237, 241, 327–329, 587, 588, 590, 591];

Entity Recognition (CLEF-ER, 2013) was a brainstorming workshop on the multilingual annotation of named entities and terminology resource acquisition with a focus on entity recognition in biomedical text, in different languages and on a large scale [530];

Biodiversity Identification and Prediction (LifeCLEF, 2014–2024) aimed at evaluating multimedia analysis and retrieval techniques on biodiversity data for species identification, namely images for plants, audio for birds, and video for fishes [290–293, 295–301];

News Recommendation Evaluation (NewsREEL, 2014–2017) focused on evaluation of news recommender systems in real-time by offering access to the APIs of a commercial system [272, 337, 338, 382]

Living Labs (LL4IR, 2015–2016) dealt with evaluation of ranking systems in a live setting with real users in their natural task environments, acting as a proxy between commercial organizations (live environments) and lab participants (experimental systems) [565];

Social Book Search (SBS, 2015–2016) investigated techniques to support users in complex book search tasks that involve more than just a query and results list [347, 348].

---

[8] https://inex.mmci.uni-saarland.de/.

Microblog Cultural Contextualization (MC2, 2016–2017) investigated techniques to support users in complex book search tasks that involve more than just a query and results list [175, 238].

Dynamic Search for Complex Tasks (CLEF DynSE, 2017–2018) promoted the development of both algorithms which interact dynamically with user (or other algorithms) towards solving a task and of evaluation methodologies to quantify their effectiveness [309, 310].

Multimodal Spatial Role Labeling (MSRL, 2017) explored the extraction of spatial information from two information resources that is image and text, which is importa in various applications such as semantic search, question answering, geographical information systems and even in robotics for machine understanding of navigational instructions or instructions for grabbing and manipulating objects [351].

Early Risk Prediction on the Internet (eRisk, 2017–2024) explored the evaluation methodology, effectiveness metrics and practical applications (particularly those related to health and safety) of early risk detection on the Internet [387–390, 459–462].

Personalised Information Retrieval (PIR-CLEF, 2017–2019) provided a framework for evaluation of *Personalized Information Retrieval (PIR)* by developing a methodology for evaluation PIR which enables repeatable experiments to enable the detailed exploration of personal models and their exploitation in IR [464–466].

Reproducibility (CENTRE@CLEF, 2018–2019) run a joint task across CLEF, NTCIR, and TREC on challenging participants to reproduce best results of the most interesting systems submitted in previous editions of CLEF/NTCIR/TREC and to contribute back to the community the additional components and resources developed to reproduce the results [197, 199].

Identification and Verification of Political Claims (CheckThat!, 2018–2024) aimed to foster the development of technology capable of spotting check-worthy claims in English political debates in addition to providing evidence-supported verification of Arabic claims [53, 54, 56, 171, 431–433].

Extracting Protests from News (ProtestNews, 2019) aimed to test and improve state-of-the-art generalizable machine learning and natural language processing methods for text classification and information extraction on English news from multiple countries such as India and China for creating comparative databases of contentious politics events (riots, social movements), i.e. the repertoire of contention that can enable large scale comparative social and political science studies [275].

Biomedical Semantic Indexing and Question Answering (BioASQ, 2020–2024) aimed to push the research frontier towards systems that use the diverse and voluminous information available online to respond directly to the information needs of biomedical scientists [436–440].

Answer Retrieval for Questions on Math (ARQMath, 2020–2022)) aimed to advance math-aware search and the semantic analysis of mathematical notation and texts [401, 402, 638].

Cheminformatics Elsevier Melbourne University lab (ChEMU, 2020–2022)) aimed to provide a unique opportunity for the development of information extraction tools over chemical patents [268, 376, 377].

Identifying Historical People, Places and other Entities (HIPE, 2020 and 2022) aimed to promote named entity recognition and linking in historical documents, with the objective of assessing and advancing the development of robust, adaptable, and transferable named entity processing systems [168, 169].

Living Labs for Academic Search (LiLAS, 2020–2021) aimed to advance online evaluation of academic search systems by improving the search for academic resources like literature (ranging from short bibliographic records to full-text papers), research data, and the interlinking between these resources [558, 561].

Argument Retrieval (Touché, 2020–2024) aimed to establish a collaborative platform for researchers in the field of argument retrieval and to provide tools for developing and evaluating argument retrieval approaches [81–84, 335].

Automatic Simplification of Scientific Texts (SimpleText, 2021–2024) aimed to make science more open and accessible via automatic generation of simplified summaries of scientific documents [173, 178–180].

Intelligent Disease Progression Prediction (iDPP, 2022–2024) aimed to design and develop an evaluation infrastructure for AI algorithms able to: (1) better describe disease mechanisms of Amyotrophic Lateral Sclerosis (ALS) and Multiple Sclerosis (MS); (2) stratify patients according to their phenotype assessed all over the disease evolution; and (3) predict disease progression in a probabilistic, time dependent fashion [78, 189, 253].

Automatic Wordplay and Humour Translation (JokeR, 2022–2024) aimed to to bring together translators and computer scientists to work on an evaluation framework for creative language, including data and metric development, and to foster work on automatic methods for wordplay translation [174, 176, 177].

Learning to Quantify (LeQua, 2022) aimed to allow the comparative evaluation of methods for "learning to quantify" in textual datasets; i.e. methods for training predictors of the relative frequencies of the classes of interest in sets of unlabelled textual documents [181].

Document Information Localization and Extraction (DocILE, 2023) aimed to benchmark Key Information Localization and Extraction (KILE) and Line Item Recognition (LIR) from business documents like invoices [575].

sEXism Identification in Social neTworks (EXIST, 2023–2024) aimed to capture and categorize sexism, from explicit misogyny to other subtle behaviors, in social networks [510, 511].

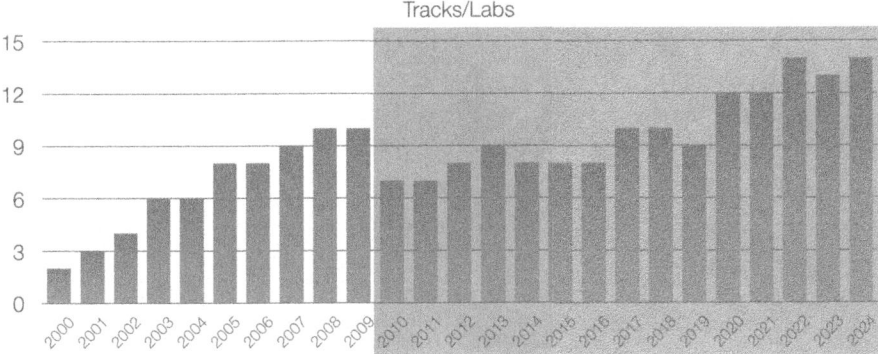

Fig. 5. Number of labs offered by CLEF over the years (CLEF "Classic" period unshaded; the CLEF Initiative period shaded).

Longitudinal Evaluation of Model Performance (LongEval, 2023–2024) aimed to evaluate the temporal persistence of information retrieval systems and text classifiers [20, 21].

Evaluating Generative Language Models (ELOQUENT, 2024) aimed to evaluate the quality of generative language models under various angles like topical competence, veracity and hallucination, and robustness [319].

Quantum Computing at CLEF (QuantumCLEF, 2024) aimed to advance the design and development of *Quantum Computing (QC)* algorithms and, in particular, for *Quantum Annealing (QA)* algorithms, for IR and *Recommender Systems (RS)* by providing access to real quantum computers [468].

Figure 5 shows the number of Labs offered by CLEF over the years. It can be noted how the new mechanism introduced for selecting labs is proving effective in restricting the number of Labs run annually, with an average of about 8 Labs per year which allows CLEF to continue successful activities for more than one cycle, typically three years, but also to introduce new activities every year. Also note that we put a cap on a maximum of 14 labs per edition, in order to avoid dispersion into too many activities. Finally, in the recent years, the lab selection process has become more and more competitive since CLEF is receiving around 20–25 lab proposals per year versus the 14 labs that can be accepted.

Figure 6 shows the number of papers published in the Working Notes papers, which represent the main output of the lab activities. We can observe has the number of papers published has a steady and consistent grow over the years, reaching quite a substantial number of papers in the last 2–3 years as a consequence of the increased participation in CLEF and the number of labs offered.

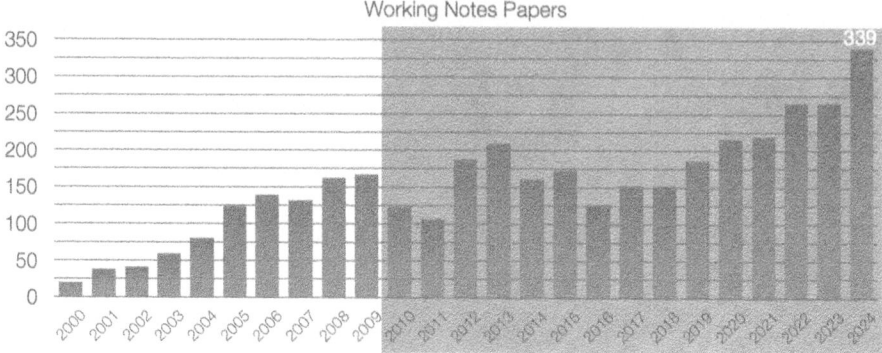

Fig. 6. Papers published in the CLEF Working Notes over the years (CLEF "Classic" period un-shaded; CLEF Initiative period shaded).

6 Impact of CLEF

Assessing the impact of an evaluation activity is a very demanding task and it can be done from multiple points of view, e.g. economic impact, industrial impact, scholarly impact, and so on.

In 2010, TREC conducted a deep study on its economic impact [539]. One of goals of CLEF has been to impact not only academia but also industrial research and society in a broader sense. Indeed, IR research can never be considered only at the theoretical level, clearly the overriding factors are the requirements of society at large. An important step in this direction, which began in "CLEF Classic" with ImageCLEF medical retrieval experiments but has certainly been increasingly reinforced in the "CLEF Initiative", is the involvement of real world user communities. Thus, just to cite a few examples, we have seen collaborations with the intellectual property and patent search domain in CLEF-IP, with health specialists in E-Health, with news portals in the NewsREEL project, until the very recent developments for early risk detection in social media as well as fact checking and trustworthiness.

When it comes to the scientific and scholarly impact, we enter the realm of bibliometrics: *TREC Video Retrieval Evaluation (TRECVID)* conducted a study on its scholarly impact [598] and some steps in this direction have been performed for CLEF as well [32,603,606]. However, analysing the impact of evaluation activities on system performances longitudinally over the years is still a research challenge, even if some attempts have been made for both TREC [40,331] and CLEF [204,206].

Such rigorous studies are beyond the scope of the present report, here we concentrate on identifying rough indicators with respect to the maturity and liveliness of the scientific production originated by CLEF. Therefore, as proxy for a more rigorous scholarly impact study, we can look at some statistics gathered from Google Scholar.

Figure 7 shows the trends of the h5-index (the largest number h such that at least h articles in that publication were cited at least h times each, only those of its articles that were published in the last five complete calendar years) and h5-median (the median number of citations for the articles that make up the h5-index), from 2016 to 2024 taken from Google Scholar Metrics[9]. We can observe a steady increase trend for both indicators, suggesting a positive scholarly impact for the research outcomes of the CLEF community. In Google Scholar Metrics 2024, CLEF achieved h5-index = 47, h5-median = 65; for a comparison: SIGIR h5-index = 103, h5-median = 149; CIKM h5-index = 91, h5-median = 133; ECIR h5-index = 42, h5-median = 60; TREC h5-index = 17, h5-median = 30.

Figure 8 reports the top-20 venues according to Google Scholar Metrics 2024 and CLEF is among them, as it happens since some years. In particular, Fig. 8a reports the top-20 venues for the "Database and Information Systems" category[10] while Fig. 8b reports the top-20 venues according to a query[11] more targeted to IR and RS

```
"AIRS" OR "WWW" OR "information retrieval" OR
"Information and Knowledge Management" OR "SIGIR" OR
"information science and technology" OR "web search" OR
"TOIS" OR "information processing & Management" OR
"Transactions on Knowledge and Data Engineering" OR
"TWEB" OR "digital libraries" OR "cross language" OR
"recommender"
```

prepared by Mark Sanderson with further refinements suggested by Damiano Spina and Martin Tomko.

As far as maturity is concerned, an indicator might be found in publications critically analysing, systematizing, and digesting the achievements, outcomes and experience; this has been done both for TREC [264, 265, 542, 617] and CLEF [97, 421, 481], with a special publication which was prepared for the 20th anniversary of CLEF, as discussed in the next section.

---

[9] Despite the expansion of the acronym for CLEF changed from *Cross-Language Evaluation Forum* to Conference and Labs of the Evaluation Forum in 2010, Google Scholar Metrics still indexes CLEF as "Cross-Language Evaluation Forum". Maybe, this is also due to the fact that also Springer still calls CLEF 'Cross-Language Evaluation Forum"; see https://link.springer.com/conference/clef.

[10] https://scholar.google.com/citations?view_op=top_venues&hl=en&vq=eng_databasesinformationsystems.

[11] https://scholar.google.com/citations?hl=en&view_op=search_venues&vq=%22AIRS%22+OR+%22WWW%22+OR+%22information+retrieval%22+OR+%22Information+and+Knowledge+Management%22+OR+%22SIGIR%22+OR+%22information+science+and+technology%22+OR+%22web+search%22+OR+%22TOIS%22+OR+%22information+processing+%26+Management%22+OR+%22Transactions+on+Knowledge+and+Data+Engineering%22+OR+%22TWEB%22+OR+%22cross+language%22+OR+%22recommender%22+OR+%22digital+libraries%22.

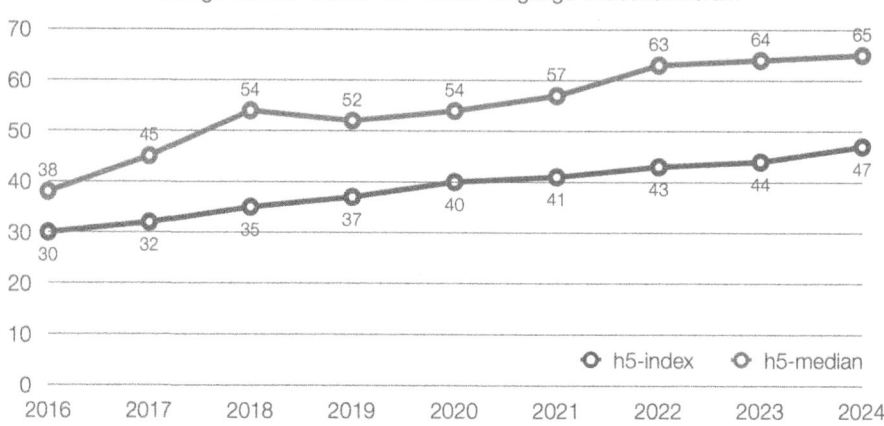

Fig. 7. Google Scholar Metrics for "Cross-Language Evaluation Forum": h5-index (blue) and h5-median (green). (Color figure online)

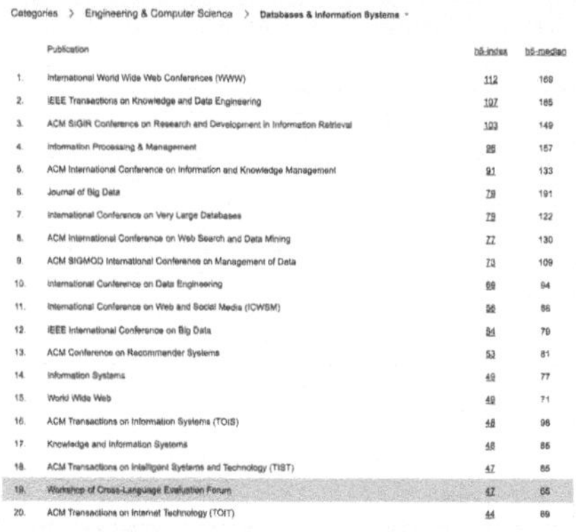

(a) Top-20 venues in the category "Database and Information Systems".

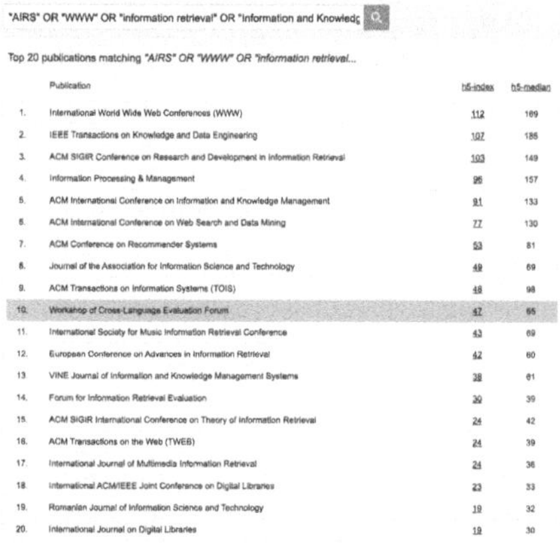

(b) Top-20 venues for a query targeted to IR and RS venues.

Fig. 8. Top venues according to Google Scholar Metrics 2024.

7 The CLEF Book

In occasion of the past 20th anniversary of CLEF, we prepared a book [202] which accounts for the evolution of CLEF over the years, its contribution to the advancement of research in multilingual and multimodal information access, and its perspectives for the future.

In order to do this, the volume is divided into six parts. The first three chapters in Part I "Experimental Evaluation and CLEF" explain what is intended by experimental evaluation and the underlying theory [618], describing how this has been interpreted in CLEF and in other internationally recognized evaluation initiatives [541]. In addition, the introductory chapter illustrates the activity and results of CLEF over the years in some detail [201]. Part II "Evaluation Infrastructure" presents research architectures and infrastructures that have been developed to manage experimental data [9] and to provide evaluation services in CLEF and elsewhere [424, 518].

Parts III, IV and V represent the core of the volume, consisting of a series of chapters presenting some of the most significant evaluation activities in CLEF, ranging from the early multilingual text processing exercises to the later, more sophisticated experiments on multimodal collections in diverse genre and media. In all cases, the focus has not only been on describing "what has been achieved" but most of all on "what has been learnt". Part III "Multilingual and Multimedia Information Retrieval" focuses on multilinguality [555] and the impact of languages on information access [321]; it then addresses multimodality from the perspective of both images [130, 428, 503, 621] and sound and vision [302]. Part IV "Retrieval in New Domains" deals with the medical domain [589], the intellectual property and patent domain [505], the biodiversity domain [294], and the structured data and semantic search domains [308]. Part V "Beyond Retrieval" covers information access tasks other than pure retrieval, namely question answering [476], digital text forensics [537], online reputation management [11], and continuous evaluation and living labs [271].

The final Part VI "Impact and Future Challenges" is dedicated to examining the impact CLEF has had on the research world and to discussing current and future challenges, both academic and industrial. We conduct a proper scholarly impact analysis [367] and we discuss open issues and areas for future development, such as reproducibility and validity [217] and *Visual Analytics (VA)* for experimental evaluation [203]. In particular, the concluding chapter discusses the relevance of IR benchmarking in an industrial setting [316].

8 The CLEF Association

The CLEF Association[12] is an independent no-profit legal entity, established in October 2013 as a result of activity of the PROMISE[13] Network of Excellence which backed CLEF from 2010 to 2013.

The CLEF Association has scientific, cultural and educational objectives and operates in the field of information access systems and their evaluation. Its mission is:

- to promote access to information and use evaluation;
- to foster critical thinking about advancing information access and use from a technical, economic and societal perspective.

Within these two areas of interest, the CLEF Association aims at a better understanding of the use and access to information and how to improve this. The

---

[12] http://www.clef-initiative.eu/association.
[13] http://www.promise-noe.eu/.

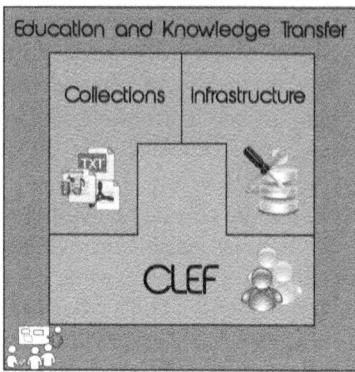

Fig. 9. Pillar activities of the CLEF Association.

two areas of interest stated in the the above mission translate into the following objectives:

- *clustering stakeholders* with multidisciplinary competences and different needs, including academia, industry, education and other societal institutions;
- *facilitating medium/long-term research* in information access and use and its evaluation;
- increasing, *transferring* and applying *expertise*.

As Fig. 9 shows, the CLEF Association pursues its mission and objectives via four pillar activities:

- *CLEF*: sustains and promotes the popular CLEF evaluation series as well as providing support for its coordination, organisation, and running;
- *Collections and Experimental Data*: fosters the adoption and exploitation of large-scale shared experimental collections, makes them available under appropriate conditions and trusted channels, and shares experimental results and scientific data for comparison with state-of-the-art and for reuse;
- *Infrastructure*: supports the adoption and deployment of software and hardware infrastructures which facilitate the experimental evaluation process, the sharing of experimental collections and results, and interaction with and understanding of experimental data;
- *Education and knowledge transfer*: organises educational events, such as summer schools, and knowledge transfer activities, such as workshops, aimed not only at spreading know-how about information access and use but also at raising awareness and stimulating alternative viewpoints about the technical, economic, and societal implications.

In it initial phase, the CLEF Association has been focused mainly on the first pillar, i.e. ensuring the continuity and self-sustainability of CLEF. CLEF 2014 was the first edition of CLEF not supported by a main European project, but run on a totally volunteer basis with only the support of the CLEF association membership fees paid by its multidisciplinary research community.

Moreover, the CLEF association plans to continue the already initiated activities for promoting and developing shared infrastructures and formats in IR evaluation [574] by also joining forces with relevant stakeholders in the fields as well as stimulating and contributing critical thinking about large-scale evaluation initiative and IR evaluation more in general.

Support for the Central Coordination of CLEF

CLEF 2000 and 2001 were supported by the European Commission under the Information Society Technologies programme and within the framework of the DELOS Network of Excellence for Digital Libraries (contract no. IST-1999-12262).

CLEF 2002 and 2003 were funded as an independent project (contract no. IST-2000-31002) under the 5th Framework Programme of the European Commission.

CLEF 2004 to 2007 were sponsored by the DELOS Network of Excellence for Digital Libraries (contract no. G038-507618) under the 6th Framework Programme of the European Commission.

Under the 7th Framework Programme of the European Commission, CLEF 2008 and 2009 were supported by TrebleCLEF Coordination Action (contract no. 215231) and CLEF 2010 to 2013 were funded by the PROMISE Network of Excellence (contract no. 258191).

CLEF 2011 to 2014 also received support from the ELIAS network (contract no. 09-RNP-085) of the European Science Foundation (ESF).

Over the years CLEF has also attracted industrial sponsorship: from 2010 onwards, CLEF has received the support of Google, Microsoft, Yandex, Xerox, Celi as well as publishers in the field such as Springer and Now Publishers.

Note that, beyond receiving the support of all the volunteer work of its community, CLEF tracks and labs have often received the support of many other projects and organisations; unfortunately, it is impossible to list them all here.

Acknowledgements. CLEF would not be possible without all the effort, enthusiasm, and passion of its community: lab organizers, lab participants, and attendees are the core and the real success of CLEF.

Many friends and colleagues – too many to mention them all but I sincerely thank all of them – have shared with me this journey through CLEF and their work, passion, ideas, expertise and wisdom have shaped what CLEF is today.

However, all of this would have not even been possible without Carol Peters, who established CLEF back in 2000, made it grown over the years with constant care, and put into CLEF her secret ingredient which makes it so special: a very friendly environment where everybody feels to be welcome and comfortable in sharing ideas and contributions. Carol has had the generosity of sharing her experience with me and teaching me a lot about how to run an evaluation initiative and grow a healthy research community: I will never thank her enough for this.

References

1. Abnar, S., Dehghani, M., Shakery, A.: Meta text aligner: text alignment based on predicted plagiarism relation. In: Mothe et al. [419], pp. 193–199 (2015)
2. Achilles, L., et al.: "Meanspo Please, I Want to Lose Weight": a characterization study of meanspiration content on tumblr based on images and texts. In: Barrón-Cedeño et al. [55], pp. 3–17 (2022)
3. Adams, J., Bedrick, S.: Automatic indexing of journal abstracts with latent semantic analysis. In: Mothe et al. [419], pp. 200–208 (2015)
4. Afzal, Z., Akhondi, S.A., van Haagen, H.H.H.B.M., van Mulligen, E.M., Kors, J.A.: Concept recognition in French biomedical text using automatic translation. In: Fuhr et al. [218], pp. 162–173 (2016)
5. Agirre, E., Di Nunzio, G.M., Ferro, N., Mandl, T., Peters, C.: CLEF 2008: ad hoc track overview. In: Peters et al. [487], pp. 15–37 (2008)
6. Agirre, E., Di Nunzio, G.M., Mandl, T., Otegi, A.: CLEF 2009 ad hoc track overview: robust-WSD task. In: Peters et al. [488], pp. 36–49 (2009)
7. Agirre, E., de Lacalle, O.L., Magnini, B., Otegi, A., Rigau, G., Vossen, P.: SemEval-2007 Task 01: evaluating WSD on cross-language information retrieval. In: Peters et al. [494], pp. 908–917 (2007)
8. Agosti, M., Di Buccio, E., Ferro, N., Masiero, I., Peruzzo, S., Silvello, G.: DIRECTions: design and specification of an ir evaluation infrastructure. In: Catarci et al. [117], pp. 88–99 (2012)
9. Agosti, M., Di Nunzio, G.M., Ferro, N., Silvello, G.: An innovative approach to data management and curation of experimental data generated through IR test collections. In: Ferro and Peters [202], pp. 105–122 (2019)
10. Agosti, M., Ferro, N., Peters, C., de Rijke, M., Smeaton, A. (eds.): CLEF 2010. LNCS, vol. 6360. Springer, Heidelberg (2010). https://doi.org/10.1007/978-3-642-15998-5
11. Carrillo-de Albornoz, J., Gonzalo, J., Amigó, E.: RepLab: an evaluation campaign for online monitoring systems. In: Ferro and Peters [202] (2019)
12. Alexander, A., Mars, M., Tingey, J.C., Yu, H., Backhouse, C., Reddy, S., Karlgren, J.: Audio features, precomputed for podcast retrieval and information access experiments. In: Candan et al. [104], pp. 3–14 (2021)
13. Alfalahi, A., Eriksson, G., Sneiders, E.: Shadow answers as an intermediary in email answer retrieval. In: Mothe et al. [419], pp. 209–214 (2015)
14. Alhamzeh, A., Bouhaouel, M., Egyed-Zsigmond, E., Mitrović, J., Brunie, L., Kosch, H.: Query expansion, argument mining and document scoring for an efficient question answering system. In: Barrón-Cedeño et al. [55], pp. 162–174 (2022)
15. Alhamzeh, A., Mukhopadhaya, S., Hafid, S., Bremard, A., Egyed-Zsigmond, E.H.K., Brunie, L.: A hybrid approach for stock market prediction using financial news and stocktwits. In: Candan et al. [104], pp. 15–26 (2021)
16. Alharbi, A., Stevenson, M.: Improving ranking for systematic reviews using query adaptation. In: Crestani et al. [138], pp. 141–148 (2019)
17. Ali, E., Caputo, A., Lawless, S., Conlan, O.: Dataset creation framework for personalized type-based facet ranking tasks evaluation. In: Candan et al. [104], pp. 27–39 (2021)
18. Aliannejadi, M., Faggioli, G., Ferro, N., Vlachos, M. (eds.): CLEF 2023 Working Notes. CEUR Workshop Proceedings (CEUR-WS.org) (2023). ISSN 1613-0073. https://ceur-ws.org/Vol-3497/

19. Aliannejadi, M., Huibers, T., Landoni, M., Murgia, E., Pera, M.S.: The effect of prolonged exposure to online education on a classroom search companion. In: Barrón-Cedeño et al. [55], pp. 62–78 (2022)
20. Alkhalifa, R., et al.: Overview of the CLEF-2023 longeval lab on longitudinal evaluation of model performance. In: Arampatzis et al. [36], pp. 440–458 (2023)
21. Alkhalifa, R., et al.: Overview of the CLEF 2024 LongEval lab on longitudinal evaluation of model performance. In: Goeuriot et al. [240] (2024)
22. Alkhawaldeh, R.S., Jose, J.M.: Experimental study on semi-structured peer-to-peer information retrieval network. In: Mothe et al. [419], pp. 3–14 (2015)
23. Almeida, J.R., Fajarda, O., Oliveira, J.L.: File forgery detection using a weighted rule-based system. In: Arampatzis et al. [37], pp. 85–96 (2020)
24. Amigó, E., et al.: Overview of RepLab 2013: evaluating online reputation monitoring systems. In: Forner et al. [210], pp. 333–352 (2013)
25. Amigó, E., et al.: Overview of RepLab 2014: author profiling and reputation dimensions for online reputation management. In: Kanoulas et al. [311], pp. 307–322 (2014)
26. Amigó, E., Artiles, J., Gonzalo, J., Spina, D., Liu, B., Corujo, A.: WePS3 evaluation campaign: overview of the on-line reputation management task. In: Braschler et al. [96] (2010)
27. Amigó, E., Corujo, A., Gonzalo, J., Meij, E., de Rijke, M.: Overview of RepLab 2012: evaluating online reputation management systems. In: Forner et al. [209] (2012)
28. Amigó, E., Gonzalo, J., Verdejo, M.F.: A comparison of evaluation metrics for document filtering. In: Forner et al. [208], pp. 38–49 (2011)
29. Anderka, M., Stein, B.: Overview of the 1th international competition on quality flaw prediction in wikipedia. In: Forner et al. [209] (2012)
30. Andrearczyk, V., Müller, H.: Deep multimodal classification of image types in biomedical journal figures. In: Bellot et al. [64], pp. 3–14 (2018)
31. Angelini, M., Ferro, N., Järvelin, K., Keskustalo, H., Pirkola, A., Santucci, G., Silvello, G.: Cumulated relative position: a metric for ranking evaluation. In: Catarci et al. [117], pp. 112–123 (2012)
32. Angelini, M., et al.: Measuring and analyzing the scholarly impact of experimental evaluation initiatives. Procedia Comput. Sci. **38**, 133–137 (2014)
33. Angelini, M., Ferro, N., Santucci, G., Silvello, G.: Improving ranking evaluation employing visual analytics. In: Forner et al. [210], pp. 29–40 (2013)
34. Antici, F., Bolognini, K., Inajetovic, M.A., Ivasiuk, B., Galassi, A., Ruggeri, F.: SubjectivITA: an Italian corpus for subjectivity detection in newspapers. In: Candan et al. [104], pp. 40–52 (2021)
35. Antunes, H., Lopes, C.: Analyzing the adequacy of readability indicators to a non-English language. In: Crestani et al. [138], pp. 149–155 (2019)
36. Mothe, J., et al. (eds.): CLEF 2015. LNCS, vol. 14163. Springer, Cham (2015). https://doi.org/10.1007/978-3-319-24027-5
37. Arampatzis, A., et al. (eds.): Experimental IR Meets Multilinguality, Multimodality, and Interaction. Proceedings of the Eleventh International Conference of the CLEF Association (CLEF 2020). Lecture Notes in Computer Science (LNCS) 12260, Springer, Heidelberg (2020). https://doi.org/10.1007/978-3-031-13643-6
38. Arcos, I., Rosso, P.: Sexism identification on tiktok: a multimodal AI approach with text, audio, and video. In: Goeuriot et al. [239] (2024)
39. Argamon, S., Juola, P.: Overview of the international authorship identification competition at PAN-2011. In: Petras et al. [501] (2011)

40. Armstrong, T.G., Moffat, A., Webber, W., Zobel, J.: Improvements that don't add up: ad-hoc retrieval results since 1998. In: Cheung, D.W.L., Song, I.Y., Chu, W.W., Hu, X., Lin, J.J. (eds.) Proceedings 18th International Conference on Information and Knowledge Management (CIKM 2009), pp. 601–610. ACM Press, New York (2009)
41. Arni, T., Clough, P., Sanderson, M., Grubinger, M.: Overview of the ImageCLEF-photo 2008 photographic retrieval task. In: Peters et al. [487], pp. 500–511 (2008)
42. Arora, P., Foster, J., Jones, G.J.F.: Query expansion for sentence retrieval using pseudo relevance feedback and word embedding. In: Jones et al. [304], pp. 97–103 (2017)
43. Artiles, J., Borthwick, A., Gonzalo, J., Sekine, S., Amigó, E.: WePS-3 evaluation campaign: overview of the web people search clustering and attribute extraction tasks. In: Braschler et al. [96] (2010)
44. Avila, J., Rodrigo, A., Centeno, R.: Best of Touché 2023 task 4: testing data augmentation and label propagation for multilingual multi-target stance detection. In: Goeuriot et al. [239] (2023)
45. Ayele, A.A., et al.: Overview of PAN 2024: multi-author writing style analysis, multilingual text detoxification, oppositional thinking analysis, and generative AI authorship verification. In: Goeuriot et al. [240] (2024)
46. Azarbonyad, H., Marx, M.: How many labels? determining the number of labels in multi-label text classification. In: Crestani et al. [138], pp. 156–163 (2019)
47. Azarbonyad, H., Saan, F., Dehghani, D., Marx, M., Kamps, J.: Are topically diverse documents also interesting? In: Mothe et al. [419], pp. 215–221 (2015)
48. Azarbonyad, H., Shakery, A., Faili, H.: Exploiting multiple translation resources for english-persian cross language information retrieval. In: Forner et al. [210], pp. 93–99 (2013)
49. Azzopardi, L., Balog, K.: Towards a living lab for information retrieval research and development - a proposal for a living lab for product search tasks. In: Forner et al. [208], pp. 26–37 (2011)
50. Balog, K., Azzopardi, L., Kamps, J., de Rijke, M.: Overview of WebCLEF 2006. In: Peters et al. [485], pp. 803–819 (2006)
51. Balog, K., Cappellato, L., Ferro, N., Macdonald, C. (eds.): CLEF 2016 Working Notes. CEUR Workshop Proceedings (CEUR-WS.org) (2016). ISSN 1613-0073. http://ceur-ws.org/Vol-1609/ (2016)
52. Baradaran Hashemi, H., Shakery, A., Feili, H.: Creating a Persian-English comparable corpus. In: Agosti et al. [10], pp. 27–39 (2010)
53. Barrón-Cedeño, A., et al.: Overview of the CLEF–2023 CheckThat! lab on checkworthiness, subjectivity, political bias, factuality, and authority of news articles and their source. In: Arampatzis et al. [36], pp. 251–275 (2023)
54. Barrón-Cedeño, A., et al.: Overview of the CLEF-2024 CheckThat! lab: checkworthiness, subjectivity, persuasion, roles, authorities, and adversarial robustness. In: Goeuriot et al. [240] (2024)
55. Barron-Cedeno, A., et al. (eds.): CLEF 2015. LNCS, vol. 13390. Springer, Cham (2015). https://doi.org/10.1007/978-3-319-24027-5
56. Barrón-Cedeño, A., et al.: Overview of CheckThat! 2020: automatic identification and verification of claims in social media. In: Arampatzis et al. [37], pp. 215–236 (2020)
57. Basile, A., Caselli, T.: Protest event detection: when task-specific models outperform an event-driven method. In: Arampatzis et al. [37], pp. 97–111 (2020)

58. Basile, A., Dwyer, G., Medvedeva, M., Rawee, J., Haagsma, H., Nissim, M.: Simply the best: minimalist system trumps complex models in author profiling. In: Bellot et al. [64], pp. 143–156 (2018)
59. Baudis, P., Sedivý, J.: Modeling of the question answering task in the YodaQA system. In: Mothe et al. [419], pp. 222–228 (2015)
60. Bauer, C., et al.: Report on the Dagstuhl seminar on frontiers of information access experimentation for research and education. SIGIR Forum **57**(1), 7:1–7:28 (2023)
61. Bauer, C., Carterette, B.A., Ferro, N., Fuhr, N., Faggioli, G. (eds.): Report from Dagstuhl Seminar 23031: Frontiers of Information Access Experimentation for Research and Education. Dagstuhl Reports, vol. 13, no. 1. Schloss Dagstuhl–Leibniz-Zentrum für Informatik, Germany (2023)
62. Bellot, P., et al.: Overview of INEX 2014. In: Kanoulas et al. [311], pp. 212–228 (2014)
63. Bellot, P., et al.: Overview of INEX 2013. In: Forner et al. [210], pp. 269–281 (2013)
64. Bellot, P., et al. (eds.): CLEF 2018. LNCS, vol. 11018. Springer, Cham (2018). https://doi.org/10.1007/978-3-319-98932-7
65. Beloborodov, A., Braslavski, P., Driker, M.: Towards automatic evaluation of health-related CQA data. In: Kanoulas et al. [311], pp. 7–18 (2014)
66. Bensalem, I., Rosso, P., Chikhi, S.: A new corpus for the evaluation of arabic intrinsic plagiarism detection. In: Forner et al. [210], pp. 53–58 (2013)
67. Berendsen, R., Tsagkias, M., de Rijke, M., Meij, E.: Generating pseudo test collections for learning to rank scientific articles. In: Catarci et al. [117], pp. 42–53 (2012)
68. Berlanga Llavori, R., Jimeno-Yepes, A., Pérez Catalán, M., Rebholz-Schuhmann, D.: Context-dependent semantic annotation in cross-lingual biomedical resources. In: Forner et al. [210], pp. 120–123 (2013)
69. Berlanga Llavori, R., Pérez Catalán, M., Museros Cabedo, L., Forcada, R.: Semantic discovery of resources in cloud-based PACS/RIS systems. In: Forner et al. [210], pp. 167–178 (2013)
70. Bernard, G., Suire, C., Faucher, C., Doucet, A., Rosso, P.: Tracking news stories in short messages in the era of infodemic. In: Barrón-Cedeño et al. [55], pp. 18–32 (2022)
71. Besançon, R., Chaudiron, S., Mostefa, D., Hamon, O., I., T., Choukri, K.: Overview of CLEF 2008 INFILE pilot track. In: Peters et al. [487], pp. 939–946 (2008)
72. Besançon, R., Chaudiron, S., Mostefa, D., I., T., Choukri, K., Laïb, M.: Information filtering evaluation: overview of CLEF 2009 INFILE track. In: Peters et al. [488], pp. 342–353 (2009)
73. Bevendorff, J., et al.: Overview of PAN 2023: authorship verification, multi-author writing style analysis, profiling cryptocurrency influencers, and trigger detection. In: Arampatzis et al. [36], pp. 459–481 (2023)
74. Bevendorff, J., et al.: Overview of PAN 2021: authorship verification, profiling hate speech spreaders on twitter, and style change detection. In: Candan et al. [104], pp. 419–431 (2021)
75. Bevendorff, J., et al.: Overview of PAN 2022: authorship verification, profiling irony and stereotype spreaders, and style change detection. In: Barrón-Cedeño et al. [55], pp. 382–394 (2022)

76. Bevendorff, J., et al.: Overview of PAN 2020: authorship verification, celebrity profiling, profiling fake news spreaders on twitter, and style change detection. In: Arampatzis et al. [37], pp. 372–383 (2020)
77. Bhaskar, P., Bandyopadhyay, S.: Language independent query focused snippet generation. In: Catarci et al. [117], pp. 138–140 (2012)
78. Birolo, G., et al.: Intelligent disease progression prediction: overview of iDPP@CLEF 2024. In: Goeuriot et al. [240] (2024)
79. Blinov, V., Mishchenko, K., Bolotova, V., Braslavski, P.: A pinch of humor for short-text conversation: an information retrieval approach. In: Jones et al. [304], pp. 3–15 (2017)
80. Boenninghoff, B., Kolossa, D., Nickel, R.M.: Self-calibrating neural-probabilistic model for authorship verification under covariate shift. In: Candan et al. [104], pp. 145–158 (2021)
81. Bondarenko, A., et al.: Overview of Touché 2020: argument retrieval - extended abstract. In: Arampatzis et al. [37], pp. 384–395 (2020)
82. Bondarenko, A., et al.: Overview of Touché 2023: argument and causal retrieval. In: Arampatzis et al. [36], pp. 507–530 (2023)
83. Bondarenko, A., et al.: Overview of Touché 2022: argument retrieval. In: Barrón-Cedeño et al. [55], pp. 311–336 (2022)
84. Bondarenko, A., et al.: Overview of Touché 2021: argument retrieval. In: Candan et al. [104], pp. 450–467 (2021)
85. Borchert, F., Llorca, I., Schapranow, M.P.: Cross-lingual candidate retrieval and re-ranking for biomedical entity linking. In: Arampatzis et al. [36], pp. 135–147 (2023)
86. Bordea, G., Thiessard, F., Hamon, T., Mougin, F.: Automatic query selection for acquisition and discovery of food-drug interactions. In: Bellot et al. [64], pp. 115–120 (2018)
87. Borri, F., Nardi, A., Peters, C., Ferro, N. (eds.): CLEF 2008 Working Notes. CEUR Workshop Proceedings (CEUR-WS.org) (2008). ISSN 1613-0073. http://ceur-ws.org/Vol-1174/
88. Borri, F., Nardi, A., Peters, C., Ferro, N. (eds.): CLEF 2009 Working Notes. CEUR Workshop Proceedings (CEUR-WS.org) (2009). ISSN 1613-0073. http://ceur-ws.org/Vol-1175/
89. Borri, F., Peters, C., Ferro, N. (eds.): CLEF 2004 Working Notes. CEUR Workshop Proceedings (CEUR-WS.org) (2004). ISSN 1613-0073. http://ceur-ws.org/Vol-1170/
90. Braschler, M.: CLEF 2000 – overview of results. In: Peters [479], pp. 89–101 (2000)
91. Braschler, M.: CLEF 2001 – overview of results. In: Peters et al. [482], pp. 9–26 (2001)
92. Braschler, M.: CLEF 2002 – overview of results. In: Peters et al. [483], pp. 9–27 (2002)
93. Braschler, M.: CLEF 2003 – overview of results. In: Peters et al. [484], pp. 44–63 (2003)
94. Braschler, M., et al.: A PROMISE for experimental evaluation. In: Agosti et al. [10], pp. 140–144 (2010)
95. Braschler, M., Di Nunzio, G.M., Ferro, N., Peters, C.: CLEF 2004: ad hoc track overview and results analysis. In: Peters et al. [486], pp. 10–26 (2004)
96. Braschler, M., Harman, D.K., Pianta, E., Ferro, N. (eds.): CLEF 2010 Working Notes. CEUR Workshop Proceedings (CEUR-WS.org) (2010). ISSN 1613-0073. http://ceur-ws.org/Vol-1176/

97. Braschler, M., Peters, C.: Cross-language evaluation forum: objectives, results, achievements. Inf. Retr. **7**(1–2), 7–31 (2004)
98. Braslavski, P., Verberne, S., Talipov, R.: Show me how to tie a tie: evaluation of cross-lingual video retrieval. In: Fuhr et al. [218], pp. 3–15 (2016)
99. Bucur, A.M.: Leveraging LLM-generated data for detecting depression symptoms on social media. In: Goeuriot et al. [239] (2024)
100. Budíková, P., Batko, M., Botorek, J., Zezula, P.: Search-based image annotation: extracting semantics from similar images. In: Mothe et al. [419], pp. 327–339 (2015)
101. Buraya, K., Farseev, A., Filchenkov, A.: Multi-view personality profiling based on longitudinal data. In: Bellot et al. [64], pp. 15–27 (2018)
102. Buscaldi, D., Rosso, P., Chulvi, B., Wang, T.: Classification of social media hateful screenshots inciting violence and discrimination. In: Goeuriot et al. [239] (2024)
103. Cabanac, G., Hubert, G., Boughanem, M., Chrisment, C.: Tie-breaking bias: effect of an uncontrolled parameter on information retrieval evaluation. In: Agosti et al. [10], pp. 112–123 (2010)
104. Cand, I., et al. (eds.): CLEF 2021. LNCS, vol. 12880. Springer, Cham (2021). https://doi.org/10.1007/978-3-030-85251-1
105. Capari, A., Azarbonyad, H., Tsatsaronis, G., Afzal, Z., Dunham, J., Kamps, J.: Knowledge acquisition passage retrieval: corpus, ranking models, and evaluation resources. In: Goeuriot et al. [239] (2024)
106. Cappellato, L., Eickhoff, C., Ferro, N., Névéol, A. (eds.): CLEF 2020 Working Notes. CEUR Workshop Proceedings (CEUR-WS.org) (2020). ISSN 1613-0073. http://ceur-ws.org/Vol-2696/
107. Cappellato, L., Ferro, N., Goeuriot, L., Mandl, T. (eds.): CLEF 2017 Working Notes. CEUR Workshop Proceedings (CEUR-WS.org) (2017). ISSN 1613-0073. http://ceur-ws.org/Vol-1866/
108. Cappellato, L., Ferro, N., Halvey, M., Kraaij, W. (eds.): CLEF 2014 Working Notes. CEUR Workshop Proceedings (CEUR-WS.org) (2014). ISSN 1613-0073. http://ceur-ws.org/Vol-1180/
109. Cappellato, L., Ferro, N., Jones, G.J.F., SanJuan, E. (eds.): CLEF 2015 Working Notes. CEUR Workshop Proceedings (CEUR-WS.org) (2015). ISSN 1613-0073. http://ceur-ws.org/Vol-1391/
110. Cappellato, L., Ferro, N., Losada, D.E., Müller, H. (eds.): CLEF 2019 Working Notes. CEUR Workshop Proceedings (CEUR-WS.org) (2019). ISSN 1613-0073, http://ceur-ws.org/Vol-2380/
111. Cappellato, L., Ferro, N., Nie, J.Y., Soulier, L. (eds.): CLEF 2018 Working Notes. CEUR Workshop Proceedings (CEUR-WS.org) (2018). ISSN 1613-0073. http://ceur-ws.org/Vol-2125/
112. Caputo, B., et al.: ImageCLEF 2014: overview and analysis of the results. In: Kanoulas et al. [311], pp. 192–211 (2014)
113. Caputo, B., et al.: ImageCLEF 2013: the vision, the data and the open challenges. In: Forner et al. [210], pp. 250–268 (2013)
114. Cardoso, R., Marinho, Z., Mendes, A., S., M.: Priberam at MESINESP multi-label classification of medical texts task. In: Candan et al. [104], pp. 159–172 (2021)
115. Cassani, L., Livraga, G., Viviani, M.: Assessing document sanitization for controlled information release and retrieval in data marketplaces. In: Goeuriot et al. [239] (2024)
116. Cassidy, T., Ji, H., Deng, H., Zheng, J., Han, J.: Analysis and refinement of cross-lingual entity linking. In: Catarci et al. [117], pp. 1–12 (2012)

117. Catarci, T., Forner, P., Hiemstra, D., Peñas, A., Santucci, G. (eds.): CLEF 2012. LNCS, vol. 7488. Springer, Heidelberg (2012). https://doi.org/10.1007/978-3-642-33247-0
118. Chaa, M., Nouali, O., Bellot, P.: Combining tags and reviews to improve social book search performance. In: Bellot et al. [64], pp. 64–75 (2018)
119. Chappell, T., Geva, S.: Overview of the INEX 2012 relevance feedback track. In: Forner et al. [209] (2012)
120. Chidlovskii, B., Csurka, G., Clinchant, S.: Evaluating stacked marginalised denoising autoencoders within domain adaptation methods. In: Mothe et al. [419], pp. 15–27 (2015)
121. Chikka, V.R., Mariyasagayam, N., Niwa, Y., Karlapalem, K.: Information extraction from clinical documents: towards disease/disorder template filling. In: Mothe et al. [419], pp. 389–401 (2015)
122. Chuklin, A., Severyn, A., Trippas, J., Alfonseca, E., Silen, H., Spina, D.: Using audio transformations to improve comprehension in voice question answering. In: Crestani et al. [138], pp. 164–170 (2019)
123. Chulif, S., Loong Chang, Y.: Herbarium-field triplet network for cross-domain plant identification. In: Candan et al. [104], pp. 173–188 (2021)
124. Cimiano, P., Lopez, V., Unger, C., Cabrio, E., Ngonga Ngomo, A.C., Walter, S.: Multilingual question answering over linked data (QALD-3): lab overview. In: Forner et al. [210], pp. 321–332 (2013)
125. Clough, P., Goodale, P.: Selecting success criteria: experiences with an academic library catalogue. In: Forner et al. [209], pp. 59–70 (2013)
126. Clough, P., Grubinger, M., Deselaers, T., Hanbury, A., Müller, H.: Overview of the ImageCLEF 2006 photographic retrieval and object annotation tasks. In: Peters et al. [485], pp. 223–256 (2007)
127. Clough, P., et al.: The CLEF 2005 cross-language image retrieval track. In: Peters et al. [493], pp. 535–557 (2005)
128. Clough, P., Müller, H., Sanderson, M.: The CLEF 2004 cross-language image retrieval track. In: Peters et al. [486], pp. 597–613 (2004)
129. Clough, P., Sanderson, M.: The CLEF 2003 cross language image retrieval track. In: Peters et al. [484], pp. 581–593 (2003)
130. Clough, P., Tsikrika, T.: Multi-lingual retrieval of pictures in ImageCLEF. In: Ferro and Peters [202] (2019)
131. Clough, P., Willett, P., Lim, J.: Unfair means: use cases beyond plagiarism. In: Mothe et al. [419], pp. 229–234 (2015)
132. Coello-Guilarte, D.L., Ortega Mendoza, R.M., Villaseñor-Pineda, L., Montes-y Gómez, M.: Cross-lingual depression detection in Twitter using bilingual word-level alignment. In: Crestani et al. [138], pp. 49–61 (2019)
133. Collovini, S., de Bairros P. Filho, M., Vieira, R.: Analysing the role of representation choices in portuguese relation extraction. In: Mothe et al. [419], pp. 105–116 (2015)
134. Conlan, O., Fraser, K., Kelly, L., Yousuf, B.: A user modeling shared challenge proposal. In: Crestani et al. [138], pp. 171–177 (2019)
135. Corbara, S., Chulvi, B., Rosso, P., Moreo, A.: Rhythmic and psycholinguistic features for authorship tasks in the Spanish parliament: evaluation and analysis. In: Barrón-Cedeño et al. [55], pp. 79–92 (2022)
136. Corezola Pereira, R., Pereira Moreira, V., Galante, R.: A new approach for cross-language plagiarism analysis. In: Agosti et al. [10], pp. 15–26 (2010)

137. Cossu, J.V., Ferreira, E., Janod, K., Gaillard, J., El-Bèze, M.: NLP-based classifiers to generalize expert assessments in e-reputation. In: Mothe et al. [419], pp. 340–351 (2015)
138. Crestani, F., et al. (eds.): CLEF 2019. LNCS, vol. 11696. Springer, Cham (2019). https://doi.org/10.1007/978-3-030-28577-7
139. Custódio, J., Paraboni, I.: An ensemble approach to cross-domain authorship attribution. In: Crestani et al. [138], pp. 201–212 (2019)
140. Dadashkarimi, J., Esfahani, H.N., Faili, H., Shakery, A.: SS4MCT: a statistical stemmer for morphologically complex texts. In: Fuhr et al. [218], pp. 201–207 (2016)
141. Daelemans, W., et al.: Overview of PAN 2019: bots and gender profiling, celebrity profiling, cross-domain authorship attribution and style change detection. In: Crestani et al. [138], pp. 402–416 (2019)
142. Dehghani, M., Azarbonyad, H., Kamps, J., Marx, M.: Two-way parsimonious classification models for evolving hierarchies. In: Fuhr et al. [218], pp. 69–82 (2016)
143. Déjean, S., Mothe, J., Ullah, M.Z.: Studying the variability of system setting effectiveness by data analytics and visualization. In: Crestani et al. [138], pp. 62–74 (2019)
144. Del Moro, M., Tudosie, S.C., Vannoni, F., Galassi, A., Ruggeri, F.: Inception models for fashion image captioning: an extensive study on multiple datasets. In: Arampatzis et al. [36], pp. 3–14 (2023)
145. Deneu, B., Servajean, M., Botella, C., Joly, A.: Evaluation of deep species distribution models using environment and co-occurrences. In: Crestani et al. [138], pp. 213–225 (2019)
146. Deselaers, T., Deserno, T.M.: Medical image annotation in ImageCLEF 2008. In: Peters et al. [487], pp. 523–530 (2008)
147. Deselaers, T., Hanbury, A.: The visual concept detection task in ImageCLEF 2008. In: Peters et al. [487], pp. 531–538 (2008)
148. Deselaers, T., et al.: Overview of the ImageCLEF 2007 object retrieval task. In: Peters et al. [494], pp. 445–471 (2007)
149. Devezas, J., Nunes, S.: Index-based semantic tagging for efficient query interpretation. In: Fuhr et al. [218], pp. 208–213 (2016)
150. Dhanani, F., Rafi, M., Atif Tahir, M.: Humour translation with transformers. In: Arampatzis et al. [36], pp. 148–160 (2023)
151. Dhrangadhariya, A., Aguilar, G., Solorio, T., Hilfiker, R., Müller, H.: End-to-End fine-grained neural entity recognition of patients, interventions, outcomes. In: Candan et al. [104], pp. 65–77 (2021)
152. Di Buccio, E., Dussin, M., Ferro, N., Masiero, I., Santucci, G., Tino, G.: To Rerank or to Re-query: can visual analytics solve this dilemma? In: Forner et al. [208], pp. 119–130 (2011)
153. Di Nunzio, G.M.: A study on a stopping strategy for systematic reviews based on a distributed effort approach. In: Arampatzis et al. [37], pp. 112–123 (2020)
154. Di Nunzio, G.M., Ferro, N., Jones, G.J.F., Peters, C.: CLEF 2005: ad hoc track overview. In: Peters et al. [493], pp. 11–36 (2005)
155. Di Nunzio, G.M., Ferro, N., Mandl, T., Peters, C.: CLEF 2006: ad hoc track overview. In: Peters et al. [487], pp. 21–34 (2006)
156. Di Nunzio, G.M., Ferro, N., Mandl, T., Peters, C.: CLEF 2007: ad hoc track overview. In: Peters et al. [494], pp. 13–32 (2007)
157. Di Nunzio, G.M., Leveling, J., Mandl, T.: LogCLEF 2011 multilingual log file analysis: language identification, query classification, and success of a query. In: Petras et al. [501] (2011)

158. Di Nunzio, G.M., Vezzani, F.: Using R markdown for replicable experiments in evidence based medicine. In: Bellot et al. [64], pp. 28–39 (2018)
159. Di Nunzio, G.M., Vezzani, F.: Did i miss anything? a study on ranking fusion and manual query rewriting in consumer health search. In: Barrón-Cedeño et al. [55], pp. 217–229 (2022)
160. Di Nunzio, G.M., Vezzani, F.: The best is yet to come: a reproducible analysis of CLEF eHealth TAR experiments. In: Arampatzis et al. [36], pp. 15–20 (2023)
161. Dicente Cid, Y., Batmanghelich, K., Müller, H.: Textured graph-based model of the lungs: application on tuberculosis type classification and multi-drug resistance detection. In: Bellot et al. [64], pp. 157–168 (2018)
162. Dietz, F.: The curious case of session identification. In: Arampatzis et al. [37], pp. 69–74 (2020)
163. Dietz, F., Petras, V.: A component-level analysis of an academic search test collection. - part i: system and collection configurations. In: Jones et al. [304], pp. 16–28 (2017)
164. Dietz, F., Petras, V.: A component-level analysis of an academic search test collection. - part ii: query analysis. In: Jones et al. [304], pp. 29–42 (2017)
165. Domann, J., Lommatzsch, A.: A highly available real-time news recommender based on apache spark. In: Jones et al. [304], pp. 161–172 (2017)
166. Dsilva, R.R.: From sentence embeddings to large language models to detect and understand wordplay. In: Goeuriot et al. [239] (2024)
167. Efimov, P., Chertok, A., Boytsov, L., Braslavski, P.: SberQuAD – Russian reading comprehension dataset: description and analysis. In: Arampatzis et al. [37], pp. 3–15 (2020)
168. Ehrmann, M., Romanello, M., Flückiger, A., Clematide, S.: Overview of CLEF HIPE 2020: named entity recognition and linking on historical newspapers. In: Arampatzis et al. [37], pp. 288–310 (2020)
169. Ehrmann, M., Romanello, M., Najem-Meyer, S., Doucet, A., Clematide, S.: Overview of HIPE-2022: named entity recognition and linking in multilingual historical documents. In: Barrón-Cedeño et al. [55], pp. 423–446 (2022)
170. El-Ebshihy, A., et al.: Predicting retrieval performance changes in evolving evaluation environments. In: Arampatzis et al. [36], pp. 21–33 (2023)
171. Elsayed, T., et al.: Overview of the CLEF-2019 CheckThat!: automatic identification and verification of claims. In: Crestani et al. [138], pp. 301–321 (2019)
172. Ermakova, K.: A method for short message contextualization: experiments at CLEF/INEX. In: Mothe et al. [419], pp. 352–363 (2015)
173. 21 Ermakova, L., et al.: Overview of SimpleText 2021 - CLEF workshop on text simplification for scientific information access. In: Candan et al. [104], pp. 432–449 (2021)
174. Ermakova, L., Bosser, A.G., Miller, T., Palma Preciado, V.M., Sidorov, G., Jatowt, A.: Overview of JOKER @ CLEF-2024: automatic humour analysis. In: Goeuriot et al. [240] (2024)
175. Ermakova, L., Goeuriot, L., Mothe, J., Mulhem, P., Nie, J.Y., SanJuan, E.: CLEF 2017 microblog cultural contextualization lab overview. In: Jones et al. [304], pp. 304–314 (2017)
176. Ermakova, L., Miller, T., Bosser, A.G., Palma Preciado, V.M., Sidorov, G., Jatowt, A.: Overview of JOKER – CLEF-2023 track on automatic wordplay analysis. In: Arampatzis et al. [36], pp. 397–415 (2023)
177. Ermakova, L., et al.: Overview of JOKER@CLEF 2022: automatic wordplay and humour translation workshop. In: Barrón-Cedeño et al. [55], pp. 447–469 (2022)

178. Ermakova, L., SanJuan, E., Huet, S., Azarbonyad, H., Augereau, O., Kamps, J.: Overview of the CLEF 2023 simpletext lab: automatic simplification of scientific texts. In: Arampatzis et al. [36], pp. 482–506 (2023)
179. Ermakova, L., et al.: Overview of the CLEF 2024 simpletext track: improving access to scientific texts for everyone. In: Goeuriot et al. [240] (2024)
180. Ermakova, L., .: Overview of the CLEF 2022 simpletext lab: automatic simplification of scientific texts. In: Barrón-Cedeño et al. [55], pp. 470–494 (2022)
181. Esuli, A., Moreo, A., Sebastiani, F., Sperduti, G.: A concise overview of LeQua@CLEF 2022: learning to quantify. In: Barrón-Cedeño et al. [55], pp. 362–381 (2022)
182. Esuli, F., Sebastiani, F.: Evaluating information extraction. In: Agosti et al. [10], pp. 100–111 (2010)
183. Eyuboglu, A.B., Altun, B., Arslan, M.B., Sonmezer, E., Kutlu, M.: Fight against misinformation on social media: detecting attention-worthy and harmful tweets and verifiable and check-worthy claims. In: Arampatzis et al. [36], pp. 161–173 (2023)
184. Ezzeldin, A.M., Kholief, M.H., El-Sonbaty, Y.: ALQASIM: arabic language question answer selection in machines. In: Forner et al. [210], pp. 100–103 (2013)
185. Fabregat, H., Duque, A., Araujo, L., Martinez-Romo, J.: A Re-labeling approach based on approximate nearest neighbors for identifying gambling disorders in social media. In: Arampatzis et al. [36], pp. 174–185 (2023)
186. Faggioli, G., Ferro, N., Galuščáková, P., Garcia Seco de Herrera, A. (eds.): CLEF 2024 Working Notes. CEUR Workshop Proceedings (CEUR-WS.org) (2017). ISSN 1613-0073
187. Faggioli, G., Ferro, N., Hanbury, A., Potthast, M. (eds.): CLEF 2022 Working Notes. CEUR Workshop Proceedings (CEUR-WS.org) (2022). ISSN 1613-0073. http://ceur-ws.org/Vol-3180/
188. Faggioli, G., Ferro, N., Joly, A., Maistro, M., Piroi, F. (eds.): CLEF 2021 Working Notes. CEUR Workshop Proceedings (CEUR-WS.org) (2021). ISSN 1613-0073. http://ceur-ws.org/Vol-2936/
189. Faggioli, G., et al.: Intelligent disease progression prediction: overview of iDPP@CLEF 2023. In: Arampatzis et al. [36], pp. 343–369 (2023)
190. Federico, M., Bertoldi, N., Levow, G.A., Jones, G.J.F.: CLEF 2004 cross-language spoken document retrieval track. In: Peters et al. [486], pp. 816–820 (2004)
191. Federico, M., Jones, G.J.F.: The CLEF 2003 cross-language spoken document retrieval track. In: Peters et al. [486], p. 646 (2003)
192. Fernández-Pichel, M., Losada, D.E., Pichel, J.C., Elsweiler, D.: Comparing traditional and neural approaches for detecting health-related misinformation. In: Candan et al. [104], pp. 78–90 (2021)
193. Ferrante, M., Ferro, N., Maistro, M.: Rethinking how to extend average precision to graded relevance. In: Kanoulas et al. [311], pp. 19–30 (2014)
194. Ferrante, M., Ferro, N., Piazzon, L.: s-AWARE: supervised measure-based methods for crowd-assessors combination. In: Arampatzis et al. [37], pp. 16–27 (2020)
195. Ferro, N.: What Happened in CLEF... For a While? In: Crestani et al. [138], pp. 3–45 (2019)
196. Ferro, N.: What happened in CLEF... for another while? In: Goeuriot et al. [239] (2024)
197. Ferro, N., Fuhr, N., Maistro, M., Sakai, T., Soboroff, I.: CENTRE@CLEF 2019. In: Azzopardi, L., Stein, B., Fuhr, N., Mayr, P., Hauff, C., Hiemstra, D. (eds.) ECIR 2019. LNCS, vol. 11438, pp. 283–290. Springer, Cham (2019). https://doi.org/10.1007/978-3-030-15719-7_38

198. Ferro, N., Harman, D.: CLEF 2009: Grid@CLEF pilot track overview. In: Peters et al. [490], pp. 552–565 (2009)
199. Ferro, N., Maistro, M., Sakai, T., Soboroff, I.: CENTRE@CLEF2018: overview of the replicability task. In: Cappellato et al. [111] (2018)
200. Ferro, N., Peters, C.: CLEF 2009 ad hoc track overview: TEL & persian tasks. In: Peters et al. [488], pp. 13–35 (2009)
201. Ferro, N., Peters, C.: From multilingual to multimodal: the evolution of clef over two decades. In: Information Retrieval Evaluation in a Changing World – Lessons Learned from 20 Years of CLEF [202], pp. 3–44 (2019)
202. Ferro, N., Peters, C. (eds.): Information Retrieval Evaluation in a Changing World - Lessons Learned from 20 Years of CLEF, The Information Retrieval Series, vol. 41. Springer, Heidelberg (2019). https://doi.org/10.1007/978-3-030-22948-1
203. Ferro, N., Santucci, G.: Visual analytics and IR experimental evaluation. In: Ferro and Peters [202], pp. 565–582 (2019)
204. Ferro, N., Silvello, G.: CLEF 15th birthday: what can we learn from ad hoc retrieval? In: Kanoulas et al. [311], pp. 31–43 (2014)
205. Ferro, N., Silvello, G.: The CLEF monolingual grid of points. In: Fuhr et al. [218], pp. 16–27 (2016)
206. Ferro, N., Silvello, G.: 3.5K runs, 5K topics, 3M assessments and 70M measures: what trends in 10 years of Adhoc-*ish* CLEF? Inf. Process. Manag. **53**(1), 175–202 (2017)
207. Fontanella, S., Rodríguez-Sánchez, A.J., Piater, J., Szedmak, S.: Kronecker decomposition for image classification. In: Fuhr et al. [218], pp. 137–149 (2016)
208. Forner, P., Gonzalo, J., Kekäläinen, J., Lalmas, M., de Rijke, M. (eds.): CLEF 2011. LNCS, vol. 6941. Springer, Heidelberg (2011). https://doi.org/10.1007/978-3-642-23708-9
209. Forner, P., Karlgren, J., Womser-Hacker, C., Ferro, N. (eds.): CLEF 2012 Working Notes. CEUR Workshop Proceedings (CEUR-WS.org) (2012). ISSN 1613-0073, http://ceur-ws.org/Vol-1178/
210. Forner, P., Müller, H., Paredes, R., Rosso, P., Stein, B. (eds.): Information Access Evaluation meets Multilinguality, Multimodality, and Visualization. Proceedings of the Fourth International Conference of the CLEF Initiative (CLEF 2013). Lecture Notes in Computer Science (LNCS) 8138, Springer, Heidelberg, Germany (2013)
211. Forner, P., Navigli, R., Tufis, D., Ferro, N. (eds.): CLEF 2013 Working Notes. CEUR Workshop Proceedings (CEUR-WS.org) (2013). ISSN 1613-0073. http://ceur-ws.org/Vol-1179/
212. Forner, P., et al.: Overview of the Clef 2008 multilingual question answering track. In: Peters et al. [487], pp. 262–295 (2008)
213. Franco-Salvador, M., Rangel Pardo, F.M., Rosso, P., Taulé, M., Martí, M.A.: Language variety identification using distributed representations of words and documents. In: Mothe et al. [419], pp. 28–40 (2015)
214. Frayling, E., Macdonald, C., McDonald, G., Ounis, I.: Using entities in knowledge graph hierarchies to classify sensitive information. In: Barrón-Cedeño et al. [55], pp. 125–132 (2022)
215. Frenda, S.and Patti, V., Rosso, P.: When sarcasm hurts: irony-aware models for abusive language detection. In: Arampatzis et al. [36], pp. 34–47 (2023)
216. Fröbe, M., Akiki, C., Potthast, M., Hagen, M.: Noise-reduction for automatically transferred relevance judgments. In: Barrón-Cedeño et al. [55], pp. 48–61 (2022)
217. Fuhr, N.: Reproducibility and validity in CLEF. In: Ferro and Peters [202] (2019)

218. Fuhr, N., et al. (eds.): CLEF 2016. LNCS, vol. 9822. Springer, Cham (2016). https://doi.org/10.1007/978-3-319-44564-9
219. Gäde, M., Ferro, N., Lestari Paramita, M.: CHiC 2011 – cultural heritage in CLEF: from use cases to evaluation in practice for multilingual information access to cultural heritage. In: Petras et al. [501] (2011)
220. Gäde, M., Stiller, J., Petras, V.: Which Log for Which Information? Gathering Multilingual Data from Different Log File Types. In: Agosti et al. [10], pp. 70–81
221. Galuscáková, P., Pecina, P., Hajic, J.: Penalty functions for evaluation measures of unsegmented speech retrieval. In: Catarci et al. [117], pp. 100–111 (2015)
222. Ganguly, D., Jones, G.J.F.: A gamified approach to relevance judgement. In: Fuhr et al. [218], pp. 214–220 (2016)
223. Ganguly, D., Leveling, J., Jones, G.J.F.: Simulation of within-session query variations using a text segmentation approach. In: Forner et al. [208], pp. 89–94 (2011)
224. Ganguly, D., Leveling, J., Jones, G.J.F.: A case study in decompounding for bengali information retrieval. In: Forner et al. [210], pp. 108–119 (2013)
225. Garmash, E., et al.: Cem mil podcasts: a spoken portuguese document corpus for multi-modal, multi-lingual and multi-dialect information access research. In: Arampatzis et al. [36], pp. 48–59 (2003)
226. Gebremeskel, G.G., de Vries, A.P.: Random performance differences between online recommender system algorithms. In: Fuhr et al. [218], pp. 187–200 (2016)
227. Gey, F., et al.: GeoCLEF 2006: the CLEF 2006 cross-language geographic information retrieval track overview. In: Peters et al. [485], pp. 852–876 (2006)
228. Gey, F.C., Larson, R.R., Sanderson, M., Joho, H., Clough, P., Petras, V.: GeoCLEF: the CLEF 2005 cross-language geographic information retrieval track overview. In: Peters et al. [493], pp. 908–919 (2005)
229. Ghosh, S., Singhania, P., Singh, S., Rudra, K., Ghosh, S.: Stance detection in web and social media: a comparative study. In: Crestani et al. [138], pp. 75–87 (2019)
230. Giampiccolo, D., et al.: Overview of the CLEF 2007 multilingual question answering track. In: Peters et al. [494], pp. 200–236 (2007)
231. Gînsca, A.L., Popescu, A., Lupu, M., Iftene, A., Kanellos, I.: Evaluating user image tagging credibility. In: Mothe et al. [419], pp. 41–52 (2015)
232. Glinos, D.G.: Discovering similar passages within large text documents. In: Kanoulas et al. [311], pp. 98–109 (2014)
233. Gobeill, J., Gaudinat, A., Ruch, P.: Instance-based learning for tweet monitoring and categorization. In: Mothe et al. [419], pp. 235–240 (2015)
234. Goëau, H., et al.: The CLEF 2011 plant images classification task. In: Petras et al. [501] (2011)
235. Goëau, H., et al.: The ImageCLEF 2012 plant identification task. In: Forner et al. [209] (2012)
236. Goeuriot, L., et al.: Overview of the CLEF eHealth evaluation lab 2015. In: Mothe et al. [419], pp. 429–443 (2015)
237. Goeuriot, L., et al.: CLEF 2017 eHealth evaluation lab overview. In: Jones et al. [304], pp. 291–303 (2015)
238. Goeuriot, L., Mothe, J., Mulhem, P., Murtagh, F., SanJuan, E.: Overview of the CLEF 2016 cultural micro-blog contextualization workshop. In: Fuhr et al. [218], pp. 371–378 (2016)
239. Goeuriot, L., et al. (eds.): Experimental IR Meets Multilinguality, Multimodality, and Interaction. Proceedings of the Fifteenth International Conference of the CLEF Association (CLEF 2024) – Part 1. Lecture Notes in Computer Science (LNCS) 14958, Springer, Heidelberg (2024)

240. Goeuriot, L., et al. (eds.): Experimental IR Meets Multilinguality, Multimodality, and Interaction. Proceedings of the Fifteenth International Conference of the CLEF Association (CLEF 2024) – Part 2. Lecture Notes in Computer Science (LNCS) 14959, Springer, Heidelberg (2024)
241. Goeuriot, L., et al.: Overview of the CLEF eHealth evaluation lab 2020. In: Arampatzis et al. [37], pp. 255–271 (2020)
242. Gollub, T., et al.: Recent trends in digital text forensics and its evaluation - plagiarism detection, author identification, and author profiling. In: Forner et al. [210], pp. 282–302 (2013)
243. Gómez-Adorno, H., et al.: Hierarchical clustering analysis: the best-performing approach at PAN 2017 author clustering task. In: Bellot et al. [64], pp. 216–223 (2017)
244. González-Sáez, G.N., Mulhem, P., Goeuriot, L.: Towards the evaluation of information retrieval systems on evolving datasets with pivot systems. In: Candan et al. [104], pp. 91–102 (2021)
245. Gonzalo, J., Clough, P., Karlgren, J.: Overview of iCLEF 2008: search log analysis for multilingual image retrieval. In: Peters et al. [487], pp. 227–235 (2008)
246. Gonzalo, J., Clough, P., Vallin, A.: Overview of the CLEF 2005 interactive track. In: Peters et al. [493], pp. 251–262 (2005)
247. Gonzalo, J., Oard, D.W.: The CLEF 2002 interactive track. In: Peters et al. [483], pp. 372–382 (2002)
248. Gonzalo, J., Oard, D.W.: iCLEF 2004 track overview: pilot experiments in interactive cross-language question answering. In: Peters et al. [486], pp. 310–322 (2004)
249. Gonzalo, J., Peinado, V., Clough, P., Karlgren, J.: Overview of iCLEF 2009: exploring search behaviour in a multilingual folksonomy environment. In: Peters et al. [496], pp. 13–20 (2009)
250. Goodwin, T., Harabagiu, S.M.: The impact of belief values on the identification of patient cohorts. In: Forner et al. [210], pp. 155–166 (2013)
251. Grotov, A., Chuklin, A., Markov, I., Stout, L., Xumara, F., de Rijke, M.: A comparative study of click models for web search. In: Mothe et al. [419], pp. 78–90 (2015)
252. Grubinger, M., Clough, P., Hanbury, A., Müller, H.: Overview of the ImageCLEFphoto 2007 photographic retrieval task. In: Peters et al. [496], pp. 433–444 (2007)
253. Guazzo, A., et al.: Intelligent disease progression prediction: overview of iDPP@CLEF 2022. In: Barrón-Cedeño et al. [55], pp. 395–422 (2022)
254. Gupta, P., Barrón-Cedeño, A., Rosso, P.: Cross-language high similarity search using a conceptual thesaurus. In: Catarci et al. [117], pp. 67–75 (2012)
255. Gupta, S., Reda Bouadjenek, M., Robles-Kelly, A.: An analysis of logic rule dissemination in sentiment classifiers. In: Barrón-Cedeño et al. [55], pp. 118–124 (2022)
256. Hagen, M., Glimm, C.: Supporting more-like-this information needs: finding similar web content in different scenarios. In: Kanoulas et al. [310], pp. 50–61 (2014)
257. Hall, M., Toms, E.: Building a common framework for IIR evaluation. In: Forner et al. [210], pp. 17–28 (2013)
258. Halvani, O., Graner, L.: Rethinking the evaluation methodology of authorship verification methods. In: Bellot et al. [64], pp. 40–51 (2018)
259. Hammarström, H.: Automatic annotation of bibliographical references for descriptive language materials. In: Forner et al. [208], pp. 62–73 (2011)
260. Han Lee, S., Loong Chang, Y., Seng Chan, S., Alexis, J., Bonnet, P., Goëau, H.: Plant classification based on gated recurrent unit. In: Bellot et al. [64], pp. 169–180 (2018)

261. Hanbury, A., Müller, H.: Automated component-level evaluation: present and future. In: Agosti et al. [10], pp. 124–135 (2010)
262. Hanbury, A., Müller, H., Langs, G., Weber, M.A., Menze, B.H., Salas Fernandez, T.: Bringing the algorithms to the data: cloud-based benchmarking for medical image analysis. In: Catarci et al. [117], pp. 24–29 (2012)
263. Hansen, C., Hansen, C., Grue Simonsen, J., Lioma, C.: Fact check-worthiness detection with contrastive ranking. In: Arampatzis et al. [37], pp. 124–130 (2020)
264. Harman, D.K.: Information Retrieval Evaluation. Morgan & Claypool Publishers, USA (2011)
265. Harman, D.K., Voorhees, E.M. (eds.): TREC. Experiment and Evaluation in Information Retrieval. MIT Press, Cambridge (2005)
266. Harris, C.G., Xu, T.: The importance of visual context clues in multimedia translation. In: Forner et al. [208], pp. 107–118 (2011)
267. Hasan, S.A., et al.: Attention-based medical caption generation with image modality classification and clinical concept mapping. In: Bellot et al. [64], pp. 224–230 (2018)
268. He, J., et al.: Overview of ChEMU 2020: named entity recognition and event extraction of chemical reactions from patents. In: Arampatzis et al. [37], pp. 237–254 (2020)
269. Hiemstra, D., Hauff, C.: MapReduce for information retrieval evaluation: "Let's quickly test this on 12 TB of data". In: Agosti et al. [10], pp. 64–69 (2010)
270. Hoang, T.B.N., Mothe, J., Baillon, M.: TwitCID: a collection of data sets for studies on information diffusion on social networks. In: Crestani et al. [138], pp. 88–100 (2019)
271. Hopfgartner, F., et al.: Continuous evaluation of large-scale information access systems: a case for living labs. In: Ferro and Peters [202] (2019)
272. Hopfgartner, F., Kille, B., Lommatzsch, A., Plumbaum, T., Brodt, T., Heintz, T.: Benchmarking news recommendations in a living lab. In: Kanoulas et al. [311], pp. 250–267 (2014)
273. Huertas-Tato, J., Martín, A., Camacho, D.: Using authorship embeddings to understand writing style in social media. In: Arampatzis et al. [36], pp. 60–71 (2023)
274. Hull, D.A., Oard, D.W.: Cross-Language Text and Speech Retrieval – Papers from the AAAI Spring Symposium. Association for the Advancement of Artificial Intelligence (AAAI), Technical Report SS-97-05 (2017). http://www.aaai.org/Press/Reports/Symposia/Spring/ss-97-05.phpX
275. Hürriyetoğlu, A., et al.: Overview of CLEF 2019 lab protestnews: extracting protests from news in a cross-context setting. In: Crestani et al. [138], pp. 425–432 (2019)
276. Huurnink, B., Hofmann, K., de Rijke, M., Bron, M.: Validating query simulators: an experiment using commercial searches and purchases. In: Agosti et al. [10], pp. 40–51 (2010)
277. Imhof, M., Braschler, M.: Are test collections "real"? mirroring real-world complexity in ir test collections. In: Mothe et al. [419], pp. 241–247 (2015)
278. Inches, G., Crestani, F.: Overview of the international sexual predator identification competition at PAN-2012. In: Forner et al. [209] (2012)
279. Ionescu, B., et al.: Overview of the ImageCLEF 2024: multimedia retrieval in medical applications. In: Goeuriot et al. [240] (2024)
280. Ionescu, B., et al.: Overview of the ImageCLEF 2023: multimedia retrieval in medical, social media and internet applications. In: Arampatzis et al. [36], pp. 370–396 (2023)

281. Ionescu, B., et al.: Overview of the ImageCLEF 2021: multimedia retrieval in medical, nature, internet and social media applications. In: Candan et al. [104], pp. 345–370 (2021)
282. Ionescu, B., et al.: Overview of the ImageCLEF 2020: multimedia retrieval in medical, lifelogging, nature, and internet applications. In: Arampatzis et al. [37], pp. 311–341 (2020)
283. Ionescu, B, et al.: ImageCLEF 2019: multimedia retrieval in medicine, lifelogging, security and nature. In: Crestani et al. [138], pp. 358–386 (2019)
284. Ionescu, B., et al.: Overview of the ImageCLEF 2022: multimedia retrieval in medical, social media and nature applications. In: Barrón-Cedeño et al. [55], pp. 541–564 (2022)
285. Ionescu, B., et al.: Overview of ImageCLEF 2017: information extraction from images. In: Jones et al. [304], pp. 315–337 (2017)
286. Ionescu, B., et al.: Overview of ImageCLEF 2018: challenges, datasets and evaluation. In: Bellot et al. [64], pp. 309–334 (2018)
287. Jabeur, L.B., Soulier, L., Tamine, L., Mousset, P.: A product feature-based user-centric Ranking model for e-commerce search. In: Fuhr et al. [218], pp. 174–186 (2016)
288. Jijkoun, V., de Rijke, M.: Overview of WebCLEF 2007. In: Peters et al. [494], pp. 725–731 (2007)
289. Jijkoun, V., de Rijke, M.: Overview of WebCLEF 2008. In: Peters et al. [487], pp. 787–793 (2008)
290. Joly, A., et al.: Overview of LifeCLEF 2023: evaluation of AI models for the identification and prediction of birds, plants, snakes and fungi. In: Arampatzis et al. [36], pp. 416–439 (2023)
291. Joly, A., et a;,.: Overview of LifeCLEF 2018: a large-scale evaluation of species identification and recommendation algorithms in the era of AI. In: Bellot et al. [64], pp. 247–266 (2018)
292. Joly, A., et al.: Overview of LifeCLEF 2019: identification of amazonian plants, south & north american birds, and niche prediction. In: Crestani et al. [138], pp. 387–401 (2019)
293. Joly, A., et al.: LifeCLEF 2017 lab overview: multimedia species identification challenges. In: Jones et al. [304], pp. 255–274 (2017)
294. Joly, A., et al.: Biodiversity information retrieval through large scale content-based identification: a long-term evaluation. In: Ferro and Peters [202] (2019)
295. Joly, A., et al.: LifeCLEF 2014: multimedia life species identification challenges. In: Kanoulas et al. [311], pp. 229–249 (2014)
296. Joly, A., et al.: LifeCLEF 2015: multimedia life species identification challenges. In: Mothe et al. [419], pp. 462–483 (2015)
297. Joly, A., et al.: Overview of LifeCLEF 2020: a system-oriented evaluation of automated species identification and species distribution prediction. In: Arampatzis et al. [37], pp. 342–363 (2020)
298. Joly, A., et al.: Overview of LifeCLEF 2021: an evaluation of machine-learning based species identification and species distribution prediction. In: Candan et al. [104], pp. 371–393 (2021)
299. Joly, A., et al.: Overview of LifeCLEF 2022: an evaluation of machine-learning based species identification and species distribution prediction. In: Barrón-Cedeño et al. [55], pp. 257–285 (2022)
300. Joly, A., et al.: LifeCLEF 2016: multimedia life species identification challenges. In: Fuhr et al. [218], pp. 286–310 (2016)

301. Joly, A., et al.: Overview of LifeCLEF 2024: challenges on species distribution prediction and identification. In: Goeuriot et al. [240] (2024)
302. Jones, G.J.F.: Bout sound and vision: CLEF beyond text retrieval tasks. In: Ferro and Peters [202] (2024)
303. Jones, G.J.F., Federico, M.: CLEF 2002 cross-language spoken document retrieval pilot track report. In: Peters et al. [483], pp. 446–457 (2002)
304. Jones, G.J.F., et al. (eds.): CLEF 2017. LNCS, vol. 10456. Springer, Cham (2017). https://doi.org/10.1007/978-3-319-65813-1
305. Juola, P.: An overview of the traditional authorship attribution subtask. In: Forner et al. [209] (2012)
306. Jürgens, J., Hansen, P., Womser-Hacker, C.: Going beyond CLEF-IP: the 'Reality' for patent searchers. In: Catarci et al. [117], pp. 30–35 (2012)
307. Kalpathy-Cramer, J., Müller, H., Bedrick, S., Eggel, I., Garcia Seco de Herrera, A., Tsikrika, T.: Overview of the CLEF 2011 medical image classification and retrieval tasks. In: Petras et al. [501] (2011)
308. Kamps, J., Koolen, M., Geva, S., Schenkel, R., SanJuan, E., Bogers, T.: From XML retrieval to semantic search and beyond. In: Ferro and Peters [202] (2019)
309. Kanoulas, E., Azzopardi, L.: CLEF 2017 Dynamic Search Evaluation Lab Overview. In: Jones et al. [304], pp. 361–366 (2017)
310. Kanoulas, E., Azzopardi, L., Hui Yang, G.: Overview of the CLEF dynamic search evaluation lab 2018. In: Bellot et al. [64], pp. 362–371 (2018)
311. Kanoulas, E., Lupu, M., Clough, P., Sanderson, M., Hall, M., Hanbury, A., Toms, E. (eds.): CLEF 2014. LNCS, vol. 8685. Springer, Cham (2014). https://doi.org/10.1007/978-3-319-11382-1
312. Karadzhov, G., Mihaylova, T., Kiprov, Y., Georgiev, G., Koychev, Y., Nakov, P.: The case for being average: a mediocrity approach to style masking and author obfuscation. In: Jones et al. [304], pp. 173–185 (2017)
313. Karan, M., Snajder, J.: Evaluation of manual query expansion rules on a domain specific FAQ collection. In: Mothe et al. [419], pp. 248–253 (2015)
314. Karimi, M.: SessionPrint: accelerating kNN via locality-sensitive hashing for session-based news recommendation. In: Goeuriot et al. [239] (2024)
315. Karisani, P., Oroumchian, F., Rahgozar, M.: Tweet expansion method for filtering task in twitter. In: Mothe et al. [419], pp. 55–64 (2015)
316. Karlgren, J.: Adopting systematic evaluation benchmarks in operational settings. In: Ferro and Peters [202] (2019)
317. Karlgren, J.: How lexical gold standards have effects on the usefulness of text analysis tools for digital scholarship. In: Crestani et al. [138], pp. 178–184 (2019)
318. Karlgren, J., et al.: Evaluating Learning Language Representations. In: Mothe et al. [419], pp. 254–260 (2015)
319. Karlgren, J., et al.: Overview of ELOQUENT 2024 — shared tasks for evaluating generative language model quality. In: Goeuriot et al. [240] (2024)
320. Karlgren, J., Gonzalo, J., Clough, P.: iCLEF 2006 overview: searching the flickr WWW photo-sharing repository. In: Peters et al. [485], pp. 186–194 (2006)
321. Karlgren, J., Hedlund, T., Järvelin, K., Keskustalo, H., Kettunen, K.: The challenges of language variation in information access. In: Ferro and Peters [202] (2019)
322. Karlsson, V., Herman, P., Karlgren, J.: Evaluating categorisation in real life – an argument against simple but impractical metrics. In: Fuhr et al. [218], pp. 221–226 (2016)

323. Kavallieratou, E., del Blanco, C.R., Cuevas, C., García, N.: Interactive learning-based retrieval technique for visual lifelogging. In: Crestani et al. [138], pp. 226–237 (2019)
324. Kazlouski, S.: Tuberculosis CT image analysis using image features extracted by 3D Autoencoder. In: Arampatzis et al. [37], pp. 131–140 (2020)
325. Keller, J., Breuer, T., Schaer, P.: Replicability measures for longitudinal information retrieval evaluation. In: Goeuriot et al. [239] (2024)
326. Keller, J., Paul, L., Munz, M.: Evaluating research dataset recommendations in a living lab. In: Barrón-Cedeño et al. [55], pp. 135–148 (2022)
327. Kelly, L., Goeuriot, L., Suominen, H., Névéol, A., Palotti, J., Zuccon, G.: Overview of the CLEF eHealth Evaluation Lab 2016. In: Fuhr et al. [218], pp. 255–266 (2016)
328. Kelly, L., et al.: Overview of the ShARe/CLEF eHealth evaluation lab 2014. In: Kanoulas et al. [311], pp. 172–191 (2014)
329. Kelly, L., et al.: Overview of the CLEF eHealth evaluation lab 2019. In: Crestani et al. [138], pp. 322–339 (2019)
330. Keszler, A., Kovács, L., Szirányi, T.: The appearance of the giant component in descriptor graphs and its application for descriptor selection. In: Catarci et al. [117], pp. 76–81 (2012)
331. Kharazmi, S., Scholer, F., Vallet, D., Sanderson, M.: Examining additivity and weak baselines. ACM Trans. Inf. Syst. (TOIS) **34**(4), 23:1–23:18 (2016)
332. Khwileh, A., Ganguly, D., Jones, G.J.F.: An investigation of cross-language information retrieval for user-generated internet video. In: Mothe et al. [419], pp. 117–129 (2015)
333. Khwileh, A., Way, A., Jones, G.J.F.: Improving the reliability of query expansion for user-generated speech retrieval using query performance prediction. In: Jones et al. [304], pp. 43–56 (2017)
334. Ki Ng, Y., Fraser, D.J., Kassaie, B., Tompa, F.W.: Dowsing for math answers. In: Candan et al. [104], pp. 201–212 (2021)
335. Kiesel, J., et al.: Overview of Touché 2024: argumentation systems. In: Goeuriot et al. [240] (2024)
336. Kiesel, J., Gohsen, M., Mirzakhmedova, N., Hagen, M., Stein, B.: Who will evaluate the evaluators? exploring the gen-IR user simulation space. In: Goeuriot et al. [239] (2024)
337. Kille, B., et al.: Overview of NewsREEL'16: multi-dimensional evaluation of real-time stream-recommendation algorithms. In: Fuhr et al. [218], pp. 311–331 (2016)
338. Kille, B., et al.: Stream-based recommendations: online and offline evaluation as a service. In: Mothe et al. [419], pp. 497–517 (2024)
339. Kim, S.J., Lee, J.H.: Subtopic mining based on head-modifier relation and co-occurrence of intents using web documents. In: Forner et al. [210], pp. 179–191 (2013)
340. Kliegr, T., Kuchar, J.: Benchmark of rule-based classifiers in the news recommendation task. In: Mothe et al. [419], pp. 130–141 (2024)
341. Kluck, M.: The domain-specific track in CLEF 2004: overview of the results and remarks on the assessment process. In: Peters et al. [486], pp. 260–270 (2004)
342. Kluck, M., Gey, F.C.: The domain-specific task of CLEF – specific evaluation strategies in cross-language information retrieval. In: Peters [479], pp. 48–56 (2001)
343. Kluck, M., Stempfhuber, M.: Domain-specific track CLEF 2005: overview of results and approaches, remarks on the assessment analysis. In: Peters et al. [493], pp. 212–221 (2005)

344. Kocher, M., Savoy, J.: Author clustering with an adaptive threshold. In: Jones et al. [304], pp. 186–198 (2017)
345. Koitka, S., Friedrich, C.M.: Optimized convolutional neural network ensembles for medical subfigure classification. In: Jones et al. [304], pp. 57–68 (2017)
346. Konstantinou, A., Chatzakou, D., Theodosiadou, O., Tsikrika, T., Vrochidis, S., Kompatsiaris, I.: Trend detection in crime-related time series with change point detection methods. In: Arampatzis et al. [36], pp. 72–84 (2023)
347. Koolen, M., et al.: Overview of the CLEF 2016 social book search lab. In: Fuhr et al. [218], pp. 351–370 (2016)
348. Koolen, M., et al.: Overview of the CLEF 2015 social book search lab. In: Mothe et al. [419], pp. 545–564 (2015)
349. Koolen, M., Kazai, G., Kamps, J., Preminger, M., Doucet, A., Landoni, M.: Overview of the INEX 2012 social book search track. In: Forner et al. [209] (2012)
350. Koops, H.V., Van Balen, J., Wiering, F.: automatic segmentation and deep learning of bird sounds. In: Mothe et al. [419], pp. 261–267 (2015)
351. Kordjamshidi, P., Rahgooy, T., Moens, M.F., Pustejovsky, J., Manzoor, U., Roberts, K.: CLEF 2017: multimodal spatial role labeling (msprl) task overview. In: Jones et al. [304], pp. 367–376 (2017)
352. Kosmajac, D., Keselj, V.: Twitter user profiling: bot and gender identification. In: Arampatzis et al. [37], pp. 141–153 (2020)
353. Kosmopoulos, A., Paliouras, G., Androutsopoulos, I.: The effect of dimensionality reduction on large scale hierarchical classification. In: Kanoulas et al. [310], pp. 160–171 (2014)
354. Kougia, V., Pavlopoulos, J., Androutsopoulos, I.: Medical image tagging by deep learning and retrieval. In: Arampatzis et al. [37], pp. 154–166 (2020)
355. Kumar, N.K., Santosh, G.S.K., Varma, V.: A language-independent approach to identify the named entities in under-resourced languages and clustering multilingual documents. In: Forner et al. [208], pp. 74–82 (2011)
356. Kurimo, M., Creutz, M., Varjokallio, M.: Morpho challenge evaluation using a linguistic gold standard. In: Peters et al. [494], pp. 864–872 (2008)
357. Kurimo, M., Turunen, V.T., Varjokallio, M.: Overview of morpho challenge 2008. In: Peters et al. [487], pp. 951–966 (2008)
358. Kurimo, M., Virpioja, S., Turunen, V.T., Blackwood, G.W., Byrne, W.: Overview and results of morpho challenge 2009. In: Peters et al. [488], pp. 587–597 (2009)
359. Kürsten, J., Eibl, M.: Comparing IR system components using beanplots. In: Catarci et al. [117], pp. 136–137 (2012)
360. Kvist, M., Velupillai, S.: SCAN: A Swedish Clinical Abbreviation Normalizer - Further Development and Adaptation to Radiology. In: Kanoulas et al. [312], pp. 62–73
361. de L. Pertile, S., Pereira Moreira, V.: A test collection to evaluate plagiarism by missing or incorrect references. In: Catarci et al. [117], pp. 141–143 (2012)
362. de L. Pertile, S., Rosso, P., Pereira Moreira, V.: Counting co-occurrences in citations to identify plagiarised text fragments. In: Forner et al. [210], pp. 150–154 (2013)
363. Lagopoulos, A., Anagnostou, A., Minas, A., Tsoumakas, G.: Learning-to-rank and relevance feedback for literature appraisal in empirical medicine. In: Bellot et al. [64], pp. 52–63 (2018)
364. Lai, M., Stranisci, M.A., Bosco, C., Damiano, R., Patti, V.: Analysing moral beliefs for detecting hate speech spreaders on twitter. In: Barrón-Cedeño et al. [55], pp. 149–161 (2022)

365. Lai, M., Tambuscio, M., Patti, V., Ruffo, G., Rosso, P.: Extracting graph topological information and users' opinion. In: Jones et al. [304], pp. 112–118 (2017)
366. Landoni, M., Matteri, D., Murgia, E., Huibers, T., Soledad Pera, M.: Sonny, cerca! evaluating the impact of using a vocal assistant to search at school. In: Crestani et al. [138], pp. 101–113 (2019)
367. Larsen, B.: The scholarly impact of CLEF 2010–2017. In: Ferro and Peters [202], pp. 547–554 (2019)
368. Larson, M., Newman, E., Jones, G.J.F.: Overview of VideoCLEF 2008: automatic generation of topic-based feeds for dual language audio-visual content. In: Peters et al. [487], pp. 906–917 (2008)
369. Larson, M., Newman, E., Jones, G.J.F.: Overview of VideoCLEF 2009: new perspectives on speech-based multimedia content enrichment. In: Peters et al. [496], pp. 354–368(2009)
370. Lasseck, M.: Towards automatic large-scale identification of birds in audio recordings. In: Mothe et al. [419], pp. 364–375 (2015)
371. Leiva, L.A., Villegas, M., Paredes, R.: Relevant clouds: leveraging relevance feedback to build tag clouds for image search. In: Forner et al. [210], pp. 143–149 (2013)
372. Leong, C.W., Hassan, S., Ruiz, M.E., Rada, M.: Improving query expansion for image retrieval via saliency and picturability. In: Forner et al. [208], pp. 137–142 (2011)
373. Lestari Paramita, M., Sanderson, M., Clough, P.: Diversity in photo retrieval: overview of the ImageCLEFPhoto Task 2009. In: Peters et al. [496], pp. 45–59 (2009)
374. Li, P., Jiang, X., Kambhamettu, C., Shatkay, H.: Segmenting compound biomedical figures into their constituent panels. In: Jones et al. [304], pp. 199–210 (2017)
375. Li, W., Jones, G.J.F.: Enhancing medical information retrieval by exploiting a content-based recommender method. In: Mothe et al. [419], pp. 142–153 (2015)
376. Li, Y., .: Overview of ChEMU 2021: reaction reference resolution and anaphora resolution in chemical patents. In: Candan et al. [104], pp. 292–307 (2021)
377. Li, Y., et al.: Overview of ChEMU 2022 evaluation campaign: information extraction in chemical patents. In: Barrón-Cedeño et al. [55], pp. 521–540 (2022)
378. Linhares Pontes, E., Huet, S., Torres-Moreno, J.M.: Microblog Contextualization: advantages and limitations of a multi-sentence compression approach. In: Bellot et al. [64], pp. 181–190 (2018)
379. Lipani, A., Piroi, F., Andersson, L., Hanbury, A.: An information retrieval ontology for information retrieval nanopublications. In: Kanoulas et al. [311], pp. 44–49 (2014)
380. Litvinova, T., Seredin, P., Litvinova, O., Ryzhkova, E.: Estimating the similarities between texts of right-handed and left-handed males and females. In: Jones et al. [304], pp. 119–124 (2017)
381. Liu, F., Peng, Y., Rosen, M.P.: An effective deep transfer learning and information fusion framework for medical visual question answering. In: Crestani et al. [138], pp. 238–247 (2019)
382. Lommatzsch, A., et al.: CLEF 2017 NewsREEL overview: a stream-based recommender task for evaluation and education. In: Jones et al. [303], pp. 239–254 (2017)
383. Lommatzsch, A., Werner, S.: Optimizing and evaluating stream-based news recommendation algorithms. In: Mothe et al. [419], pp. 376–388 (2015)
384. Longhin, F., Guazzo, A., Longato, E., Ferro, N., Di Camillo, B.: DAVI: a dataset for automatic variant interpretation. In: Arampatzis et al. [36], pp. 85–96 (2023)

385. Loponen, A., Järvelin, K.: A dictionary- and corpus-independent statistical lemmatizer for information retrieval in low resource languages. In: Agosti et al. [10], pp. 3–14 (2010)
386. Losada, D.E., Crestani, F.: A test collection for research on depression and language use. In: Fuhr et al. [218], pp. 28–39 (2016)
387. Losada, D.E., Crestani, F., Parapar, J.: eRISK 2017: CLEF lab on early risk prediction on the internet: experimental foundations. In: Jones et al. [304], pp. 346–360 (2017)
388. Losada, D.E., Crestani, F., Parapar, J.: Overview of eRisk: early risk prediction on the internet. In: Bellot et al. [64], pp. 343–361 (2019)
389. Losada, D.E., Crestani, F., Parapar, J.: Overview of eRisk 2019: early risk prediction on the internet. In: Crestani et al. [138], pp. 340–357 (2019)
390. Losada, D.E., Crestani, F., Parapar, J.: Overview of eRisk 2020: early risk prediction on the internet. In: Arampatzis et al. [37], pp. 272–287 (2020)
391. Mackie, S., McCreadie, R., Macdonald, C., Ounis, I.: Comparing algorithms for microblog summarisation. In: Kanoulas et al. [311], pp. 153–159 (2014)
392. Magdy, W., Jones, G.J.F.: Examining the robustness of evaluation metrics for patent retrieval with incomplete relevance judgements. In: Agosti et al. [10], pp. 82–93 (2010)
393. Magnini, B., et al.: Overview of the CLEF 2006 multilingual question answering track. In: Peters et al. [485], pp. 223–256 (2006)
394. Magnini, B., et al.: The multiple language question answering track at CLEF 2003. In: Peters et al. [484], pp. 471–486 (2003)
395. Magnini, B., et al.: Overview of the CLEF 2004 multilingual question answering track. In: Peters et al. [486], pp. 371–391 (2004)
396. Mandl, T., et al.: LogCLEF 2009: the CLEF 2009 multilingual logfile analysis track overview. In: Peters et al. [488], pp. 508–517 (2009)
397. Mandl, T., et al.: GeoCLEF 2008: the CLEF 2008 cross-language geographic information retrieval track overview. In: Peters et al. [487], pp. 808–821 (2008)
398. Mandl, T., Di Nunzio, G.M., Schulz, J.M.: LogCLEF 2010: the CLEF 2010 multilingual logfile analysis track overview. In: Braschler et al. [96] (2010)
399. Mandl, T., et al.: GeoCLEF 2007: the CLEF 2007 cross-language geographic information retrieval track overview. In: Peters et al. [494], pp. 745–772 (2007)
400. Manotumruksa, J., Macdonald, C., Ounis, I.: Predicting contextually appropriate venues in location-based social networks. In: Fuhr et al. [218], pp. 96–109 (2016)
401. Mansouri, B., Novotný, V., Agarwal, A., Oard, D.W., Zanibbi, R.: Overview of ARQMath-3 (2022): third CLEF lab on answer retrieval for questions on math. In: Barrón-Cedeño et al. [55], pp. 286–310 (2022)
402. Mansouri, B., Zanibbi, R., Oard, D.W., Agarwal, A.: Overview of ARQMath-2 (2021): second CLEF lab on answer retrieval for questions on math. In: Candan et al. [104], pp. 215–238 (2021)
403. Martínez-Castaño, R., Htait, A., Azzopardi, L., Moshfeghi, Y.: BERT-based transformers for early detection of mental health illnesses. In: Candan et al. [104], pp. 189–200 (2021)
404. Martínez-Gómez, J., García-Varea, I., Caputo, B.: Overview of the ImageCLEF 2012 robot vision task. In: Forner et al. [209] (2012)
405. Mayfield, J., Lawrie, D., McNamee, P., Oard, D.W.: Building a cross-language entity linking collection in twenty-one languages. In: Forner et al. [208], pp. 3–13 (2011)
406. McCreadie, R., Macdonald, C., Ounis, I., Brassey, J.: A study of personalised medical literature search. In: Kanoulas et al. [311], pp. 74–85 (2014)

407. McMinn, A.J., Jose, J.M.: Real-time entity-based event detection for twitter. In: Mothe et al. [419], pp. 65–77 (2015)
408. Medvedeva, M., Haagsma, H., Nissim, M.: An analysis of cross-genre and in-genre performance for author profiling in social media. In: Jones et al. [304], pp. 211–223 (2017)
409. Merker, J.H., Merker, L., Bondarenko, A.: The impact of web search result quality on decision-making. In: Goeuriot et al. [239]
410. Michail, A., Andermatt, P.S., , Fankhauser, T.: SimpleText best of labs in CLEF-2023: scientific text simplification using multi-prompt minimum bayes risk decoding. In: Goeuriot et al. [239] (2023)
411. Miftahutdinov, Z., Tutubalina, E.: Deep learning for ICD coding: looking for medical concepts in clinical documents in English and in French. In: Bellot et al. [64], pp. 203–215 (2018)
412. Mirsarraf, M.R., Dehghani, N.: A Dependency-inspired semantic evaluation of machine translation systems. In: Forner et al. [210], pp. 71–74 (2013)
413. Mitrovic, S., Müller, H.: Summarizing citation contexts of scientific publications. In: Mothe et al. [419], pp. 154–165 (2015)
414. Mohtaj, S.and Möller, S.: The impact of pre-processing on the performance of automated fake news detection. In: Barrón-Cedeño et al. [55], pp. 93–102 (2022)
415. 3 Molina, A., SanJuan, E., Torres-Moreno, J.M.: A turing test to evaluate a complex summarization task. In: Forner et al. [210], pp. 75–80 (201)
416. Molina, S., Mothe, J., Roques, D., Tanguy, L., Ullah, M.Z.: IRIT-QFR: IRIT query feature resource. In: Jones et al. [304], pp. 69–81 (2017)
417. Morante, R., Daelemans, W.: Overview of the QA4MRE pilot task: annotating modality and negation for a machine reading evaluation. In: Petras et al. [501] (2012)
418. Moreno, R., Huáng, W., Younus, A., O'Mahony, M.P., Hurley, N.J.: Evaluation of hierarchical clustering via markov decision processes for efficient navigation and search. In: Jones et al. [304], pp. 125–131 (2017)
419. Mothe, J., et al. (eds.): CLEF 2015. LNCS, vol. 9283. Springer, Cham (2015). https://doi.org/10.1007/978-3-319-24027-5
420. Mulhem, P., Goeuriot, L., Dogra, N., Amer, N.O.: TimeLine illustration based on microblogs: when diversification meets metadata re-ranking. In: Jones et al. [303], pp. 224–235 (2017)
421. Müller, H., Clough, P., Deselaers, T., Caputo, B. (eds.): ImageCLEF - Experimental Evaluation in Visual Information Retrieval. Springer, Heidelberg (2010). https://doi.org/10.1007/978-3-642-15181-1
422. Müller, H., Deselaers, T., Deserno, T.M., Clough, P., Kim, E., Hersh, W.R.: overview of the ImageCLEFmed 2006 medical retrieval and medical annotation tasks. In: Peters et al. [485], pp. 595–608 (2006)
423. Müller, H., Deselaers, T., Deserno, T.M., Kalpathy-Cramer, J., Kim, E., Hersh, W.R.: Overview of the ImageCLEFmed 2007 medical retrieval and medical annotation tasks. In: Peters et al. [494], pp. 472–491 (2007)
424. Müller, H., Hanbury, A.: EaaS: evaluation-as-a-service and experiences from the VISCERAL project. In: Ferro and Peters [202] (2019)
425. Müller, H., Garcia Seco de Herrera, A., Kalpathy-Cramer, J., Demner-Fushman, D., Antani, S., Eggel, I.: Overview of the ImageCLEF 2012 medical image retrieval and classification tasks. In: Forner et al. [209] (2012)
426. Müller, H., et al.: Overview of the CLEF 2009 medical image retrieval track. In: Peters et al. [496], pp. 72–84 (2009)

427. Müller, H., Kalpathy-Cramer, J., Eggel, I., Bedrick, S., Reisetter, J., Khan Jr., C.E., Hersh, W.R.: Overview of the CLEF 2010 medical image retrieval track. In: Braschler et al. [96] (2010)
428. Müller, H., Kalpathy-Cramer, J., Garcia Seco de Herrera, A.: Experiences from the ImageCLEF medical retrieval and annotation tasks. In: Ferro and Peters [202] (2019)
429. Müller, H., Kalpathy-Cramer, J., Kahn, C.E., Hatt, W., Bedrick, S., Hersh, W.: Overview of the ImageCLEFmed 2008 medical image retrieval task. In: Peters et al. [488], pp. 512–522 (2008)
430. Murauer, B., Specht, G.: Generating cross-domain text corpora from social media comments. In: Crestani et al. [138], pp. 114–125 (2019)
431. Nakov, P., et al.: Overview of the CLEF–2022 CheckThat! lab on fighting the COVID-19 infodemic and fake news detection. In: Barrón-Cedeño et al. [55], pp. 495–520 (2022)
432. Nakov, P., et al.: Overview of the CLEF-2018 CheckThat! lab on automatic identification and verification of political claims. In: Bellot et al. [64], pp. 372–387 (2018)
433. Nakov, P., et al.: Overview of the CLEF–2021 CheckThat! lab on detecting check-worthy claims, previously fact-checked claims, and fake news. In: Candan et al. [104], pp. 264–291 (2021)
434. Nardi, A., Peters, C., Ferro, N. (eds.): CLEF 2007 working notes. CEUR Workshop Proceedings (CEUR-WS.org) (2007). ISSN 1613-0073. http://ceur-ws.org/Vol-1173/
435. Nardi, A., Peters, C., Vicedo, J.L., Ferro, N. (eds.): CLEF 2006 Working Notes. CEUR Workshop Proceedings (CEUR-WS.org) (2007). ISSN 1613-0073, http://ceur-ws.org/Vol-1172/
436. Nentidis, A., et al.: Overview of BioASQ 2023: the eleventh BioASQ challenge on large-scale biomedical semantic indexing and question answering. In: Arampatzis et al. [36], pp. 227–250 (2023)
437. Nentidis, A., et al.: Overview of BioASQ 2024: the twelfth BioASQ challenge on large-scale biomedical semantic indexing and question answering. In: Goeuriot et al. [240] (2024)
438. Nentidis, A., et al.: Overview of BioASQ 2021: the ninth BioASQ challenge on large-scale biomedical semantic indexing and question answering. In: Candan et al. [104], pp. 239–263 (2021)
439. Nentidis, A., et al.: Overview of BioASQ 2022: the tenth BioASQ challenge on large-scale biomedical semantic indexing and question answering. In: Barrón-Cedeño et al. [55], pp. 337–361 (2022)
440. Nentidis, A., et al.: Overview of BioASQ 2020: the eighth BioASQ challenge on large-scale biomedical semantic indexing and question answering. In: Arampatzis et al. [37], pp. 194–214 (2020)
441. Nicolson, A., Dowling, J., Koopman, B.: ImageCLEF 2021 best of labs: the curious case of caption generation for medical images. In: Barrón-Cedeño et al. [55], pp. 175–189 (2021)
442. Ningtyas, A.M., El-Ebshihy, A., Budi Herwanto, G., Piroi, F., Hanbury, A.: Leveraging wikipedia knowledge for distant supervision in medical concept normalization. In: Barrón-Cedeño et al. [55], pp. 33–47 (2021)
443. Ningtyas, A.M., El-Ebshihy, A., Piroi, F., Hanbury, A.: Improving laypeople familiarity with medical terms by informal medical entity linking. In: Goeuriot et al. [239] (2024)

444. Nordlie, R., Pharo, N.: Seven years of INEX interactive retrieval experiments - lessons and challenges. In: Catarci et al. [117], pp. 13–23 (2012)
445. Nowak, S., Dunker, P.: Overview of the CLEF 2009 large-scale visual concept detection and annotation task. In: Peters et al. [496], pp. 94–109 (2009)
446. Nowak, S., Huiskes, M.J.: New strategies for image annotation: overview of the photo annotation task at ImageCLEF 2010. In: Braschler et al. [96] (2010)
447. Nowak, S., Nagel, K., Liebetrau, J.: The CLEF 2011 photo annotation and concept-based retrieval tasks. In: Petras et al. [501] (2011)
448. Oard, D.W., Gonzalo, J.: The CLEF 2001 interactive track. In: Peters et al. [482], pp. 308–319 (2001)
449. Oard, D.W., Gonzalo, J.: The CLEF 2003 interactive track. In: Peters et al. [484], pp. 425–434 (2003)
450. Oard, D.W., et al: Overview of the CLEF-2006 cross-language speech retrieval track. In: Peters et al. [485], pp. 744–758 (2006)
451. Oh, H.S., Jung, Y., Kim, K.Y.: A multiple-stage approach to re-ranking medical documents. In: Mothe et al. [419], pp. 166–177 (2015)
452. Olvera-Lobo, M.D., Gutiérrez-Artacho, J.: Multilingual question-answering system in biomedical domain on the web: an evaluation. In: Forner et al. [208], pp. 83–88 (2011)
453. Orio, N., Liem, C.C.S., Peeters, G., Schedl, M.: MusiClef: multimodal music tagging task. In: Catarci et al. [117], pp. 36–41 (2012)
454. Orio, N., Rizo, D.: Overview of MusiCLEF 2011. In: Petras et al. [501] (2011)
455. Ortega-Mendoza, M., Franco-Arcega, A., López-Monroy, A.P., Montes-y Gómez, M.: I, me, mine: the role of personal phrases in author profiling. In: Fuhr et al. [218], pp. 110–122 (2016)
456. Otterbacher, J.: Addressing social bias in information retrieval. In: Bellot et al. [64], pp. 121–127 (2018)
457. Pääkkönen, T., et al.: Exploring behavioral dimensions in session effectiveness. In: Mothe et al. [419], pp. 178–189 (2015)
458. Palotti, J., Zuccon, G., Bernhardt, J., Hanbury, A., Goeuriot, L.: Assessors agreement: a case study across assessor type, payment levels, query variations and relevance dimensions. In: Fuhr et al. [218], pp. 40–53 (2016)
459. Parapar, J., Martín-Rodilla, P., Losada, D.E., Crestani, F.: Overview of eRisk 2021: early risk prediction on the internet. In: Candan et al. [104], pp. 324–344 (2021)
460. Parapar, J., Martín-Rodilla, P., Losada, D.E., Crestani, F.: Overview of eRisk 2022: early risk prediction on the internet. In: Barrón-Cedeño et al. [55], pp. 233–256 (2022)
461. Parapar, J., Martín-Rodilla, P., Losada, D.E., Crestani, F.: Overview of eRisk 2023: early risk prediction on the internet. In: Arampatzis et al. [36], pp. 294–315 (2023)
462. Parapar, J., Martín-Rodilla, P., Losada, D.E., Crestani, F.: Overview of eRisk 2024: early risk prediction on the internet. In: Goeuriot et al. [240] (2024)
463. Parks, M., Karlgren, J., Stymne, S.: Plausibility testing for lexical resources. In: Jones et al. [304], pp. 132–137 (2017)
464. Pasi, G., Jones, G.J.F., Curtis, K., Marrara, S., Sanvitto, C., Ganguly, D., Sen, P.: Evaluation of personalised information retrieval at CLEF 2018 (PIR-CLEF). In: Bellot et al. [64], pp. 335–342 (2018)
465. Pasi, G., Jones, G.J.F., Goeuriot, L., Kelly, L., Marrara, S., Sanvitto, C.: Overview of the CLEF 2019 personalised information retrieval lab (PIR-CLEF 2019). In: Crestani et al. [138], pp. 417–424 (2019)

466. Pasi, G., Jones, G.J.F., Marrara, S., Sanvitto, C., Ganguly, D., Sen, P.: Overview of the CLEF 2017 personalised information retrieval pilot lab (PIR-CLEF 2017). In: Jones et al. [304], pp. 338–345 (2017)
467. Pasin, A., Ferrari Dacrema, M., Cremonesi, P., Ferro, N.: qCLEF: a proposal to evaluate quantum annealing for information retrieval and recommender systems. In: Arampatzis et al. [36], pp. 97–108 (2023)
468. Pasin, A., Ferrari Dacrema, M., Cremonesi, P., Ferro, N.: Overview of quantum CLEF 2024: the quantum computing challenge for information retrieval and recommender systems at CLEF. In: Goeuriot et al. [240] (2024)
469. Pecina, P., Hoffmannová, P., Jones, G.J.F., Zhang, Y., Oard, D.W.: Overview of the CLEF-2007 cross-language speech retrieval track. In: Peters et al. [494], pp. 674–686 (2007)
470. Pellegrin, L., et al.: A two-step retrieval method for image captioning. In: Fuhr et al. [218], pp. 150–161 (2016)
471. Peñas, A., Forner, P., Rodrigo, A., Sutcliffe, R.F.E., Forascu, C., Mota, C.: Overview of ResPubliQA 2010: question answering evaluation over european legislation. In: Braschler et al. [96] (2010)
472. Peñas, A., et al.: Overview of ResPubliQA 2009: question answering evaluation over european legislation. In: Peters et al. [488], pp. 174–196 (2009)
473. Peñas, A., et al.: Overview of QA4MRE at CLEF 2011: question answering for machine reading evaluation. In: Petras et al. [501] (2011)
474. Peñas, A., Hovy, E.H., Forner, P., Rodrigo, A., Sutcliffe, R.F.E., Morante, R.: QA4MRE 2011-2013: overview of question answering for machine reading Evaluation. In: Forner et al. [210], pp. 303–320 (2013)
475. Peñas, A., et al.: Overview of QA4MRE at CLEF 2012: question answering for machine reading evaluation. In: Forner et al. [209] (2012)
476. Peñas, A., et al.: Results and lessons of the question answering track at CLEF. In: Ferro and Peters [202] (2019)
477. Peñas, A., Unger, C., Ngonga Ngomo, A.C.A.: Overview of CLEF question answering track 2014. In: Kanoulas et al. [311], pp. 300–306 (2014)
478. Peñas, A., Unger, C., Paliouras, P., Kakadiaris, I.A.: Overview of the clef question answering track 2015. In: Mothe et al. [419], pp. 539–544 (2015)
479. Peters, C. (ed.): CLEF 2000. LNCS, vol. 2069. Springer, Heidelberg (2001). https://doi.org/10.1007/3-540-44645-1
480. Peters, C.: Introduction. In: Cross-Language Information Retrieval and Evaluation: Workshop of Cross-Language Evaluation Forum (CLEF 2000) [479], pp. 1–6 (2000)
481. Peters, C., Braschler, M., Clough, P.: Multilingual Information Retrieval. Springer, Heidelberg (2011). https://doi.org/10.1007/978-3-642-23008-0
482. Peters, C., Braschler, M., Gonzalo, J., Kluck, M. (eds.): CLEF 2001. LNCS, vol. 2406. Springer, Heidelberg (2002). https://doi.org/10.1007/3-540-45691-0
483. Peters, C., Braschler, M., Gonzalo, J., Kluck, M. (eds.): CLEF 2002. LNCS, vol. 2785. Springer, Heidelberg (2003). https://doi.org/10.1007/b12018
484. Peters, C., Gonzalo, J., Braschler, M., Kluck, M. (eds.): CLEF 2003. LNCS, vol. 3237. Springer, Heidelberg (2004). https://doi.org/10.1007/b102261
485. Peters, C., et al. (eds.): CLEF 2006. LNCS, vol. 4730. Springer, Heidelberg (2007). https://doi.org/10.1007/978-3-540-74999-8
486. Peters, C., Clough, P., Gonzalo, J., Jones, G.J.F., Kluck, M., Magnini, B. (eds.): CLEF 2004. LNCS, vol. 3491. Springer, Heidelberg (2005). https://doi.org/10.1007/b138934

487. Peters, C., Deselaers, T., Ferro, N., Gonzalo, J., Jones, G.J.F., Kurimo, M., Mandl, T., Peñas, A., Petras, V. (eds.): CLEF 2008. LNCS, vol. 5706. Springer, Heidelberg (2009). https://doi.org/10.1007/978-3-642-04447-2
488. Peters, C., Di Nunzio, G.M., Kurimo, M., Mandl, T., Mostefa, D., Peñas, A., Roda, G. (eds.): CLEF 2009. LNCS, vol. 6241. Springer, Heidelberg (2010). https://doi.org/10.1007/978-3-642-15754-7
489. Peters, C., Ferro, N. (eds.): CLEF 2000 Working Notes. CEUR Workshop Proceedings (CEUR-WS.org) (2000). ISSN 1613-0073. http://ceur-ws.org/Vol-1166/
490. Peters, C., Ferro, N. (eds.): CLEF 2001 Working Notes. CEUR Workshop Proceedings (CEUR-WS.org) (2001). ISSN 1613-0073. http://ceur-ws.org/Vol-1167/
491. Peters, C., Ferro, N. (eds.): CLEF 2002 Working Notes. CEUR Workshop Proceedings (CEUR-WS.org) (2002). ISSN 1613-0073. http://ceur-ws.org/Vol-1168/ (2002)
492. Peters, C., Ferro, N. (eds.): CLEF 2003 Working Notes. CEUR Workshop Proceedings (CEUR-WS.org) (2000). ISSN 1613-0073. http://ceur-ws.org/Vol-1169/
493. Peters, C., Gey, F.C., Gonzalo, J., Müller, H., Jones, G.J.F., Kluck, M., Magnini, B., de Rijke, M. (eds.): CLEF 2005. LNCS, vol. 4022. Springer, Heidelberg (2006). https://doi.org/10.1007/11878773
494. Peters, C., Jijkoun, V., Mandl, T., Müller, H., Oard, D.W., Peñas, A., Petras, V., Santos, D. (eds.): CLEF 2007. LNCS, vol. 5152. Springer, Heidelberg (2008). https://doi.org/10.1007/978-3-540-85760-0
495. Peters, C., Quochi, V., Ferro, N. (eds.): CLEF 2005 Working Notes. CEUR Workshop Proceedings (CEUR-WS.org) (2005). ISSN 1613-0073. http://ceur-ws.org/Vol-1171/
496. Peters, C., Caputo, B., Gonzalo, J., Jones, G.J.F., Kalpathy-Cramer, J., Müller, H., Tsikrika, T. (eds.): CLEF 2009. LNCS, vol. 6242. Springer, Heidelberg (2010). https://doi.org/10.1007/978-3-642-15751-6
497. Petras, V., Baerisch, S.: The domain-specific track at CLEF 2008. In: Peters et al. [487], pp. 186–198 (2008)
498. Petras, V., Baerisch, S., Stempfhuber, M.: The domain-specific track at CLEF 2007. In: Peters et al. [494], pp. 160–173 (2007)
499. Petras, V., et al.: Cultural Heritage in CLEF (CHiC) 2013. In: Forner et al. [210], pp. 192–211 (2013)
500. Petras, V., et al.: Cultural Heritage in CLEF (CHiC) overview 2012. In: Forner et al. [209] (2012)
501. Petras, V., Forner, P., Clough, P., Ferro, N. (eds.): CLEF 2011 Working Notes. CEUR Workshop Proceedings (CEUR-WS.org) (2011). ISSN 1613-0073, http://ceur-ws.org/Vol-1177/
502. Petras, V., Lüschow, A., Ramthun, R., Stiller, J., España-Bonet, C., Henning, S.: Query or document translation for academic search – what's the real difference? In: Arampatzis et al. [37], pp. 28–42 (2020)
503. Piras, L., Caputo, B., Dang-Nguyen, D.T., Riegler, M., Halvorsen, P.: image retrieval evaluation in specific domains. In: Ferro and Peters [202] (2019)
504. Piroi, F.: CLEF-IP 2010: retrieval experiments in the intellectual property domain . In: Braschler et al. [96] (2010)
505. Piroi, F., Hanbury, A.: Multilingual patent text retrieval evaluation: CLEF-IP. In: Ferro and Peters [202] (2019)
506. Piroi, F., Lupu, M., Hanbury, A.: Effects of language and topic size in patent IR: an empirical study. In: Catarci et al. [117], pp. 54–66 (2012)
507. Piroi, F., Lupu, M., Hanbury, A.: Overview of CLEF-IP 2013 lab - information retrieval in the patent domain. In: Forner et al. [210], pp. 232–249 (2013)

508. Piroi, F., Lupu, M., Hanbury, A., Sexton, A.P., Magdy, W., Filippov, I.V.: CLEF-IP 2012: retrieval experiments in the intellectual property Domain. In: Forner et al. [209] (2012)
509. Piroi, F., Lupu, M., Hanbury, A., Zenz, V.: CLEF-IP 2011: retrieval in the intellectual property domain. In: Petras et al. [501] (2011)
510. Plaza, L., et al.: Overview of EXIST 2023 – learning with disagreement for sexism identification and characterization. In: Arampatzis et al. [36], pp. 316–342 (2023)
511. Plaza, L., et al.: Overview of EXIST 2024 – learning with disagreement for sexism identification and characterization in tweets and memes. In: Goeuriot et al. [240] (2024)
512. Popescu, A., Tsikrika, T., Kludas, J.: Overview of the wikipedia retrieval task at ImageCLEF 2010. In: Braschler et al. [96] (2010)
513. Potha, N., Stamatatos, E.: An improved impostors method for authorship verification. In: Jones et al. [304], pp. 138–144 (2017)
514. Potthast, M., Barrón-Cedeño, A., Eiselt, A., Stein, B., Rosso, P.: Overview of the 2nd international competition on plagiarism detection. In: Braschler et al. [96] (2010)
515. Potthast, M., Eiselt, A., Barrón-Cedeño, A., Stein, B., Rosso, P.: Overview of the 3rd international competition on plagiarism detection. In: Petras et al. [501] (2011)
516. Potthast, M., et al.: Overview of the 4th international competition on plagiarism detection. In: Forner et al. [209] (2015)
517. Potthast, M., Gollub, T., Rangel Pardo, F., Rosso, P., Stamatatos, E., Stein, B.: Improving the reproducibility of PAN's shared tasks: plagiarism detection, author identification, and author profiling. In: Kanoulas et al. [311], pp. 268–299 (2014)
518. Potthast, M., Gollub, T., Wiegmann, M., Stein, b.: TIRA integrated research architecture. In: Ferro and Peters [202], pp. 123–160 (2019)
519. Potthast, M., Holfeld, T.: Overview of the 2nd international competition on wikipedia vandalism detection. In: Petras et al. [501] (2011)
520. Potthast, M., et al.: Overview of PAN'17 - author identification, author profiling, and author obfuscation. In: Jones et al. [304], pp. 275–290 (2017)
521. Potthast, M., Stein, B., Holfeld, T.: Overview of the 1st international competition on wikipedia vandalism detection. In: Braschler et al. [96] (2010)
522. Pradel, C., Sileo, D., Rodrigo, A., Peñas, A., Agirre, E.: Question answering when knowledge bases are incomplete. In: Arampatzis et al. [37], pp. 43–54 (2020)
523. Prasetyo Putri, D.G., Viviani, M., Pasi, G.: A multi-task learning model for multidimensional relevance assessment. In: Candan et al. [104], pp. 103–115 (2021)
524. Pritsos, D.A., Stamatatos, E.: The impact of noise in web genre identification. In: Mothe et al. [419], pp. 268–273 (2015)
525. Pronobis, A., Fornoni, M., Christensen, H.I., Caputo, B.: The robot vision track at ImageCLEF 2010. In: Braschler et al. [96] (2010)
526. Pronobis, A., Xing, L., Caputo, B.: Overview of the CLEF 2009 robot vision track. In: Peters et al. [496], pp. 110–119 (2009)
527. Raghavi, K.C., Chinnakotla, M.K., Black, A.W., Shrivastava, M.: WebShodh: a code mixed factoid question answering system for web. In: Jones et al. [304], pp. 104–111 (2017)
528. Ragheb, W., Azé, J., Bringay, S., Servajean, M.: Language modeling in temporal mood variation models for early risk detection on the internet. In: Crestani et al. [138], pp. 248–259 (2019)

529. Rangel Pardo, F., Rosso, P.: On the multilingual and genre robustness of emographs for author profiling in social media. In: Mothe et al. [419], pp. 274–280 (2019)
530. Rebholz-Schuhmann, D., et al.: Entity recognition in parallel multi-lingual biomedical corpora: the CLEF-ER laboratory overview. In: Forner et al. [210], pp. 353–367 (2013)
531. Rekabsaz, N., Lupu, M.: A real-world framework for translator as expert retrieval. In: Kanoulas et al. [311], pp. 141–152 (2014)
532. Reusch, A., Thiele, M., Lehner, W.: Transformer-encoder-based mathematical information retrieval. In: Barrón-Cedeño et al. [55], pp. 175–189 (2022)
533. de Rijke, M., Balog, K., Bogers, T., van den Bosch, A.: On the evaluation of entity profiles. In: Agosti et al. [10], pp. 94–99 (2010)
534. Roda, G., Tait, J., Piroi, F., Zenz, V.: CLEF-IP 2009: retrieval experiments in the intellectual property domain. In: Peters et al. [488], pp. 385–409 (2009)
535. Rodrigo, A., Peñas, A., Verdejo, M.F.: Overview of the answer validation exercise 2008. In: Peters et al. [487], pp. 296–313 (2008)
536. Roller, R., Stevenson, M.: Self-supervised relation extraction using UMLS. In: Kanoulas et al. [311], pp. 116–127 (2014)
537. Rosso, P., Potthast, M., Stein, B., Stamatatos, E., Rangel Pardo, F.M., Daelemans, W.: Evolution of the PAN lab on digital text forensics. In: Ferro and Peters [202] (2019)
538. Rosso, P., Rangel, F., Potthast, M., Stamatatos, E., Tschuggnall, M., Stein, B.: Overview of PAN 2016. In: Fuhr et al. [218], pp. 332–350 (2016)
539. Rowe, B.R., Wood, D.W., Link, A.L., Simoni, D.A.: Economic Impact Assessment of NIST's Text REtrieval Conference (TREC) Program. RTI Project Number 0211875, RTI International, USA (2010). http://trec.nist.gov/pubs/2010.economic.impact.pdf
540. Sabetghadam, S., Bierig, R., Rauber, A.: A hybrid approach for multi-faceted ir in multimodal domain. In: Kanoulas et al. [311], pp. 86–97 (2014)
541. Sakai, T.: How to run an evaluation task. In: Ferro and Peters [202], pp. 71–102 (2019)
542. Sakai, T., Oard, D.W., Kando, N. (eds.): Evaluating Information Retrieval and Access Tasks - NTCIR's Legacy of Research Impact, The Information Retrieval Series, vol. 43. Springer, Germany (2021). https://doi.org/10.1007/978-981-15-5554-1
543. Sakhovskiy, A., Semenova, N., Kadurin, A., Tutubalina, E.: Graph-enriched biomedical entity representation transformer. In: Arampatzis et al. [36], pp. 109–120 (2023)
544. Saleh, S., Pecina, P.: Reranking hypotheses of machine-translated queries for cross-lingual information retrieval. In: Fuhr et al. [218], pp. 54–68 (2016)
545. Samuel, J.: Analyzing and visualizing translation patterns of wikidata properties. In: Bellot et al. [64], pp. 128–134 (2018)
546. Sánchez-Cortés, D., Burdisso, S., Villatoro-Tello, E., Motlicek, P.: Mapping the media landscape: predicting factual reporting and political bias through web interactions. In: Goeuriot et al. [239] (2024)
547. Sánchez-Junquera, J., Villaseñor-Pineda, L., Montes-y Gómez, M., Rosso, P.: Character N-grams for detecting deceptive controversial opinions. In: Bellot et al. [64], pp. 135–140 (2018)
548. Sanchez-Perez, M.A., Gelbukh, A.F., Sidorov, G.: Adaptive algorithm for plagiarism detection: the best-performing approach at PAN 2014 text alignment competition. In: Mothe et al. [419], pp. 402–413 (2014)

549. Sanchez-Perez, M.A., Markov, I., Gómez-Adorno, H., Sidorov, G.: Comparison of character n-grams and lexical features on author, gender, and language variety identification on the same spanish news corpus. In: Jones et al. [304], pp. 145–151 (2017)
550. SanJuan, E., Moriceau, V., Tannier, X., Bellot, P., Mothe, J.: Overview of the INEX 2012 tweet contextualization track. In: Forner et al. [209] (2012)
551. Santos, D., Cabral, L.M.: GikiCLEF: expectations and lessons learned. In: Peters et al. [487], pp. 212–222 (2009)
552. Ramos dos Santos, W., Paraboni, I.: Personality facets recognition from text. In: Crestani et al. [138], pp. 185–190 (2019)
553. Sarvazyan, A.M., González, J.A., Rosso, P., Franco-Salvador, M.: Supervised machine-generated text detectors: family and scale matters. In: Arampatzis et al. [36], pp. 121–132 (2023)
554. Savenkov, D., Braslavski, P., Lebedev, M.: Search snippet evaluation at yandex: lessons learned and future directions. In: Forner et al. [208], pp. 14–25 (2011)
555. Savoy, J., Braschler, M.: Lessons learnt from experiments on the ad-hoc multilingual test collections at CLEF. In: Ferro and Peters [202] (2019)
556. Sawinski, M., Wecel, K., Księżniak, E.: Under-sampling strategies for better transformer-based classifications models. In: Goeuriot et al. [239] (2024)
557. Schaer, P.: Better than their reputation? on the reliability of relevance assessments with students. In: Catarci et al. [117], pp. 124–135 (2012)
558. Schaer, P., Breuer, T., Castro, L.J., Wolff, B., Schaible, J., Tavakolpoursaleh, N.: Overview of LiLAS 2021 – living labs for academic search. In: Candan et al. [104], pp. 394–418 (2021)
559. Schaer, P., Mayr, P., Sünkler, S., Lewandowski, D.: How relevant is the long tail?: a relevance assessment study on million short. In: Fuhr et al. [218], pp. 227–233 (2016)
560. Schaer, P., Neumann, M.: Enriching existing test collections with OXPath. In: Jones et al. [304], pp. 152–158 (2017)
561. Schaer, P., Schaible, J., Jael García Castro, L.: Overview of LiLAS 2020 - living labs for academic search. In: Arampatzis et al. [37], pp. 364–371 (2020)
562. Schaüble, P., Sheridan, P.: Cross-language information retrieval (CLIR) track overview. In: Voorhees, E.M., Harman, D.K. (eds.) The Sixth Text REtrieval Conference (TREC-6), pp. 31–44. National Institute of Standards and Technology (NIST), Special Publication 500-240, Washington, USA (1997)
563. Schreieder, T., Braker, J.: Touché 2022 best of labs: neural image retrieval for argumentation. In: Arampatzis et al. [36], pp. 186–197 (2022)
564. Schubotz, M., Krämer, L., Meuschke, N., Hamborg, F., Gipp, B.: Evaluating and improving the extraction of mathematical identifier definitions. In: Jones et al. [304], pp. 82–94 (2017)
565. Schuth, A., Balog, K., Kelly, L.: Overview of the living labs for information retrieval evaluation (LL4IR) CLEF Lab 2015. In: Mothe et al. [419], pp. 484–496 (2015)
566. Schuth, A., Marx, M.: Evaluation methods for rankings of facetvalues for faceted search. In: Forner et al. [208], pp. 131–136 (2011)
567. Shahshahani, M.S., Kamps, J.: Argument retrieval from web. In: Arampatzis et al. [37], pp. 75–81 (2020)
568. Shen, W., Nie, J.Y.: Is concept mapping useful for biomedical information retrieval. In: Mothe et al. [419], pp. 281–286 (2015)
569. Shepeleva, N., Balog, K.: Towards an understanding of transactional tasks. In: Fuhr et al. [218], pp. 234–240 (2016)

570. Sherkat, E., Velcin, J., Milios, E.E.: Fast and simple deterministic seeding of kmeans for text document clustering. In: Bellot et al. [64], pp. 76–88 (2018)
571. Shing, H.S., Barrow, J., Galuščáková, P., Oard, D.W., Resnik, P.: Unsupervised system combination for set-based retrieval with expectation maximization. In: Crestani et al. [138], pp. 191–197 (2019)
572. Sierek, T., Hanbury, A.: Using health statistics to improve medical and health search. In: Mothe et al. [419], pp. 287–292 (2015)
573. Sigurbjörnsson, B., Kamps, J., de Rijke, M.: Overview of WebCLEF 2005. In: Peters et al. [493], pp. 810–824 (2005)
574. Silvello, G., Bordea, G., Ferro, N., Buitelaar, P., Bogers, T.: Semantic representation and enrichment of information retrieval experimental data. Int. J. Dig. Libr. (IJDL) **18**(2), 145–172 (2017)
575. Šimsa, S., et al.: Overview of DocILE 2023: document information localization and extraction. In: Arampatzis et al. [36], pp. 276–293 (2023)
576. Singh, G., Mantrach, A., Silvestri, F.: Improving profiles of weakly-engaged users: with applications to recommender systems. In: Fuhr et al. [218], pp. 123–136 (2016)
577. Skalický, M., Šimsa, S., Uřičář, M., Šulc, M.: Business document information extraction: towards practical benchmarks. In: Barrón-Cedeño et al. [55], pp. 105–117 (2022)
578. Smith, E., Weiler, A., Braschler, M.: Skill extraction for domain-specific text retrieval in a job-matching platform. In: Candan et al. [104], pp. 116–128 (2021)
579. Sorg, P., Cimiano, P., Schultz, A., Sizov, S.: Overview of the cross-lingual expert search (CriES) pilot challenge. In: Braschler et al. [96] (2010)
580. Spina, D., Amigó, E., Gonzalo, J.: Filter keywords and majority class strategies for company name disambiguation in twitter. In: Forner et al. [208], pp. 38–49 (2011)
581. Stamatatos, E., Potthast, M., Rangel Pardo, F.M., Rosso, P., Stein, B.: Overview of the PAN/CLEF 2015 evaluation lab. In: Mothe et al. [419], pp. 518–538 (2015)
582. Stamatatos, E., et al.: Overview of PAN 2018. In: Bellot et al. [64], pp. 267–285 (2018)
583. Stathopoulos, S., Kalamboukis, T.: Medical image classification with weighted latent semantic tensors and deep convolutional neural networks. In: Bellot et al. [64], pp. 89–100 (2018)
584. Stefanov, V., Sachs, A., Kritz, M., Samwald, M., Gschwandtner, M., Hanbury, A.: A Formative evaluation of a comprehensive search system for medical professionals. In: Forner et al. [210], pp. 81–92 (2013)
585. Stempfhuber, M., Baerisch, S.: The domain-specific track at CLEF 2006: overview of approaches, results and assessment. In: Peters et al. [485], pp. 163–169 (2006)
586. Suchomel, S., Brandejs, M.: Determining window size from plagiarism corpus for stylometric features. In: Mothe et al. [419], pp. 293–299 (2015)
587. Suominen, H.: CLEFeHealth2012 - The CLEF 2012 workshop on cross-language evaluation of methods, applications, and resources for ehealth document analysis. In: Forner et al. [209] (2012)
588. Suominen, H., et al.: Overview of the CLEF eHealth evaluation lab 2021. In: Candan et al. [104], pp. 308–323 (2021)
589. Suominen, H., Kelly, L., Goeuriot, L.: The scholarly impact and strategic intent of CLEF eHealth Labs from 2012–2017. In: Ferro and Peters [202] (2019)
590. Suominen, H., et al.: Overview of the CLEF eHealth evaluation lab 2018. In: Bellot et al. [64], pp. 286–301 (2018)

591. Suominen, H., et al.: Overview of the ShARe/CLEF eHealth evaluation lab 2013. In: Forner et al. [210], pp. 212–231 (2013)
592. Tannebaum, W., Mahdabi, P., Rauber, A.: Effect of log-based query term expansion on retrieval effectiveness in patent searching. In: Mothe et al. [419], pp. 300–305 (2015)
593. Tannebaum, W., Rauber, A.: Mining query logs of USPTO patent examiners. In: Forner et al. [210], pp. 136–142 (2013)
594. Teixeira Lopes, C., Almeida Fernandes, T.: Health suggestions: a chrome extension to help laypersons search for health information. In: Fuhr et al. [218], pp. 241–246 (2016)
595. Teixeira Lopes, C., Ribeiro, C.: Effects of language and terminology on the usage of health query suggestions. In: Fuhr et al. [218], pp. 83–95 (2023)
596. Teixeira Lopes, C., Ribeiro, C.: Effects of language and terminology of query suggestions on the precision of health searches. In: Bellot et al. [64], pp. 101–111 (2018)
597. Thomee, B., Popescu, A.: Overview of the ImageCLEF 2012 flickr photo annotation and retrieval task. In: Forner et al. [209] (2011)
598. Thornley, C.V., Johnson, A.C., Smeaton, A.F., Lee, H.: The scholarly impact of TRECVid (2003–2009). J. Am. Soc. Inf. Sci. Technol. (JASIST) **62**(4), 613–627 (2011)
599. Tian, L., Huang, N., Zhang, X.: Large language model cascades and persona-based in-context learning for multilingual sexism detection. In: Goeuriot et al. [239] (2024)
600. Tommasi, T., Caputo, B., Welter, P., Güld, M.O., Deserno, T.M.: Overview of the CLEF 2009 medical image annotation track. In: Peters et al. [496], pp. 85–93 (2009)
601. Trappett, M., Geva, S., Trotman, A., Scholer, F., Sanderson, M.: Overview of the INEX 2012 snippet retrieval track. In: Forner et al. [209] (2012)
602. Trotzek, M., Koitka, S., Friedrich, C.M.: Early detection of depression based on linguistic metadata augmented classifiers revisited. In: Bellot et al. [64], pp. 191–202 (2018)
603. Tsikrika, T., Garcia Seco de Herrera, A., Müller, H.: Assessing the scholarly impact of ImageCLEF. In: Forner et al. [208], pp. 95–106 (2011)
604. Tsikrika, T., Kludas, J.: Overview of the WikipediaMM task at ImageCLEF 2008. In: Peters et al. [487], pp. 539–550 (2008)
605. Tsikrika, T., Kludas, J.: Overview of the WikipediaMM task at ImageCLEF 2009. In: Peters et al. [496], pp. 60–71 (2009)
606. Tsikrika, T., Larsen, B., Müller, H., Endrullis, S., Rahm, E.: The scholarly impact of CLEF (2000–2009). In: Forner et al. [210], pp. 1–12 (2009)
607. Tsikrika, T., Popescu, A., Kludas, J.: Overview of the wikipedia image retrieval task at ImageCLEF 2011. In: Petras et al. [501] (2011)
608. Turchi, M., Steinberger, J., Alexandrov Kabadjov, M., Steinberger, R.: Using parallel corpora for multilingual (multi-document) summarisation evaluation. In: Agosti et al. [10], pp. 52–63 (2010)
609. Turmo, J., et al.: Overview of QAST 2009. In: Peters et al. [487], pp. 197–211 (2009)
610. Turmo, J., Comas, P., Rosset, S., Lamel, L., Moreau, N., Mostefa, D.: Overview of QAST 2008. In: Peters et al. [487], pp. 296–313 (2008)
611. Turner, M., Ive, J., Velupillai, S.: Linguistic uncertainty in clinical NLP: a taxonomy, dataset and approach. In: Candan et al. [104], pp. 129–141 (2021)

612. Vallin, A., et al.: Overview of the CLEF 2005 multilingual question answering track. In: Peters et al. [493], pp. 307–331 (2005)
613. Valverde-Albacete, F., Carrillo de Albornoz, J., Peláez-Moreno, C.: A Proposal for new evaluation metrics and result visualization technique for sentiment analysis tasks. In: Forner et al. [210], pp. 41–42 (2013)
614. Villegas, M., et al.: General overview of ImageCLEF at the CLEF 2015 labs. In: Mothe et al. [419], pp. 444–461 (2015)
615. Villegas, M., et al.: General overview of ImageCLEF at the CLEF 2016 labs. In: Fuhr et al. [218], pp. 267–285 (2016)
616. Villegas, M., Paredes, R.: Overview of the ImageCLEF 2012 scalable web image annotation task. In: Forner et al. [209] (2012)
617. Voorhees, E.M.: TREC: continuing information retrieval's tradition of experimentation. Commun. ACM (CACM) **50**(11), 51–54 (2007)
618. Voorhees, E.M.: The evolution of cranfield. In: Ferro and Peters [202], pp. 45–69 (2019)
619. Wakeling, S., Clough, P.: Integrating mixed-methods for evaluating information access systems. In: Mothe et al. [419], pp. 306–311 (2015)
620. Walker, A., Starkey, A., Pan, J.Z., Siddharthan, A.: Making test corpora for question answering more representative. In: Kanoulas et al. [311], pp. 1–6 (2014)
621. Wang, J., Gilbert, A., Thomee, B., Villegas, M.: Automatic image annotation at ImageCLEF. In: Ferro and Peters [202] (2019)
622. Wang, Q., et al.: Overview of the INEX 2012 linked data track. In: Forner et al. [209] (2012)
623. Wang, X., Guo, Z., Zhang, Y., Li, J.: Medical image labeling and semantic understanding for clinical applications. In: Crestani et al. [138], pp. 260–270 (2019)
624. Wang, X., X., W., Zhang, Q.: A web-based CLIR system with cross-lingual topical pseudo relevance feedback. In: Forner et al. [210], pp. 104–107 (2013)
625. Weitzel, L., Bernardini, F., Quaresma, P., Alves, C.A., Zacharski, W., de Figueiredo, L.G.: Brazilian social mood: the political dimension of emotion. In: Fuhr et al. [218], pp. 247–255 (2016)
626. White, R.W., Oard, D.W., Jones, G.J.F., Soergel, D., Huang, X.: Overview of the CLEF-2005 cross-language speech retrieval track. In: Peters et al. [493], pp. 744–759 (2006)
627. Wiegmann, M., Stein, B., Potthast, M.: De-noising document classification benchmarks via prompt-based rank pruning: a case study. In: Goeuriot et al. [239] (2024)
628. Wilhelm-Stein, T., Eibl, M.: A quantitative look at the CLEF working notes. In: Forner et al. [210], pp. 13–16 (2013)
629. Wilhelm-Stein, T., Eibl, M.: Teaching the IR process using real experiments supported by game mechanics. In: Mothe et al. [419], pp. 312–317 (2015)
630. Wilhelm-Stein, T., Herms, R., Ritter, M., Eibl, M.: Improving transcript-based video retrieval using unsupervised language model adaptation. In: Kanoulas et al. [311], pp. 110–115 (2014)
631. Wu, S.H., Huang, H.Y.: SimpleText best of labs in CLEF-2022: simplify text generation with prompt engineering. In: Arampatzis et al. [36], pp. 198–208 (2022)
632. Xu, K., Feng, Y., Huang, S., Zhao, D.: Question answering via phrasal semantic parsing. In: Mothe et al. [419], pp. 414–426 (2015)
633. Yan, X., Gao, G., Su, X., Wei, H., Zhang, X., Lu, Q.: Hidden Markov model for term weighting in verbose queries. In: Catarci et al. [117], pp. 82–87 (2012)
634. Yang, H., Gonçalves, T.: A compound model for consumer health search. In: Bellot et al. [64], pp. 231–236 (2018)

635. Yeshambel, T., Mothe, J., Assabie, Y.: 2AIRTC: the amharic adhoc information retrieval test collection. In: Arampatzis et al. [37], pp. 55–66 (2020)
636. Yoon, W., et al.: Data-centric and model-centric approaches for biomedical question answering. In: Barrón-Cedeño et al. [55], pp. 204–216 (2022)
637. Zamani, H., Esfahani, H.N., Babaie, P., Abnar, S., Dehghani, M., Shakery, A.: Authorship Identification using dynamic selection of features from probabilistic feature set. In: Kanoulas et al. [311], pp. 128–140 (2014)
638. Zanibbi, R., Oard, D.W., Agarwal, A., Mansouri, B.: Overview of ARQMath 2020: CLEF lab on answer retrieval for questions on math. In: Arampatzis et al. [37], pp. 169–193 (2020)
639. Zellhöfer, D.: Overview of the personal photo retrieval pilot task at ImageCLEF 2012. In: Forner et al. [209] (2012)
640. Zhang, L., Rettinger, A., Färber, M., Tadic, M.: A comparative evaluation of cross-lingual text annotation techniques. In: Forner et al. [210], pp. 124–135 (2013)
641. Zhong, W., Xie, Y., Lin, J.: Answer retrieval for math questions using structural and dense retrieval. In: Arampatzis et al. [36], pp. 209–223 (2023)
642. Ziak, H., Kern: evaluation of pseudo relevance feedback techniques for cross vertical aggregated search. In: Mothe et al. [419], pp. 91–102 (2015)
643. Zingla, M.A., Latiri, C., Slimani, Y.: Tweet contextualization using association rules mining and DBpedia. In: Mothe et al. [419], pp. 318–323 (2015)
644. Zlabinger, M., Rekabsaz, N., Zlabinger, S., Hanbury, A.: Efficient answer-annotation for frequent questions. In: Crestani et al. [138], pp. 126–137 (2019)
645. Zuo, C., Karakas, A., Banerjee, R.: To check or not to check: syntax, semantics, and context in the language of check-worthy claims. In: Crestani et al. [138], pp. 271–283 (2019)

Conference Papers

Sexism Identification on TikTok: A Multimodal AI Approach with Text, Audio, and Video

Iván Arcos[1](✉)[iD] and Paolo Rosso[1,2][iD]

[1] Universitat Politècnica de València, Valencia, Spain
iarcgab@etsinf.upv.es, prosso@dsic.upv.es
[2] ValgrAI-Valencian Graduate School and Research Network of Artificial Intelligence, Valencia, Spain

Abstract. Sexism persists as a pervasive issue in society, particularly evident on social media platforms like TikTok. This phenomenon encompasses a spectrum of expressions, ranging from subtle biases to explicit misogyny, posing unique challenges for detection and analysis. While previous research has predominantly focused on textual analysis, the dynamic nature of TikTok demands a more comprehensive approach. This study leverages advancements in Artificial Intelligence (AI), specifically multimodal deep learning, to establish a robust framework for identifying and interpreting sexism on TikTok. We compiled the first dataset of TikTok videos tailored for analyzing sexism in both English and Spanish. This dataset serves as an initial benchmark for comparing models or for future investigations in this area. By integrating text, linguistic features, emotions, audio, and video features, this study identifies unique indicators of sexist content. Multimodal analysis surpasses text-only methods, particularly in understanding the intentions behind sexism.

Keywords: Multimodal Sexism Identification · TikTok · Artificial Intelligence

1 Introduction

Sexism refers to multifaceted, encompassing subtle expressions that can be as insidious as explicit misogyny. Whether presented as seemingly positive remarks, jokes, or offensive comments, sexism permeates various aspects of individuals' lives, influencing domestic and parenting roles, career opportunities, sexual image, and life expectations. Recognizing the diverse forms of sexism is crucial to understanding its impact on society.

Social media platforms have become conduits for the dissemination of sexist content, perpetuating and even normalizing gender differences and biased attitudes. The Internet, with its vast reach, reflects and amplifies societal inequalities and discrimination against women. This study is particularly crucial given the significant presence of teenagers on social media platforms, urging the need for urgent investigation and societal dialogue, especially from an educational standpoint.

TikTok, a dynamic social media platform with 1.218 billion users aged 18 and above, has revolutionized content consumption. Known for its role in shaping fast-paced, short-form videos, TikTok has become a hub for diverse content creation and dissemination. Out of 5.3 billion internet users worldwide, 23% actively engage with TikTok[1]. A study conducted by The Observer revealed that TikTok's algorithm can lead users down a path of increasingly sexist content[2]. This raises concerns about the platform's potential to reinforce negative preconceptions and misogyny. If someone comes to the app already thinking negatively about a group, TikTok's algorithm shows them more content that supports and even makes those negative thoughts seem acceptable.

The rest of the paper is structured as follows. Section 2 presents some related work. Section 3 introduces the tasks of sexism detection, source intention classification and sexism categorization, as well as the dataset we compiled. Section 4 describes the models for text, audio, video and multimodal data. Section 5 presents the results and, finally, Sect. 6 draws some conclusions and discusses future work.

2 Related Work

Hate Speech (HS) is generally described as any form of communication that belittles a person or a group based on attributes such as race, ethnicity, gender, sexual orientation, nationality, religion, among others [13]. When the target of hate speech is women, it manifests as a form of misogyny. However, *misogyny*, as defined by the Oxford English Dictionary[3], refers to feelings of hatred or dislike towards women, or beliefs that devalue women compared to men. Misogyny can exist in behaviors, attitudes, or beliefs that demean women or see them as inferior to men, without the need for overt hate speech. On the other hand, *sexism* is defined as prejudice, stereotyping, or discrimination, often against women, based on sex. Unlike misogyny, sexism can manifest subtly, such as through gender stereotypes, traditional gender roles or unequal access to opportunities[4].

The field of NLP has increasingly focused on detecting hate speech and sexism, driven by their growing societal impacts, especially on social platforms. Early efforts included a foundational corpus of misogynous tweets that explored various NLP features and machine learning models for classifying misogynistic language [3]. Building on these approaches, SemEval-2019 Task 5 targeted hate speech against immigrants and women, featuring a binary classification task to identify hate speech and a more detailed classification to analyze its aggressive features and targets [5]. Following up, the AMI challenge at Evalita2020 focused on misogyny and aggressiveness in Italian tweets [10]. More recently,

---

[1] https://www.businessofapps.com/data/tiktok-report/.
[2] https://medium.com/moviente/does-tiktok-have-a-misogyny-problem-c1033fbb2cc2.
[3] https://www.oed.com/view/Entry/misogyny.
[4] https://www.oed.com/view/Entry/sexism.

SemEval-2023 Task 10 developed a hierarchical taxonomy of sexist content and curated a dataset of 20,000 social media comments to enhance the explainability of detection methods [12]. Since 2021, the EXIST task has also been addressing sexism identification in social networks, further contributing to the field's ongoing efforts [14–16].

Recent advancements in multimodal analysis have significantly enhanced the detection of hate speech in memes and images. The Multimodal Hate Speech Event Detection task organized in 2023 explored binary and target-specific detection strategies in text-embedded images, demonstrating the effectiveness of multimodal approaches in identifying hate speech [17]. A novel method introduced in 2023 utilizes pre-trained vision-language models (PVLMs) for hateful meme detection [6]. Additionally, a study conducted in 2018 demonstrated the superiority of a multimodal approach over unimodal methods in detecting sexist content in advertisements [11]. The Multimedia Automatic Misogyny Identification (MAMI) task at SemEval-2022 focused on identifying misogynous content in memes [9].

Recent studies have significantly advanced the field of hate speech detection in video content. The authors in [18] developed a method that classifies videos into normal or hateful categories by exclusively analyzing the transcriptions of spoken content. In another important study, published a dataset of Portuguese videos, which primarily includes textual features extracted from the video content [1]. Furthermore, in [7] created a dataset by manually annotating around 43 h of BitChute videos. Their multi-modal approach led to a 5.7% improvement in F1-score over systems that used only one modality.

Our work focuses on addressing sexism detection in TikTok videos by integrating text, images, and audio analysis. While previous efforts concentrated on text and images, this research expands into video content.

3 Tasks and Datasets

3.1 Tasks

Following what was done in EXIST[5], our aim is to address sexism identification in the following three tasks:

1. **Sexism Detection.** Determine if TikTok videos contain sexist content. This is a binary classification:
    - **Not Sexist.** Videos not focusing on gender-related themes.
    - **Sexist.** Videos discussing or portraying gender-related stereotypes or issues.
2. **Source Intention Classification.** Categorize sexist videos based on the creation intent:
    - **Reported Sexist.** Videos sharing experiences of encountering sexism.
    - **Direct Sexist.** Videos explicitly promoting sexist beliefs.

---

[5] http://nlp.uned.es/exist2024/.

3. **Sexism Categorization.** Classify sexist videos by the aspect of sexism they exhibit:
    - **Ideological and Inequality.** Videos undermining women's rights or contributions.
    - **Role Stereotyping and Dominance.** Videos perpetuating gender role stereotypes.
    - **Objectification.** Videos portraying women solely as objects of desire.
    - **Sexual Violence.** Videos containing or promoting sexual harassment or assault.
    - **Misogyny and Non-sexual Violence.** Videos expressing hostility or violence towards women.

3.2 Dataset

The TikTok dataset was collected using Apify's TikTok Hashtag Scraper tool[6], focusing on hashtags associated with potentially sexist content. A total of 185 Spanish hashtags and 61 English hashtags were selected.

The annotation was conducted using Servipoli's service[7] with eight students organized in pairs. Each pair, consisting of one male and one female student to avoid biases, was tasked with annotating 1000 TikTok videos in either Spanish or English. A preliminary test using 10 TikToks helped familiarize the annotators with the process. In cases of disagreement between annotator pairs, the final decision were made by a research team member. The Spanish TikTok dataset consists of a total of 1969 TikToks, with a cumulative duration of 13.86 h. Among these, 817 (41.49%) are categorized as non-sexist, while 1152 (58.51%) are considered sexist. There is a significant imbalance between the reported sexist content, which accounts for 362 (31.42%), and directed sexist content, which comprises 790 (68.58%). The English TikTok dataset comprises a total of 1773 TikToks, with a cumulative duration of 11.83 h. Out of these, 975 (55%) are non-sexist, while 798 (45%) are categorized as sexist. Similar to the Spanish dataset, there is an imbalance between the reported sexist content, which constitutes 297 (37.22%), and directed sexist content, which accounts for 501 (62.78%). Tables 1, 2 show the detailed statistics.

Table 1. Statistics of the Spanish TikTok Dataset

	Non-sexist	Reported	Direct
Count (%)	817 (41.49%)	362 (31.42%)	790 (68.58%)
Count (%) by Category			
Ideological Inequality	–	140 (38.67%)	204 (25.82%)
Stereotyping-Dominance	–	186 (51.38%)	536 (67.84%)
Objectification	–	47 (12.98%)	156 (19.75%)
Sexual Violence	–	95 (26.24%)	54 (6.84%)
Misogyny-Non-Sexual Violence	–	89 (24.59%)	134 (16.96%)

---
[6] https://apify.com/clockworks/tiktok-hashtag-scraper.
[7] https://www.servipoli.es/.

Cohen's Kappa was utilized to evaluate inter-annotator agreement (IAA) across two tasks. In the *sexism detection* task, the average Kappa value between annotator pairs was 0.499, indicating moderate agreement ($\kappa \in [0.41, 0.60]$). The *source intention classification* task showed substantial agreement with an average Kappa of 0.672 ($\kappa \in [0.61, 0.80]$).

Table 2. Statistics of the English TikTok Dataset

	Non-sexist	Reported	Direct
Count (%)	975 (54.99%)	297 (37.22%)	501 (62.78%)
Count (%) by Category			
Ideological Inequality	–	180 (60.61%)	158 (31.54%)
Stereotyping-Dominance	–	252 (84.85%)	391 (78.04%)
Objectification	–	91 (30.64%)	135 (19.75%)
Sexual Violence	–	53 (17.85%)	18 (3.59%)
Misogyny-Non-Sexual Violence	-	46 (15.49%)	57 (11.38%)

Although the dataset employed in the experiments was entirely annotated by human annotators, we were interested in investigating the Kappa values for annotations made by GPT-3.5 Turbo. The average Kappa values for the *sexism detection* task and the *source intention classification* task by GPT-3.5 Turbo were 0.282 and 0.246, respectively. These results, which show fair agreement ($\kappa \in [0.21, 0.40]$), indicate that while GPT-3.5 Turbo achieves a basic level of concordance with human annotations, it does not yet reach the agreement levels of human annotators.

While the values of IAA can be considered good in the two first tasks, the are low in the third task that involves the categorization into five categories of sexism. Here, the average Kappa values across categories were: Ideological Inequality ($\kappa = 0.306$), Stereotyping-Dominance ($\kappa = 0.409$), Objectification ($\kappa = 0.312$), Sexual Violence ($\kappa = 0.396$), and Misogyny-Non-Sexual Violence ($\kappa = 0.179$).

4 Models

4.1 Text Models

In the preprocessing phase of TikTok text data, several measures have been taken to ensure an accurate and unbiased representation of the content. Hashtags and user mentions are removed from titles to prevent biases in determining whether a video is sexist or inferring its intent or category. Emojis are retained and interpreted using the *emoji* library[8]. For the transcription of spoken content,

---

[8] https://pypi.org/project/emoji/.

the *clu-ling/whisper-large-v2-spanish* model is used for Spanish TikToks[9] and *openai/whisper-large-v3* for English TikToks[10]. Additionally, Optical Character Recognition (OCR) is conducted using the *easyocr* library, which supports over 80 languages[11]. Text is extracted from each TikTok video every 30 frames to ensure no critical information is missed. Despite these efforts, the preprocessing is not perfect and may include noise in the transcriptions or OCR, or miss certain information.

In this work, various linguistic resources were utilized for text by amalgamating title, transcription, and OCR text of the TikToks. These resources include the Linguistic Inquiry and Word Count (LIWC) and HURTLEX lexicons, which provide tools for psycholinguistic and socio-linguistic analysis. Furthermore, pre-trained transformer models were employed to extract complex language features. EmoRoBERTa[12], a model trained on the GoEmotions dataset[13] comprising 58000 Reddit comments labeled with 28 different emotions, was used to extract emotional contexts from the text. Additionally, models designed for identifying hate speech were applied, specifically the *twitter RoBERTa base* for English[14] and *BETO hate speech* for Spanish[15].

Transcriptions, OCR, and titles of TikToks were transformed into features for three tasks. Term Frequency - Inverse Document Frequency (TF-IDF) was applied to both English and Spanish TikToks. Pretrained transformers, including *RoBERTa large*[16] and *twitter RoBERTa base* for English, and *RoBERTa large*[17] and *BETO hate speech* for Spanish, were utilized. TF-IDF captures term frequency and document specificity, while pretrained embeddings learn contextualized features. These features will feed into Support Vector Machine (SVM), Multi-Layer Perceptron (MLP), Extra-Trees classifiers, and a Stacking ensemble. Additionally, linguistic variables extraction will be experimented with.

4.2 Audio Models

In the context of detecting sexism in TikTok videos, audio analysis plays a crucial role due to its ability to capture unique aspects that may not be apparent from textual analysis alone. Mel-Frequency Cepstral Coefficients (MFCC) and Pre-trained Wav2Vec2 Embeddings are employed for this purpose.

MFCC. Widely used in speech processing, MFCC effectively captures spectral characteristics [2], enabling comprehensive analysis for tasks such as classification and discrimination detection.

---

[9] https://huggingface.co/clu-ling/whisper-large-v2-spanish.
[10] https://huggingface.co/openai/whisper-large-v3.
[11] https://pypi.org/project/easyocr/.
[12] https://huggingface.co/arpanghoshal/EmoRoBERTa.
[13] https://huggingface.co/datasets/go_emotions.
[14] https://huggingface.co/cardiffnlp/twitter-roberta-base-hate-multiclass-latest.
[15] https://huggingface.co/piuba-bigdata/beto-contextualized-hate-speech.
[16] https://huggingface.co/FacebookAI/roberta-large.
[17] https://huggingface.co/PlanTL-GOB-ES/roberta-large-bne.

Pre-trained Wav2Vec2 Embeddings. Wav2Vec2, proposed in [4], extracts features from audio signals. For feature extraction, pre-trained models such as `Wav2Vec2 large` for Spanish[18] and for English[19] have been utilized.

These features are utilized with SVM, MLP, Extra-Trees, and a Stacking ensemble, similar to text models.

4.3 Video Models

Video models provide a unique perspective by capturing visual features and temporal dynamics, complementing text and audio analysis. In this study, we explore three video models for identifying sexism in TikTok content.

ResNet+LSTM

This model extracts N frames from TikTok videos and utilizes a pre-trained ResNet model for feature extraction[20]. The ResNet extracts 2048-dimensional feature vectors for each frame, which are then processed by an Long Short-Term Memory (LSTM) to model temporal relationships. Finally, fully connected layers produce predictions for sexism detection or categorization.

ViT+LSTM

Similar to the previous model, this one extracts features from each frame but utilizes Vision Transformers (ViT) instead of Convolutional Neural Networks (CNNs) like ResNet [8]. ViT extracts 768-dimensional feature vectors for each frame, which are temporally modeled by an LSTM. The LSTM output is passed through fully connected layers to generate the final prediction.

BLIP+TF-IDF

In this model, N frames from each TikTok video are processed to generate captions using the `BLIP large model`[21] model, which has been fine-tuned on the `Captions` dataset[22]. Each caption undergoes TF-IDF transformation to create feature vectors, which are then processed by fully connected layers or any other machine learning model for the final prediction.

4.4 Multimodal Models

Our comprehensive approach to detect sexism in TikTok videos involves an SVM that integrates all modalities: text, audio, and video. The text component is represented by embeddings; the audio by MFCCs or Wav2Vec2 embeddings; and the video by average pooled features from ResNet, ViT or TF-IDF vectors derived from captions generated by the BLIP model. For video, average pooled features from either ResNet or ViT. This choice was driven by the limited size of our dataset, which made the average pooling of video features more effective than sequential processing with LSTMs.

---

[18] https://huggingface.co/jonatasgrosman/wav2vec2-large-xlsr-53-spanish.
[19] https://huggingface.co/jonatasgrosman/wav2vec2-large-xlsr-53-english.
[20] https://keras.io/api/applications/resnet/#resnet50-function.
[21] https://huggingface.co/unography/blip-large-long-cap.
[22] https://huggingface.co/datasets/unography/laion-14k-GPT4V-LIVIS-Captions.

Ablation tests were conducted to determine the impact of each modality on the overall performance of the model. By excluding one type of feature (text, audio, or video) at a time, we were able to identify the critical components necessary for effective sexism detection. These tests highlight the unique contributions of each modality.

5 Results

5.1 Text Models

Our analysis reveals notable linguistic and emotional distinctions between sexist and non-sexist content on TikTok. Sexist videos typically contain language related to sexuality, affect, and derogation, including terms linked to the seven deadly sins, while non-sexist content features words associated with achievement and relationships. Emotional analysis shows that non-sexist TikToks evoke positive feelings like joy and love, whereas sexist ones provoke negative reactions such as anger and sadness. Interestingly, direct sexist content sometimes generates amusement and neutrality, possibly due to perceived humor.

Embeddings outperform TF-IDF in capturing semantic relationships, but linguistic variables do not significantly enhance embeddings' performance, suggesting redundancy. However, adding linguistic variables to TF-IDF improves performance, indicating TF-IDF's limitations in capturing all linguistic features. In English, SVM with embeddings achieved the best results for task 1 (macro F1: 0.728) and task 2 (macro F1: 0.701). For Spanish, a stacking model with linguistic features on embeddings performed best for task 1 (macro F1: 0.696), while Extra-Trees with TF-IDF features excelled in task 2 (macro F1: 0.696). Task 3 results were suboptimal due to insufficient representation of data in some categories, obtaining a macro F1 of approximately 0.5 in both languages. Performance details for the best models are summarized in Table 3.

Table 3. Summary of the top performing text models for each task

Language	Task	Model	M-F1	F1	P	R
English	Task 1	SVM (embeddings)	0.728	0.705	0.694	0.719
	Task 2	SVM (embeddings)	0.701	0.788	0.771	0.806
	Task 3	SVM + ling. (embeddings)	0.490	–	–	–
Spanish	Task 1	Stacking + ling. (embeddings)	0.696	0.757	0.742	0.773
	Task 2	Extra-Trees (TF-IDF)	0.696	0.816	0.803	0.830
	Task 3	SVM + ling. (embeddings)	0.498	–	–	–

5.2 Audio Models

For the sexism detection task, the audio variables, both MFCCs and Wav2Vec2 embeddings, do not perform as strongly, achieving an F1-macro around 0.6. This suggests that audio features might not capture the nuanced patterns of sexism as effectively as text-based features for this particular task.

However, when considering the task of detecting the intention behind sexism, the audio-based models present a more competitive performance. Notably, the Wav2Vec2 embeddings outperform the MFCCs significantly. In English, the SVM model using Wav2Vec2 achieves the best performance with an F1-macro of 0.654, while in Spanish, it reaches 0.687, also using SVM. This demonstrates that audio embeddings like Wav2Vec2 can approximate the performance of text-based models for this specific task. For task 3, the best results are again obtained with Wav2Vec2 embeddings employing SVM. In English, an F1-macro of 0.381 is achieved, while in Spanish, a value of 0.367 is reached. Performance details for the best models are summarized in Table 4.

Table 4. Summary of the top performing audio models for each task

Language	Task	Model	M-F1	F1	P	R
English	Task 1	SVM (Wav2Vec2)	0.608	0.567	0.572	0.565
	Task 2	SVM (Wav2Vec2)	0.654	0.733	0.750	0.719
	Task 3	SVM (Wav2Vec2)	0.381	–	–	–
Spanish	Task 1	SVM (MFCCs)	0.579	0.648	0.653	0.645
	Task 2	SVM (Wav2Vec2)	0.687	0.810	0.800	0.820
	Task 3	SVM (Wav2Vec2)	0.367	–	–	–

5.3 Video Models

This section analyzes the performance of three different video models used to identify sexism in TikTok videos. For the English dataset, the best-performing model for both task 1 and task 2 was found to be BLIP+TF-IDF, achieving macro F1 scores of 0.630 and 0.687, respectively. Similarly, in the Spanish dataset, BLIP+TF-IDF emerged as the top-performing model for both task 1 and task 2, with macro F1 scores of 0.592 and 0.693, respectively. However, for task 3 in English, BLIP+TF-IDF obtained the highest F1-macro score of 0.302, while in Spanish, ViT+LSTM achieved the highest F1-macro score of 0.312. Performance details for the best models are summarized in Table 5.

Table 5. Summary of the top performing video models for each task

Language	Task	Model	M-F1	F1	P	R
English	Task 1	BLIP+TF-IDF	0.630	0.578	0.611	0.553
	Task 2	BLIP+TF-IDF	0.687	0.794	0.749	0.846
	Task 3	BLIP+TF-IDF	0.302	–	–	–
Spanish	Task 1	BLIP+TF-IDF	0.592	0.708	0.650	0.778
	Task 2	BLIP+TF-IDF	0.693	0.827	0.791	0.867
	Task 3	ViT+LSTM	0.312	–	–	–

Figure 1 shows that in Spanish TikTok videos, terms like 'dark', 'purple', and 'cheerful' in video captions are associated with sexist classifications by the BLIP model, while neutral terms such as 'dog' or 'road' are linked to non-sexist content. Additionally, 'dark background' and 'smiley' relate to direct sexism, whereas 'a woman' and 'social event' are indicative of reported sexism. Similar patterns are observed in English TikToks.

5.4 Multimodal Models

In task 1, using only the text modality for detecting sexism presence outperformed multimodal approaches, indicating that textual linguistic features were more crucial than audio or video cues. Similarly, for task 3, which aimed to identify specific categories of sexism, text alone showed superior performance.

For task 2 of detecting the intent of sexism, the multimodal approach significantly improved the outcomes compared to unimodal models for both English and Spanish languages. Figure 2 illustrates these findings, demonstrating the efficacy of integrating multiple modalities. The best-performing unimodal model was using the ViT in both English and Spanish, achieving an F1-macro score of 0.709 and 0.720, respectively.

However, the highest performance was observed with the TAV model (combining Text, Audio, and Video) that excluded linguistic features (L). Specifically, for audio, the model employed MFCCs, and for video, it used ViT features. This configuration led to F1-macro scores of 0.753 and 0.768 in English and Spanish respectively, marking an improvement of 4.4% and 4.8% over the best unimodal models. The multimodal approach demonstrates the benefits of integrating text, audio, and video to achieve a more comprehensive understanding of content, leading to more accurate detection of sexist intent.

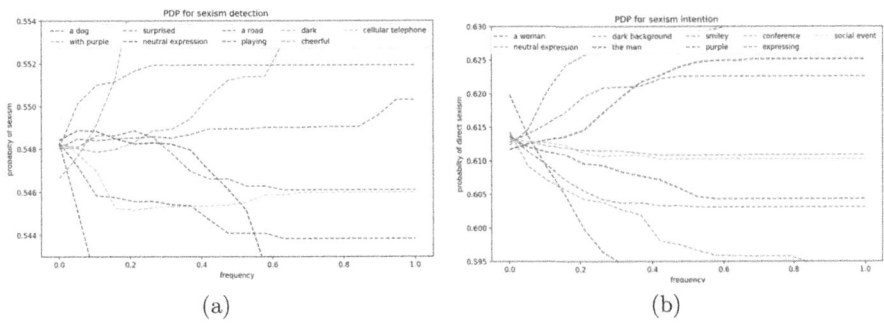

Fig. 1. Partial Dependence Plots (PDPs) for detected words by the BLIP model on the (a) sexism detection task, and (b) sexism intent task.

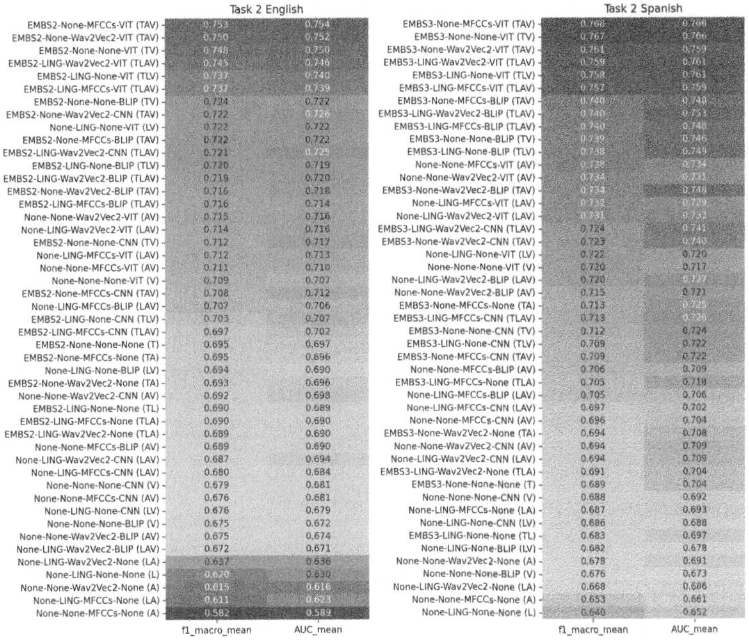

Fig. 2. Multimodality Results for Task 2

6 Conclusions and Future Work

In this work, the detection of sexist content on TikTok was explored through multimodal analysis. The study examined specific features of each modality (text, audio, and video) to ascertain their effectiveness in detecting sexism or its underlying intent. While the integration of multiple modalities did not yield improvements in some tasks, it notably surpassed unimodal models in detecting sexist intent.

This finding underscored the superiority of a multimodal approach over unimodal models, particularly in discerning the intent behind sexist content. The TAV model (Text, Audio, and Video, excluding linguistic features) emerged as the top performer, achieving remarkable results with F1-macro scores of 0.753 and 0.768 for English and Spanish, respectively. Notably, this configuration led to an improvement of 4.4% and 4.8% over the best unimodal models.

Future studies should focus on expanding datasets, particularly to better represent various categories of sexism, and integrate innovative paradigms such as Learning With Disagreement (LeWiDi) [14] to handle conflicting annotations, which address the subjective nature of sexism annotations. Improving text preprocessing to minimize information loss from transcription or optical character recognition (OCR) is also crucial. Moreover, addressing dataset imbalance through mitigation steps like class weighting, resampling techniques, and data augmentation can ensure model robustness.

Exploring more types of features like linguistic elements, user interactions and comments can offer a deeper understanding in detecting sexist content. Fine-tuning transformers or other large models, rather than just using pretrained embeddings, can significantly boost performance. It's crucial to broaden the diversity of datasets to enhance the generalizability of these models across different global contexts and demographics. This approach aims to reduce gender-based biases and promote a more inclusive online environment.

Acknowledgements. The work of Paolo Rosso was conducted within the framework of the FairTransNLP-Stereotypes project - "Fairness and Transparency for equitable NLP applications in social media: Identifying stereotypes and prejudices and developing equitable systems". This project is funded by the grant PID2021-124361OB-C31, supported by MCIN/AEI/10.13039/501100011033 and by the European Regional Development Fund (ERDF), under the banner "A way of making Europe".

References

1. Alcântara, C., Moreira, V., Feijo, D.: Offensive video detection: dataset and baseline results. In: Proceedings of the Twelfth Language Resources and Evaluation Conference, pp. 4309–4319 (2020)
2. Ali, S., Tanweer, S., Khalid, S., Rao, N.: Mel frequency cepstral coefficient: a review, January 2021
3. Anzovino, M., Fersini, E., Rosso, P.: Automatic identification and classification of misogynistic language on twitter, pp. 57–64, May 2018
4. Baevski, A., Zhou, Y., Mohamed, A., Auli, M.: wav2vec 2.0: a framework for self-supervised learning of speech representations. In: Larochelle, H., Ranzato, M., Hadsell, R., Balcan, M., Lin, H. (eds.) Advances in Neural Information Processing Systems, vol. 33, pp. 12449–12460. Curran Associates, Inc. (2020)
5. Basile, V., et al.: Semeval-2019 task 5: multilingual detection of hate speech against immigrants and women in twitter, pp. 54–63, January 2019
6. Cao, R., Hee, M.S., Kuek, A., Chong, W.H., Lee, R.K.W., Jiang, J.: Pro-cap: leveraging a frozen vision-language model for hateful meme detection, pp. 5244–5252, October 2023

7. Das, M., Raj, R., Saha, P., Mathew, B., Gupta, M., Mukherjee, A.: Hatemm: a multi-modal dataset for hate video classification. In: Proceedings of the International AAAI Conference on Web and Social Media, vol. 17, pp. 1014–1023, June 2023
8. Dosovitskiy, A., et al.: An image is worth 16 × 16 words: transformers for image recognition at scale (2021)
9. Fersini, E., et al.: Semeval-2022 task 5: multimedia automatic misogyny identification. In: Proceedings of the 16th International Workshop on Semantic Evaluation (SemEval-2022), pp. 533–549. Association for Computational Linguistics (2022)
10. Fersini, E., Nozza, D., Rosso, P.: AMI @ EVALITA2020: automatic misogyny identification, pp. 21–28, January 2020
11. Gasparini, F., Erba, I., Fersini, E., Corchs, S.: Multimodal classification of sexist advertisements, pp. 565–572, January 2018
12. Kirk, H., Yin, W., Vidgen, B., Röttger, P.: SemEval-2023 task 10: explainable detection of online sexism. In: Ojha, A.K., Doğruöz, A.S., Da San Martino, G., Tayyar Madabushi, H., Kumar, R., Sartori, E. (eds.) Proceedings of the 17th International Workshop on Semantic Evaluation (SemEval-2023), pp. 2193–2210. Association for Computational Linguistics, Toronto, Canada, July 2023
13. Nockleby, J.T.: Hate Speech, 2nd edn. Macmillan, New York (2000)
14. Plaza, L., et al.: Overview of EXIST 2023 - learning with disagreement for sexism identification and characterization, pp. 316–342, September 2023
15. Rodríguez-Sánchez, F., et al.: Overview of exist 2021: sexism identification in social networks. Procesamiento del Lenguaje Natural **67**, 195–207 (2021)
16. Rodríguez-Sánchez, F., et al.: Overview of exist 2022: sexism identification in social networks. Procesamiento del Lenguaje Natural **69**, 229–240 (2022)
17. Thapa, S., Jafri, F.A., Hürriyetoğlu, A., Vargas, F., Lee, R.K.W., Naseem, U.: Multimodal hate speech event detection - shared task 4, case 2023. In: Proceedings of the 6th Workshop on Challenges and Applications of Automated Extraction of Socio-political Events from Text, pp. 151–159. Incoma Ltd (2023)
18. Wu, C.S., Bhandary, U.: Detection of hate speech in videos using machine learning. In: 2020 International Conference on Computational Science and Computational Intelligence (CSCI), pp. 585–590. IEEE (2020)

Knowledge Acquisition Passage Retrieval: Corpus, Ranking Models, and Evaluation Resources

Artemis Capari[1](✉), Hosein Azarbonyad[1], Georgios Tsatsaronis[1], Zubair Afzal[1], Judson Dunham[1], and Jaap Kamps[2]

[1] Elsevier, Amsterdam, The Netherlands
a.capari@elsevier.com
[2] University of Amsterdam, Amsterdam, The Netherlands

Abstract. Knowledge acquisition passage retrieval is a task that captures search in a learning or educational setting, where users seek to find key educational information within their field of interest. Traditional relevance assessments used in ad-hoc retrieval tasks tend to focus on topical relevance, often overlooking other factors such as the "informativeness" of the retrieved educational content in relation to the user's knowledge acquisition needs. This paper presents a new test collection for the knowledge acquisition passage retrieval (KAPR) task, constructed using the data and production systems of a large academic publisher containing: First, a set of search requests covering key educational topics/concepts across different science domains. Second, a large corpus of passages extracted from review (survey) articles published in over 2,700 journals as well as the content of 43,000 books published in a wide range of science domains. Third, relevance assessments on both topical relevance as well as informativeness, reflecting the task-specific relevance. This resource enables direct evaluation of the user's utility of the retrieved content and provides a comparative analysis with traditional topical relevance. Our findings indicate a strong correlation between relevance and informativeness, although the distribution of these labels varies per domain.

Keywords: passage retrieval · knowledge acquisition · test collections

1 Introduction

It is often argued that traditional IR evaluation methods are not sufficient as they assess retrieved documents only on their topical relevance in relation to the query [4,9], while failing to consider what relevance actually entails from a user's perspective. Widely used test collections such as MS MARCO and early versions of TREC have been criticized for their rather simplistic definition of what is deemed 'relevant' [24,26]. Their binary relevance judgments only address topical relevance and therefore include documents that might be 'about' a certain topic, but are not useful for learning about said topic [26]. Furthermore,

Table 1. Examples of passages where the degree of relevance (Rel) and informativeness (Inf) do not align. Labels are assigned by subject matter experts.

Concept in Domain	Snippet	Rel.	Inf.	Explanation
Natural Disaster in Earth and Planetary Science	Floods are the most frequent natural disaster. They represent approximately 40% of the total number of natural disasters worldwide. Figure 1 illustrates the number of flood disasters per country from 1990 to 2007. Asian countries carry the largest burden of floods. In particular, India, Bangladesh, and China are the countries where floods affect the most people...	1	0	Is about a sub-type of the concept, mainly discussing examples of it without actually explaining it.
Poisson Distribution in Mathematics	The Poison distribution arises from situations where there is a large number of opportunities for the event under scrutiny to occur but a small chance that it will occur on any one trial. The number of cases of bubonic plague would follow Poison: A large number of patients can be found with chills, fever, tender lymph nodes...	2	1	Is about the concept, but does not exactly explain the concept.
Narcissism in Psychology	Hubris as extreme narcissism is egotism, self-centeredness, grandiosity, lack of empathy, exploitation, exaggerated self-love, recklessness, and failure to acknowledge nonmanipulative boundaries... Diminutive states of hubris reflect the classic "show-off" personality. Pretentious styles often hide insecurity. They ostentatiously proclaim wished-for minimal or nonexistent assets revealing a sense of deep-seated privation...	1	2	Not about the concept itself, but rather a sub-concept. Still, very informative for a reader with knowledge of the concept.

these datasets have also been criticized for the sparsity of their relevance labels, where queries often only have one or a few positive labels, but neural rankers often return unlabeled documents that are superior or equal in quality to the labeled documents. Arabzadeh et al. [1] therefore argues that the method used to determine MS MARCO qrels does not capture relevance in the conventional manner; rather, it merely aims to identify an answer, not necessarily the best one. Hence, we must carefully consider the different types of relevance and which types are important for the specific task under consideration. The notion of relevance has long been discussed and defined in many different, but often similar ways [2,6,9,21,22]. The main aspect used for evaluating IR systems is **relevance**, which is generally defined as *"direct matching between the overall topic of a relevant document and the overall topic of the user need"* [8]. This notion of relevance is the main driver behind most existing models and is the main proxy to evaluate such models. Topical relevance is a good criterion to assess the similarity of user queries and documents in an ad-hoc retrieval task, but there is a limited study of the generalizability of this metric to other types of search tasks. In this paper, we further consider the evaluation of IR systems from an **informativeness** aspect. Our notion of informativeness is based on a type of relevance often referred to as *pertinence*, which is defined as: *"Relation between the cognitive state of knowledge of a user and information or information objects (retrieved or in the systems file, or even in existence)"* [22]. Examples can be found in Table 1.

The specific retrieval task we consider in this work is the knowledge acquisition passage retrieval task introduced above. The primary objective of this research is to introduce a new dataset specifically designed for this retrieval task and evaluate the performance of existing lexical and semantic matching retrieval models from both topical relevance and informativeness aspects on this dataset to gain insights into the utility of these models in such a task. In doing so, we answer our main research questions:

1. To what extent do relevance and informativeness correlate with each other in the knowledge acquisition passage retrieval task?
2. Which factors contribute to an increase in the performance of retrieval models in the knowledge acquisition passage retrieval task?
   (a) Which semantic matching model is most suited for this task and how does it compare to the lexical matching models?
   (b) Does the difference in text style between different domains affect the performance of the passage retrieval models?

Our main contributions are the following. First, we create a benchmark set, which we refer to as KAPR (Knowledge Acquisition Passage Retrieval), containing 100 queries per topic across 20 different science domains and use a pooling approach to collect both relevance and informativeness scores from subject matter experts. Each passage is evaluated on a three-point scale, assigning a value of 0 for not being relevant or informative, 1 for partial relevance and informativeness, and 2 for high relevance/informativeness. To the best of our knowledge, KAPR is the first dataset containing both labels on the various science domains. Second, we evaluate the performance of different retrieval models on the collected benchmark set from both relevance and informativeness aspects and analyze their effectiveness in the knowledge acquisition retrieval setting. Third, we measure the effectiveness of different retrieval models across science domains and analyze their utility from different aspects on all domains.

2 Related Work

Dense retrieval models have been shown to outperform traditional sparse retrieval models, such as BM25 [27], in a variety of tasks, including question answering and document ranking. Two popular types of such dense retrieval models are bi-encoders [3,10] and cross-encoders [11,13,17]. Both models have the same objective, i.e. capturing the semantic meaning of queries and documents into dense vector representations, but differ in the architecture of the neural network used to learn their representations. It is argued that the majority of information retrieval methods are not sufficient as they assess retrieved documents solely based on their topical relevance to the query, without taking into account the user's perspective on what relevance truly means [4,9].

In the knowledge acquisition use-case, it is imperative that the results are highly informative to the user and not just relevant to the topic. Hence, we must carefully consider the different types of relevance and which types are important for our specific task. The notion of relevance has long been discussed and defined in many different, but often similar ways. For instance, Borlund [2] distinguished between two types of relevance: *objective/system-based relevance* and *subjective/user-based relevance*. Saracevic [21] further categorized the subjective relevance into four sub-types: "topicality", "cognitive relevance/pertinence", "situational relevance' and "intentional relevance". However, as most of these types are highly subjective per user and would require some type of user-based interaction, we define our notions of relevance that are important to this specific research. The first aspect we assess our test collection on concerns *topical relevance*, which is generally defined as *"direct topical matching or the direct matching between the overall topic of a relevant document and the overall topic of the user need"* [8]. Traditionally, most IR systems are evaluated on this type of relevance [5,24], and we will simply refer to it as **relevance**. However, this type of relevance does not necessarily help a user learn about the topic. Hence, we evaluate the snippets on how ***informative*** a snippet is as well. Our notion of **informativeness** is based on a type of relevance often referred to as *pertinence*, which is defined by Saracevic [22] as follows: *"Relation between the cognitive state of knowledge of a user and information or information objects (retrieved or in the systems file, or even in existence). Cognitive correspondence, informativeness, novelty, information quality, and the like are criteria by which cognitive relevance is inferred"*, thus better reflecting the educational level of a document.

3 Test Collection Construction

We aim to evaluate different state-of-the-art ranking models from two different perspectives: traditional topical relevance and informativeness. These ranking models will perform a KAPR task, where given a query from a user on a scientific topic, passages that are both relevant as well as informative should be retrieved. To do this, we start by building a test collection[1], containing a set of queries

---
[1] The dataset can be found here: https://github.com/acapari/KAPR.git.

and passages with relevance and informativeness labels for query-passage pairs. In this section, we describe our methodology for building this data set.

3.1 Query Selection

The first step in creating the benchmark set is to select a set of queries to annotate passages for. Considering our setting is a knowledge acquisition task, we choose a set of key educational topics/concepts in different science domains. For each of the science domains[2], five concepts were selected from an extensive science taxonomy curated by subject matter experts. Three of the concepts were chosen based on popularity (the number of times concepts are searched by users based on click logs of a large search engine on scientific documents), while the remaining two concepts were selected randomly from the science taxonomy, resulting in a collection of 100 concepts.

3.2 Document Collection

For the document collection, we use a set of review (survey) articles published in over 2,700 journals as well as the content of 43,000 books published in different science domains. Considering the targeted task is focused around knowledge acquisition, books and review journal articles could be appropriate sources for addressing the information need in this setting. We consider each section (sub-section) in such documents as a passage. Each passage is truncated to a maximum length of 500 words. As searching through all passages per concept would be very time-consuming, first, all passages have been tagged with the concept they belong to. The matching is done by finding a mention of the concept itself or any of its synonyms in the passage [14]. This matching module processes content in *XML* format, identifying mentions of concepts in articles and books. This process greatly reduces the number of passages to be encoded per concept, creating a separate corpus for each concept, rather than one large corpus containing all passages. In our setting, where users might not be very familiar with the concepts they want to learn about, an exact match of the concept in the passage can help users concentrate their attention on the most relevant pieces of information. Therefore, the selection of passages described above can help devise effective ranking models in this setting.

3.3 Ranking Models

To retrieve the most well-rounded set of passages for each of the 100 concepts, several ranking models were used and their results were combined. Two lexical

---

[2] Agricultural and Biological Sciences, Biochemistry, Genetics and Molecular Biology, Chemical Engineering, Chemistry, Computer Science, Earth and Planetary Sciences, Economics, Econometrics and Finance, Engineering, Food Science, Immunology and Microbiology, Materials Science, Mathematics, Medicine and Dentistry, Neuroscience, Nursing and Health Professions, Pharmacology, Toxicology and Pharmaceutical Science, Physics and Astronomy, Psychology, Social Science, Veterinary Science and Veterinary Medicine.

search models were included in the set: BM25, and the baseline model that is currently deployed and used on a publicly available user-facing product. Additionally, the passages were ranked using semantic search models as well. We have tested several of the top-performing models on retrieval-tasks. Through a careful selection process of manually comparing the top-3 passages of these ranking models, we have chosen two bi-encoders and one cross-encoder (see selected methods for "pooling" in Table 2).

3.4 Pooling Method

The dataset should consist of 50 passages per concept. As each of the $|M| = 5$ different ranking models has its own top-50 ranking, there are 250 snippets in total per concept. The final selection is made by ranking each passage i based on a certain weight, denoted as w_{pool}. As overlap between the rankings from different models is expected, w_{pool}^i should not only be calculated based on how highly a passage is ranked, but how many rankings it occurs in as well. Therefore, it is calculated as follows:

$$w_{freq}^i = \frac{f_i}{|M|} \tag{1}$$

$$w_{rank}^i = \frac{\left(\frac{(|N|+1) - \sum_{j=1}^{f_i} r_{ij}}{|N|}\right)}{|M|} \tag{2}$$

$$w_{pool}^i = \frac{w_{freq}^i + w_{rank}^i}{2} \tag{3}$$

where $m \in$ with $M = \{1, \ldots, 5\}$ is the number of models in the pooling set, $f_i \in M$ with $M = \{1, \ldots, 5\}$ is the number of rankings snippet i occurs in and $r_{im} \in N$ with $N = \{1, \ldots, 50\}$ is the rank position of snippet i in ranking of model m.

3.5 Annotations

Each passage is evaluated by at least one domain expert on their *relevance* and *informativeness* regarding the concept. The evaluators are in-house subject matter experts with extensive knowledge of their respective fields. While each passage was only evaluated thoroughly by one expert, the experts reviewed and commented on each other's assessments. Both aspects are judged with an ordinal three-point scale, with 0 being not relevant/informative at all, 1 being partially relevant/informative, and 2 being very relevant/informative.

Relevance. A passage is considered very relevant (2) when it only covers the concept, clearly discusses the said concept, and there is no other piece of information discussed in such passages. It becomes less relevant when other concepts are discussed in the passage as well or when it only discusses a specific aspect of the concept.

Informativeness. Assesses the amount of information about the concept contained in the passage. A very informative snippet should answer questions such as: *How much would an average user learn about the concept after reading the passage? How much "educational" or "essential" information about the concept is included in the passage? Does the snippet provide an answer to "what CONCEPT is?" like questions?*

4 Experimental Setup

We have conducted a series of experiments to evaluate various pre-trained ranking models using our test collection on the knowledge acquisition retrieval task. The semantic search models were used without any parameter tuning. We first assessed the performance of the ranking models on the entire test collection. Subsequently, the models were further evaluated per domain to determine if certain domains pose a difficulty and whether certain models are more sensitive to such domains. Finally, we selected the highest-performing semantic search models to assess the impact of the query input on performance. We report the performance of multiple IR-based metrics, namely Precision, Recall, MRR, and NDCG at different cut-off points. We consider NDCG as the strongest indicator of success as it is suitable for tasks with graded relevance judgments (unlike MRR) [25, 28]. To rank documents per concept, we first use a lightweight annotation method to find mentions of a given scientific concept (query) in documents. Then, we re-rank the documents matching the concept using ranking models. The main intuition behind this approach is that in a setting where a user is not familiar with the concept, an explicit mention of the concept in the context is necessary to help the user read the relevant context. The annotation method searches for occurrences of concepts in documents. If an abbreviation for the concept is proposed in the text, then the abbreviation is also added as an alias for the concept and searched in the article. We use the Schwartz and Hearst [23] method to detect such abbreviations.

Models. We evaluate different types of lexical- and semantic search models on the KAPR task, all of which can be found in Table 2. Lexical search models are designed to match and score passages based on lexical similarity with a given query. One example is our **Baseline Model**, which utilizes a location-aware term frequency score and is commonly used for scientific document search. We also use **BM25** [19] as it is one of the widely used lexical search models. Next to these lexical models, we use several semantic search models including a set of bi-encoders and cross-encoders.

5 Results

In this section, we evaluate the performance of different ranking models and perform several analyses on the differences between models on the KAPR task.

Table 2. Ranking Models used for evaluation

Search Type	Encoder	In Pooling Set	Name
Lexical	–	✓	Baseline Model (TF)
Lexical	–	✓	BM25 [19]
Semantic	Bi-Encoder	✓	ST msmarco-distilbert-base-tas-b [7]
Semantic	Bi-Encoder	✓	ST msmarco-distilbert-base-v4 [18]
Semantic	Bi-Encoder	✗	ST msmarco-bert-dot-v5 [18]
Semantic	Bi-Encoder	✗	ST msmarco-MiniLM-L-6-v3 [18]
Semantic	Bi-Encoder	✗	ST RoBerta-large-v1 [12]
Semantic	Bi-Encoder	✗	flax-distilRoBerta-v3 [20]
Semantic	Bi-Encoder	✗	ST T5-xl [15]
Semantic	Bi-Encoder	✗	ST gtr-t5-l [16]
Semantic	Cross-Encoder	✓	ST msmarco-MiniLM-L6-v2
Semantic	Cross-Encoder	✗	ST msmarco-Electra
Semantic	Cross-Encoder	✗	ST ms-marco-MiniLM-L-12-v2

Table 3. Distribution of labels over 5,000 passages

	Not (0)	Partially (1)	Very (2)
Relevant	32.98%	40.28%	26.74%
Informative	33.28%	43.28%	23.44%

Table 4. Total number of passages per relevance and informativeness label combination.

Info	Rel		
	0	1	2
0	1593	69	2
1	55	1892	217
2	1	53	1118

5.1 Relevance Versus Informativeness

Per Table 3, passages were most often labeled partially relevant (40.28%) and/or informative (43.28%). If we further compare the assessment of each individual document in Table 4, we find that 92.06% of passages are equally relevant and informative, and only 7.94% are not equally relevant and informative, out of which 5.67% of passages are more relevant than informative and 2.18% of passages are more informative than relevant. In total, passages that were deemed more relevant than informative made up approximately 6% of the test collection, with the highest number of such passages coming from the *Food Science* and *Psychology* domains. Annotators indicated that passages that were more informative than relevant were often highly detailed, but focused more on a subtype or specific aspect. This type of passage might not be relevant if the user wants to learn what the concept is in general, but it is nonetheless very informative as it provides the user with in-depth knowledge. Such passages most often came from *Chemical Engineering, Earth and Planetary Sciences* and *Nursing and Health Professions*.

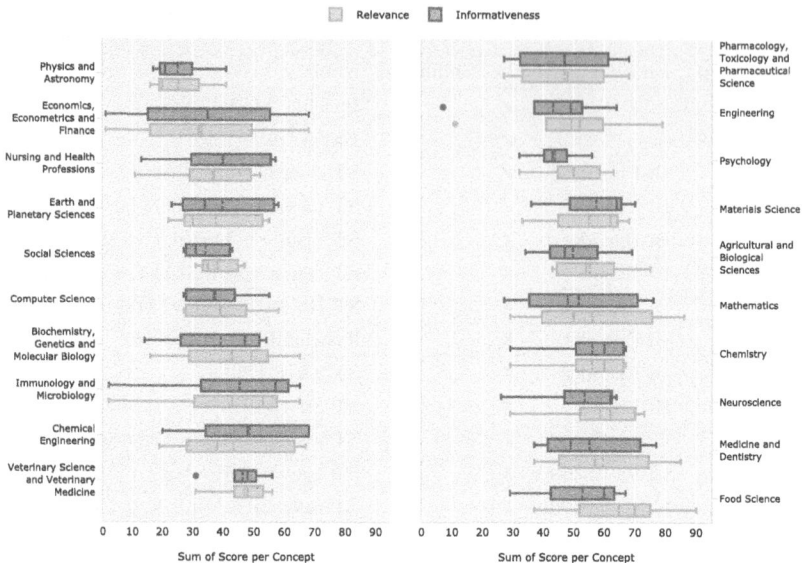

Fig. 1. Boxplot of Total Scores per Concept by Domain

Moreover, the quality of passages within our test collection varies depending on the concept and domain under consideration. As 50 passages have been evaluated for each
concept, a concept can receive a maximum score of $50 \cdot 2 = 100$ on each aspect if all passages have been labeled as very relevant or informative. For each domain, 5 concepts have been evaluated. Figure 1 shows that the average assessment of passages varies significantly across domains, in terms of total score, but the difference in score between relevance and informativeness as well. Certain domains show a wide range of scoring between concepts, such as *Economics, Econometrics and Finance* and *Immunology and Microbiology*, where certain concepts received a total score of less than 2 on both relevance and informativeness, meaning that only one out of 50 passages were relevant and informative. Other domains score more consistently across concepts, such *Veterinary Science and Veterinary Medicine* and *Social Sciences*. Furthermore, Fig. 1 presents the domains in increasing order of total concept score, i.e. the sum of scores per concept over 50 passages. We observe that as the score increases, the difference in score between relevance and informativeness shifts. Lower-scoring domains often score higher on informativeness than relevance or equally high, while relevance is higher than informativeness for the higher-scoring domains.

The discrepancies may stem from the limited availability of high-quality passages within our dataset. As described in Sect. 3.2, a separate corpus was created for each concept from which the final selection of 50 passages was then retrieved. However, the sizes of these concept-corpora differ. A domain such as *Economics, Econometrics, and Finance* has an average concept-corpus size of only 160 pas-

Table 5. Performance on relevance and informativeness of passages including domain-standard deviation (best models in bold, second best underlined)

Model	Aspect	P@10	R@10	R@50	MRR@10	nDCG@10	nDCG@50
BASELINE	rel	0.52±0.09	0.17±0.06	0.36±0.07	0.84±0.08	0.46±0.09	0.40±0.06
	info	0.51±0.09	0.17±0.06	0.36±0.07	0.85±0.08	0.45±0.08	0.40±0.06
BM25	rel	0.54±0.13	0.18±0.06	0.51±0.11	0.82±0.10	0.46±0.10	0.49±0.10
	info	0.54±0.13	0.179±0.06	0.51±0.11	0.82±0.10	0.46±0.09	0.48±0.10
BE ST msmarco-distilbert-tas-b	rel	<u>0.77 ± 0.11</u>	0.25 ± 0.06	0.78±0.06	0.91±0.11	<u>0.69 ± 0.10</u>	0.75±0.06
	info	<u>0.76 ± 0.10</u>	0.25 ± 0.06	0.78±0.06	0.92±0.11	<u>0.66 ± 0.07</u>	0.74±0.06
BE ST msmarco-distilbert-v4	rel	**0.78 ± 0.11**	**0.26 ± 0.06**	<u>0.80 ± 0.05</u>	0.92±0.10	**0.70 ± 0.09**	**0.77 ± 0.05**
	info	**0.78 ± 0.10**	**0.26 ± 0.06**	<u>0.81 ± 0.05</u>	0.91±0.09	**0.68 ± 0.07**	**0.76 ± 0.05**
CE ST msmarco-MiniLM-L6-v2	rel	0.76±0.11	0.25±0.06	**0.82 ± 0.05**	<u>0.93 ± 0.11</u>	0.68±0.09	<u>0.77 ± 0.05</u>
	info	0.75±0.10	0.25±0.05	**0.81 ± 0.04**	<u>0.93 ± 0.12</u>	0.65±0.08	<u>0.75 ± 0.05</u>
CE ST msmarco-MiniLM-L12-v2	rel	0.76±0.10	0.25±0.06	0.75±0.05	**0.94 ± 0.09**	0.69±0.08	0.73±0.05
	info	0.76±0.10	0.25±0.05	0.74±0.05	**0.93 ± 0.09**	0.67±0.07	0.72±0.04
CE ST msmarco Electra	rel	0.70±0.12	0.23±0.06	0.66±0.06	0.90±0.11	0.64±0.09	0.66±0.05
	info	0.79±0.12	0.23±0.05	0.66±0.06	0.90±0.10	0.62±0.09	0.65±0.05
BE ST T5-xl	rel	0.66±0.09	0.21±0.05	0.62±0.09	0.87±0.11	0.60±0.07	0.62±0.06
	info	0.66±0.08	0.21±0.05	0.62±0.09	0.87±0.11	0.59±0.06	0.61±0.06
BE ST gtr-t5-l	rel	0.74±0.09	0.24±0.06	0.69±0.08	0.92±0.11	0.67±0.09	0.69±0.07
	info	0.74±0.10	0.24±0.06	0.69±0.09	0.92±0.09	0.65±0.08	0.68±0.07
BE ST msmarco-bert-dot-v5	rel	0.63±0.13	0.21±0.04	0.65±0.07	0.82±0.16	0.54±0.11	0.61±0.06
	info	0.62±0.11	0.20±0.04	0.65±0.07	0.82±0.17	0.50±0.08	0.59±0.05
BE ST RoBerta-large-v1	rel	0.64±0.09	0.22±0.06	0.60±0.10	0.86±0.10	0.58±0.09	0.60±0.08
	info	0.64±0.09	0.22±0.06	0.60±0.10	0.87±0.10	0.57±0.08	0.60±0.08
BE flax-distilRoBerta-v3	rel	0.66±0.11	0.22±0.06	0.62±0.09	0.86±0.12	0.58±0.09	0.61±0.07
	info	0.66±0.10	0.22±0.06	0.63±0.09	0.85±0.11	0.57±0.07	0.60±0.07
BE ST msmarco-MiniLM-L-6-v3	rel	0.71±0.11	0.23±0.06	0.68±0.08	0.90±0.13	0.63±0.10	0.66±0.07
	info	0.69±0.11	0.23±0.06	0.68±0.08	0.89±0.13	0.59±0.07	0.65±0.05

sages, and this is reflected in the scores presented in Fig. 1 as well. Conversely, *Food Science* has an average concept-corpus size of 10,000 and receives the highest score on relevance. If more passages are available, the likelihood of finding high-quality passages is simply higher. The variance in scores could be further attributed to the inherent nature of the topics themselves as well. For instance, topics within *Physics and Astronomy* might demand more precise descriptions compared to concepts in *Food Science*, making them less likely to be classified as relevant or informative. Although *Food Science* may rank highest overall, this is partly because it contains 127 highly relevant passages (i.e., rel=2) but not necessarily highly informative ones. On the other hand, *Chemical Engineering* has 95 very informative passages but only 79 very relevant ones. Thus, while variant across domains, relevance and informativeness are largely correlated.

5.2 Model Performance

As presented in Table 5, all semantic matching models consistently outperform the lexical search models. While *BE ST msmarco-distilbert-v4* [18] shows the highest performance on both relevance, as well as informativeness across most metrics, the performance gaps between most of the semantic models are not as

Table 6. Best and worst performing domains for various models

	CE ST MiniLM-L6-v2	BE ST TAS-B	BE ST distilbert-v4	BM25
Top-3 Best Domains	Neuroscience	Agricultural and Biological Sciences	Biochemistry, Genetics and Molecular Biology	Economics, Econometrics and Finance
	Chemistry	Neuroscience	Food Science	Social Sciences
	Medicine and Dentistry	Mathematics	Neuroscience	Immunology and Microbiology
Top-3 Worst Domains	Immunology and Microbiology	Immunology and Microbiology	Chemical Engineering	Chemical Engineering
	Nursing and Health Professions	Chemical Engineering	Immunology and Microbiology	Engineering
	Earth and Planetary Sciences	Earth and Planetary Sciences	Nursing and Health Professions	Earth and Planetary Sciences

significant, nor is there a significant difference in performance between cross-encoders and bi-encoders.

As Precision and Recall do not take into account graded relevance and therefore scores 1 and 2 are both labeled as positive, while a score of 0 is labeled as negative. Table 4 thus shows that most passages were labeled as positive for both aspects or negative for both aspects, while few passages were positive on one aspect and negative on the other. We therefore do not expect significant differences in performance for Precision and Recall.

One might argue that the first five models in Table 5, which were utilized in creating KAPR, could introduce a bias that affects their performance evaluation. However, since $R@50$ signifies how many of the positive passages were retrieved, it could be seen as a reflection of this bias, or at the very least, the advantage of this bias. Notably, despite the baseline model and $BM25$ being part of the pooling set for the test collection, their $R@50$ scores fall below those of models not used in pooling. Conversely, the semantic search models from the pooling set consistently obtain the highest $R@50$ scores, indicating a slight bias toward these models. This bias should be taken into account when comparing the models, as unbiased models achieve $R@50$ scores ranging only from 0.6 to 0.75, which ultimately constrain the maximum achievable $nDCG$ score.

Moreover, when we consider our most important metric, $nDCG@10$, all models perform better on relevance compared to informativeness. While the differences in performance between the two aspects are small in general, the lexical search models display the most similar performance, with a difference of approximately 0.008. In contrast, *BE ST msmarco-bert-dot-v5* and *BE ST msmarco-MiniLM-L-6-v3* exhibit a gap in performance of approximately 0.04. Table 5 presents the standard deviations for domain performance as well, which are notably consistent but also quite high for all models. This indicates that the models do not generalize well across various domains.

5.3 Domain Performance

While most semantic models demonstrated similar performance across the entire test set, certain models may excel in particular domains. It is to be noted that 9 out of 11 semantic search models list *Immunology & Microbiology* among their three worst-performing domains, while half of them rank *Computer Science* within their top three domains. This pattern suggests that many semantic search models encounter challenges in handling the same domains. As previously mentioned, the quality of passages within the test collection exhibits variations across different domains. However, the total domain relevance- and informativeness scores do not necessarily align with the performance of ranking models in those domains. For instance, *Computer Science*, one of the top-performing domains, and *Earth & Planetary Sciences*, one of the worst-performing domains, have domain scores of 383 and 387, respectively.

If we further examine domain-specific performance for each model, as shown in Table 6, we observe that most of the overlap occurs in the worst-performing domains. Although the semantic search models do not consistently excel in precisely the same domains, their top three domains often cluster around the 'exact' sciences. However, it is particularly interesting to note that *BM25* performs best in domains where semantic models often struggle. This could be attributed to *BM25* heavily relying on term frequency. For instance, in domains like *Social Sciences* concepts such as *Imperialism* are more likely to be mentioned exactly in documents, as opposed to *Atom-Transfer Radical-Polymerization* in *Chemical Engineering*, where abbreviations are often used rather than the exact term.

6 Conclusion

The superficiality of relevance judgments in IR evaluation methods has long been criticized [1,21,24,26], as mere topicality is often not sufficient for satisfying a user's information need. For evaluating search tasks such as Knowledge Acquisition Passage Retrieval (KAPR), existing scientific IR benchmark sets might therefore not be sufficient. Furthermore, such data sets often only cover a certain scientific domain. In this research, we considered these criticisms by creating the KAPR dataset, which provides both relevance- as well as informativeness assessments on passages from scientific books and articles from a wide range of scientific domains. Despite the criticisms on topical relevance as an evaluation method, we found that relevance and informativeness appear to be highly correlated, but the degree of this correlation differs between domains. Considering the nature of the *topics* varies between domains, the standards for a passage to be deemed relevant or informative may vary as well. Certain domains might contain more complex topics that would require more exact explanations and are therefore less often considered informative. On the other hand, the nature of the *text* describing such concepts varies between domains as well. Passages in certain domains might be colored by opinions for instance, and therefore negatively impacting the informativeness. It should also be noted that a bias is present in our dataset as most assessments were made by only one annotator per document.

When evaluating various ranking models on the KAPR task, we found that all semantic search models outperformed the lexical search models. However, it is to be noted that lexical search models performed more similarly on relevance and informativeness, while semantic search models performed relatively better on relevance. Considering these semantic search models are trained on traditional IR relevance judgments, it is to be expected that they would not perform as well on informativeness. Finally, our evaluation showed a significant difference in performance between scientific domains, and most semantic search models showed difficulty with the same set of domains, highlighting the fact that the evaluated retrieval systems do not generalize well. Future work could explore the use of KAPR for fine-tuning a model that is suitable for the knowledge acquisition setting across all scientific domains.

Acknowledgments. The KAPR test collection is constructed using the data and production systems of Elsevier's ScienceDirect. The constructed dataset is available here: https://github.com/acapari/KAPR.git. Jaap Kamps is partly funded by the Netherlands Organization for Scientific Research (NWO CI # CISC.CC.016, NWO NWA # 1518.22.105), the University of Amsterdam (AI4FinTech program), and ICAI (AI for Open Government Lab). Views expressed in this paper are not necessarily shared or endorsed by those funding the research.

References

1. Arabzadeh, N., Vtyurina, A., Yan, X., Clarke, C.L.: Shallow pooling for sparse labels. Inf. Retrieval J. **25**(4), 365–385 (2022)
2. Borlund, P.: The concept of relevance in IR. J. Am. Soc. Inform. Sci. Technol. **54**(10), 913–925 (2003)
3. Devlin, J., Chang, M.W., Lee, K., Toutanova, K.: Bert: pre-training of deep bidirectional transformers for language understanding. arXiv:1810.04805 (2018)
4. Ghafourian, Y.; Knoth, P., Hanbury, A.: Information retrieval evaluation in knowledge acquisition tasks. WEPIR 2021: The 3rd Workshop on Evaluation of Personalisation in Information Retrieval at CHIIR, pp. 88–95 (2021)
5. Ghafourian, Y.: Relevance models based on the knowledge gap. In: ECIR, pp. 488–495 (2022)
6. Hjørland, B.: The foundation of the concept of relevance. JASIST **61**(2), 217–237 (2010)
7. Hofstätter, S., Lin, S.C., Yang, J.H., Lin, J., Hanbury, A.: Efficiently teaching an effective dense retriever with balanced topic aware sampling. In: SIGIR, pp. 113–122 (2021)
8. Huang, X., Soergel, D.: Relevance judges' understanding of topical relevance types: an explication of an enriched concept of topical relevance. JASIST **41**(1), 156–167 (2004)
9. Ingwersen, P., Järvelin, K.: The turn: integration of information seeking and retrieval in context, vol. 18. Springer, Cham (2006)
10. Karpukhin, V., et al.: Dense passage retrieval for open-domain question answering. arXiv:2004.04906 (2020)

11. Li, C., Yates, A., MacAvaney, S., He, B., Sun, Y.: Parade: passage representation aggregation for document reranking. preprint arXiv:2008.09093 (2020)
12. Liu, Y., et al.: Roberta: a robustly optimized BERT pretraining approach. CoRR **abs/1907.11692** (2019), http://arxiv.org/abs/1907.11692
13. MacAvaney, S., Yates, A., Cohan, A., Goharian, N.: CEDR: contextualized embeddings for document ranking. In: SIGIR, pp. 1101–1104 (2019)
14. Malaisé, V., Otten, A., Coupet, P.: Omniscience and extensions–lessons learned from designing a multi-domain, multi-use case knowledge representation system. In: European Knowledge Acquisition Workshop, pp. 228–242 (2018)
15. Ni, J., et al.: Sentence-t5: scalable sentence encoders from pre-trained text-to-text models. arXiv:2108.08877 (2021a)
16. Ni, J., et al.: Large dual encoders are generalizable retrievers. arXiv:2112.07899 (2021b)
17. Nogueira, R., Cho, K.: Passage re-ranking with bert. arXiv:1901.04085 (2019)
18. Reimers, N., Gurevych, I.: Sentence-bert: sentence embeddings using siamese bert-networks. In: EMNLP, Association for Computational Linguistics (2019)
19. Robertson, S.E., Walker, S., Beaulieu, M., Willett, P.: Okapi at TREC-7: automatic ad hoc, filtering, VLC and interactive track. Nist Special Publication SP, pp. 253–264 (1999)
20. Sanh, V., Debut, L., Chaumond, J., Wolf, T.: Distilbert, a distilled version of bert: smaller, faster, cheaper and lighter. ArXiv abs/1910.01108 (2019)
21. Saracevic, T.: Relevance reconsidered. In: Proceedings of the Second Conference on Conceptions of Library and Information Science (CoLIS 2), pp. 201–218 (1996)
22. Saracevic, T.: Relevance: a review of the literature and a framework for thinking on the notion in information science. part ii. Adv. Librarianship **30**, 3–71 (2006)
23. Schwartz, A.S., Hearst, M.A.: A simple algorithm for identifying abbreviation definitions in biomedical text. In: Biocomputing 2003, pp. 451–462, World Scientific (2002)
24. Sormunen, E.: Liberal relevance criteria of trec- counting on negligible documents? In: SIGIR, pp. 324–330 (2002)
25. Thakur, N., Reimers, N., Rücklé, A., Srivastava, A., Gurevych, I.: Beir: a heterogenous benchmark for zero-shot evaluation of information retrieval models. arXiv:2104.08663 (2021)
26. Voorhees, E.M., Craswell, N., Lin, J.: Too many relevants: Whither cranfield test collections? In: SIGIR, pp. 2970–2980 (2022)
27. Wang, X., Macdonald, C., Ounis, I.: Improving zero-shot retrieval using dense external expansion. Inf. Process. Manage. **59**(5), 103026 (2022)
28. Wang, Y., Wang, L., Li, Y., He, D., Liu, T.Y.: A theoretical analysis of NDCG type ranking measures. In: Conference on Learning Theory, pp. 25–54 (2013)

Assessing Document Sanitization for Controlled Information Release and Retrieval in Data Marketplaces

Luca Cassani[1], Giovanni Livraga[2], and Marco Viviani[1](✉)

[1] Department of Informatics, Systems, and Communication (DISCo),
University of Milano-Bicocca, Edificio U14 (ABACUS),
Viale Sarca 336, 20216 Milan, Italy
l.cassani@campus.unimib.it, marco.viviani@unimib.it

[2] Department of Computer Science "Giovanni Degli Antoni", University of Milan,
via Giovanni Celoria, 18, 20133 Milan, Italy
giovanni.livraga@unimi.it

Abstract. This study provides insights into both addressing data confidentiality concerns and enhancing document retrieval effectiveness in Data Marketplaces, which in this specific study consist of unstructured, textual documents. Through a semi-automatic sanitization process leveraging token masking with text summarization, possibly complemented by Coreference Resolution, the proposed solution mitigates the risk of inferring confidential information while maintaining search performance. Experimental results demonstrate encouraging improvements in both aspects with respect to baseline solutions.

Keywords: Text Sanitization · Confidentiality · Text Summarization · Coreference Resolution · Information Retrieval · Data Marketplaces

1 Introduction

The current online landscape is characterized by vast amounts of publicly available data, generated with different purposes, referring to multiple application domains, and characterized by a variable level of quality [2]. This often makes it complicated for users to find the most useful and reliable information for specific purposes, partly because of the *information overload* problem [3]. In this context, *Data Marketplaces* (DMs) are specialized virtual spaces that allow the exchange among users of various kinds of data that can range from highly specific and niche data to more general and broadly applicable information [1]. These exchanges typically involve monetization, as *data owners*, i.e., the proprietors of data, are willing to offer them for a fee on a DM. A *registered user* can then explore the platform to retrieve the data they need and, should they find data of interest, proceed with the purchase. Hence, DMs generate revenue usually through commissions from processed transactions [1].

The main issues associated with DM platforms are closely related to the service and data they provide: unlike marketplaces for physical items, whose products can be presented with accurate descriptions and photographs and are subject to return and warranty policies, *digital information* presents different characteristics by its nature. In particular, the data stored within DM platforms must be protected so that they are only visible to users who have purchased them. These platforms must also equip potential buyers with the tools needed to determine whether the data they find are indeed what they are looking for, without exposing the entire content before the sale is concluded.

For *structured data*, usable solutions include providing detailed descriptions of the dataset for sale, supported by overall statistics (e.g., number of rows, columns, distributions, etc.), and/or displaying a data sample, possibly after applying *sanitization techniques* [19] to information that the data owner deems inappropriate for pre-sale exposure. However, modern DMs are increasingly including *unstructured data*, for which the objective of providing an accurate description remains the same, but needs strategies tailored to the particular type of data. For images, transformations like blurring can be applied to partially obscure content; for videos, extracting key frames or important short segments is feasible. In the context of *textual documents*, which is the focus of this research, content sanitization techniques involving text masking and/or summarization can be employed. Through masking, data owners can obfuscate specific content from being shown to users [20]. Through summarization, relevant information can be extracted from textual documents, and presented concisely while avoiding exposure of the entire content [17]. The two techniques can then also be used together in a hybrid mode, as we do and assess in this article.

However, within the realm of such sanitized textual documents, challenges arise concerning the effective search within the DM platform. If the sanitized text fails to sufficiently represent buyers' needs and, at the same time, to be sufficiently representative of the original text, potential buyers would be unable to identify the most relevant content for purchase, posing a challenge to a DM model based on commissions. Therefore, in this article, we study various sanitization techniques applied to textual documents within the DM platform. We assess the effectiveness of retrieving sanitized documents to verify that data sanitization, while concealing confidential content,[1] compromises neither retrieval effectiveness nor data saleability.

2 Background and Related Work

The challenges associated with DMs are manifold [7]: beyond the aspect related to multiple data owners, as also discussed in [4], and the correctness of exchanges [13], the issue of *data confidentiality* is crucial. Like in any application scenario in which data collections are outsourced to *honest-but-curious* subjects (as DM providers are typically considered), a possible approach for protecting confidentiality requires the adoption of *owner-side encryption* (e.g., [6]). Being

---

[1] We consider as confidential any content the owner may not wish to publicly disclose (i.e., including but not limited to personal/sensitive/company confidential data).

data wrapped in a layer of encryption, unauthorized subjects who do not possess the correct encryption keys (possibly including the DM provider) cannot access their plaintext values. While effectively safeguarding data confidentiality, the adoption of encryption however clashes with the need for searchability, a paramount requirement in DM scenarios. In the context of *structured data*, approaches aimed at balancing these two contrasting requirements are often based on the creation of *metadata*, based on the original plaintext values, on which (partial) query evaluation at the provider side is performed without the need to decrypt data (e.g., [10]). Another solution is the adoption of different *encryption schemes* that support different types of queries (e.g., combining in an onion structure deterministic encryption for supporting `group by` queries and order-preserving encryption for supporting *range* queries [15]). In the context of *unstructured data*, particular attention has been lately devoted to the definition and adoption of *searchable encryption* schemes (e.g., [14]). Also in this case, metadata generated from plaintext documents can be used for indexing, but are typically stored in encrypted form along with the encrypted documents. These solutions can then permit some searches (via simple keyword-based queries to more complex queries [8,9]) over encrypted documents.

While related, these approaches are orthogonal to ours, as they are focused on the development and/or assume the adoption of specific encryption schemes, while our approach is independent on whether encryption is applied and, if so, on which scheme(s) can be used.

3 Controlled Information Release and Retrieval

The proposed solution is based on the development of a system on top of a DM platform for making available to a set of interested consumers, through a search engine, unstructured (textual) documents while enjoying protection from a confidentiality perspective. The proposed architecture is shown in Fig. 1.

- First, the data owner performs *document sanitization* in-house (Fig. 1a), and sends sanitized versions of original documents to the DM platform. At this point, the original documents can either be *encrypted* and sent to the platform, offering support for scenarios in which the provider is considered *honest-but-curious* (trusted to follow protocols and manage data, but not trusted to access plaintext data) [6], or kept securely in-house by the owner; in either case, the plaintext versions are never available to the provider;
- When a consumer formulates a query using the DM platform's *search engine*, the retrieval process operates on the sanitized documents, as the original versions are not accessible for this purpose (Fig. 1c). The search engine then returns a ranked list of sanitized documents based on the query;
- From such a ranked list, the consumer selects a subset of documents of interest; then, the consumer purchases the original versions of these documents from either the owner or the DM platform.

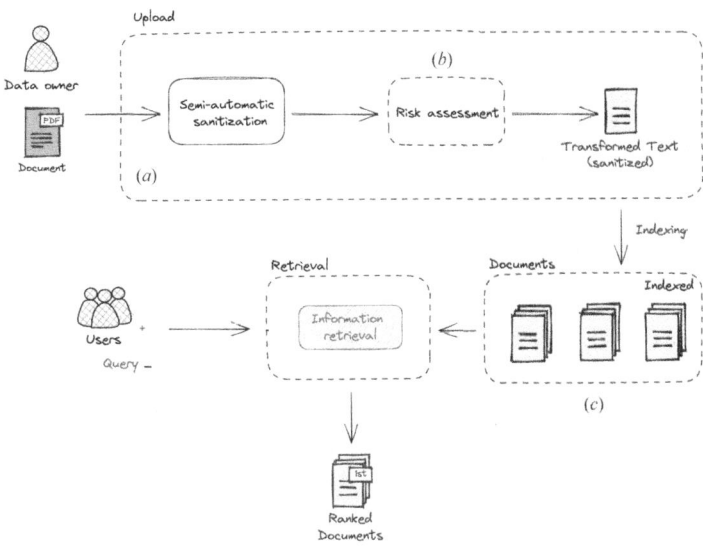

Fig. 1. The proposed architecture for document sanitization and retrieval in a DM.

In proposing such a solution, we considered the fact that the exposure of only non-confidential data does not completely eliminate confidentiality issues. This is even more challenging with the advent of *Large Language Models* (LLMs) [5], which may infer portions of text that do not appear in the sanitized version of the document. For this reason, we also propose in the system a simple and optional solution for *confidentiality risk assessment* (Fig. 1b).

3.1 Document Sanitization

An effective *document sanitization* phase should satisfy the following requirements: (*i*) avoid exposure of the document's entire informative content; (*ii*) be sufficiently representative of the original document; and (*iii*) maximize the retrieval effectiveness at the DM provider. The first requirement concerns avoiding (economic) damages to the document owner, as it limits the possibility that a malicious consumer may benefit from the simple observation of the search results. The second requirement concerns permitting consumers who observe such results – we recall, sanitized documents – to assess whether it might actually be of interest and worth paying the economic incentive to obtain the associated original documents. The last requirement concerns ensuring that search results are relevant to the queries and, hence, potentially useful for consumers.

To perform document sanitization, we leverage two solutions, i.e., *document masking* and *summarization*, which can be used either individually or together.

– *Document masking.* This implies *selectively masking* parts of the document (in terms of *tokens*) as deemed necessary by the owner. For example, if the

owner does not wish to include the word 'cat' in a masked document (e.g., as it would reveal the nature of its informative content), the original document can be sanitized by masking all occurrences of the token 'cat';
- *Document summarization.* This implies *generating a summary* of a document, by either keeping just the most important sentences in the summary, i.e., *extractive summarization* [12], or rephrasing the original documents in a shorter version, i.e., *abstractive summarization* [11]. Extractive summarization preserves the original document's representativeness by including original sentences in the summarized document. Abstractive techniques could incur the problem of presenting a summarized text that significantly diverges from the original text, which may negatively impact consumers' satisfaction with the purchase based on the retrieved sanitized documents. For this reason, in this paper we focus on extractive summarization (see Sect. 4.2).

The two solutions exhibit both pros and cons. While intuitive, document masking has to be performed with care, since – as well-known from the literature on privacy and data protection – the simple hiding of a piece of information may not offer adequate protection due to the possibility for an observer to reconstruct it through inferences [19]. Furthermore, while permitting more 'surgical' interventions, masking requires the owner for the identification of the tokens to be masked, and hence implicitly requires some technical skills from the owner themselves. Concerning summarization, it represents an approach that may be more easily enforced also by less technically skilled owners, either manually or with off-the-shelves tools, although it permits less control than identifying confidential information that should not appear in the sanitized document. For these reasons, supporting both solutions together for sanitization can provide a nice flexibility that can fit the specific needs and technical skills of the data owners. In this paper, we simply follow the summarization phase with a masking phase.

3.2 Confidentiality Risk Assessment

In this paper, we intend the confidentiality risk as the possibility of *demasking* tokens that have been obfuscated in the masking phase by the data owner. Therefore, regardless of whether we perform masking alone or summarization and masking together, we propose the following solution that makes use of *Coreference Resolution* (CR) [18] and a *demasking resistance* measure with respect to the information to be kept confidential.

Coreference Resolution. Let us consider the following text as an example:

> **The mouse** and **the elephant** are two animals, belonging to the class of mammals. **The former** has an average weight of 20 g, while **the latter** can weigh up to 6,000 kg. In addition, **the latter**, unlike **the former**, has a proboscis.

In case the data owner wanted to *mask* the tokens associated with the *animals*, the following result would be obtained:

[MASK] and [MASK] are two animals, belonging to the class of mammals. **The former** has an average weight of 20 g, while **the latter** can weigh up to 6,000 kg. In addition, **the latter**, unlike **the former**, has a proboscis.

In this case, it remains explicit that the animal that reaches 6,000 kg has a proboscis, unlike the other. Hence, it might be relatively easy for a human and/or an LLM to demask the elephant. Via CR, i.e., the task of finding all linguistic expressions (called *mentions*) in a given text that refer to the same subject, we can resolve them by replacing, in this case, pronouns with noun phrases. Hence, the use of CR leads to modifying the previous example as follows:

The mouse and **the elephant** are two animals, belonging to the class of mammals. **The mouse** has an average weight of 20 g, while **the elephant** can weigh up to 6,000 kg. In addition, **the elephant**, unlike **the mouse**, has a proboscis.

Now applying token masking on those associated with animals, the result would be as follows:

[MASK] and [MASK] are two animals, belonging to the class of mammals. [MASK] has an average weight of 20 g, while [MASK] can weigh up to 6,000 kg. In addition, [MASK], unlike [MASK], has a proboscis.

In this case, the link between the masked tokens is weakened; in the specific example, it is impossible to tell whether the presence of the proboscis relates to the lightest or the heaviest specimen: there are in fact small mammals weighing about 20 g characterized by a proboscis, such as specimens of the family *Macroscelididae*. By loosening the association among masked tokens, we hypothesize that improvements may occur in preventing the demasking risk from an LLM, as the context is more limited and references are unclear.

Demasking Resistance Measure. To assess the risk related to the possible demasking of tokens, we perform a *fill-mask* task [16], which tries to infer the tokens masked during the sanitization stage. To this aim, we employ the DistilRoBERTa model.[2] The tokens inferred by the model, with a score higher than a defined *threshold*, are compared with the original tokens in the unmasked document. It is then possible to define a *metric* for the demasking resistance, as follows:

$$dr(d) = 1 - \frac{n_{inf}}{n_{max}} \qquad (1)$$

where d is the sanitized document, n_{inf} is the number of tokens that the DistilRoBERTa model can infer from the sanitized text, and n_{max} is the number of tokens that have been masked in the non-sanitized text. The metric takes values in the [0, 1] range, with higher values implying higher resistance to demasking: when no token is demasked, $n_{inf} = 0$ and $dr(d) = 1$ (maximum resistance); when all tokens are demasked, $n_{inf} = n_{max}$ and $dr(d) = 0$ (no resistance).

---

[2] https://huggingface.co/distilbert/distilroberta-base/.

3.3 Document Retrieval

Once the documents have been sanitized, possibly taking into account the confidentiality risk considering demasking resistance, the retrieval phase is carried out by means of the search engine implemented within the DM platform. To this aim, we use standard retrieval models (see Sect. 4.2). Furthermore, *Query Expansion* (QE) techniques are applied to the users' original queries to assess their effectiveness when retrieval is done on sanitized documents and thus the performance of the search engine may deteriorate.

4 Experimental Evaluation

In this section, after illustrating the dataset employed and some implementation details, we illustrate the experiments conducted, evaluated by means of appropriate metrics, in order to select the document sanitization solutions that allow the best trade-off between data confidentiality and retrieval.

4.1 Data

As a reference scenario, we consider that of *online news*, given the fact that showing only the title or the first portion of an article may not be the best choice for a customer interested in purchasing the article itself. Specifically, the data used relate to articles from the Washington Post collected as part of TREC .[3] This collection includes 595,037 articles, stored in a JSON Lines format file, collected around 50 different *topics*. A qrels.txt file is also provided for performance evaluation in IR. This file associates a relevance score (in the range of 0 to 4) for each query-document pair in the dataset. Only documents with a length of less than 512 tokens (a limit imposed by BERT) were considered for evaluation, given the fact that we employ the DistilRoBERTa model for token demasking. Thus, a subset consisting of 3,776 articles was considered.[4]

4.2 Implementation Details

A *Python* framework has been implemented, which integrates various libraries to perform different tasks. For the masking phase, the spaCy library was utilized.[5] This library proves useful for the *Named Entity Recognition* (NER) task, carried out to simulate the identification of *confidential entities* to be masked by the data owner.[6] CR was implemented using the coreferee library.[7] Regarding

---

[3] https://trec.nist.gov/data/wapost/.
[4] The file qrels.txt was modified accordingly.
[5] https://spacy.io/.
[6] This choice is made to simulate the behavior of a data owner who decides to consider certain entities within textual documents as confidential. In this case, we assume that all entities extracted by NER algorithms constitute the tokens to be masked.
[7] https://github.com/msg-systems/coreferee/.

text summarization, the sumy library implementations of well-known extractive algorithms encompassing *Luhn*, *KLSummarizer* (a method that greedily adds sentences to a summary so long as it decreases the *Kullback-Leibler* (KL) divergence), *Latent Semantic Analysis* (LSA), *LexRank*, and *TextRank* were employed.[8] In addition, an algorithm based on *Sentence Transformers* (SBERT), namely *SBertSummarizer*,[9] was considered. For developing and evaluating the search engine, we used PyTerrier.[10] In particular, we considered the models provided by the library, i.e., TF-IDF, BM25, DLH, DPH, InL2, and MDL2.

4.3 Evaluation Metrics

Evaluations have been conducted regarding both the *effectiveness* in document retrieval by the search engine concerning the various sanitization techniques applied, and the adequacy of these techniques in terms of *demasking resistance* within the sanitized documents. The evaluation metrics considered, pertaining to the IR effectiveness, are the most commonly used ones in the literature, namely *Mean Average Precision* (MAP) and *normalized Discounted Cumulative Gain* (nDCG). As for the measure assessing demasking resistance, we consider the average of the demasking resistances defined by Eq (1), which we denote as $adr(D)$, where D is the set of documents in the collection.

It is worth to be underlined that, when employing summarization for sanitization, both IR effectiveness and demasking resistance may be impacted by the length of the generated summaries: we expect that shorter summaries induce better results for limiting demasking and worse results for IR effectiveness, and viceversa. For this reason, sanitization based on summarization is evaluated w.r.t. varying the lengths of the generated summaries, considering values equal to 10%, 20%, ..., 80% of the length of the original documents. For space constraints, for the experiments that include summarization, we report the average scores obtained over these lengths for the considered evaluation metrics, denoted in this case as MAP_{as}, nDCG_{as}, and $adr(D)_{as}$.

4.4 Results

This section illustrates the results of distinct experiments conducted with respect to: (*i*) the usage of simple token masking for performing document sanitization on the original documents by the data owner; (*ii*) the usage of various summarization techniques followed by a subsequent masking phase to perform document sanitization; and (*iii*) the usage of the best summarization technique followed by a subsequent masking phase to perform document sanitization coupled with the usage of CR and QE.

---

[8] https://github.com/miso-belica/sumy.
[9] https://github.com/dmmiller612/bert-extractive-summarizer.
[10] https://pyterrier.readthedocs.io.

Document Masking. This preliminary set of experiments evaluates the effects of adopting entity masking only (i.e., without summarization, CR, or QE).[11] MAP and nDCG values related to baselines (bl), i.e., where retrieval is performed on non-sanitized documents, are also reported. The results of these experiments are reported in Table 1.

Table 1. Evaluation metrics considering document masking for sanitization.

Model	MAP_{bl}	$nDCG_{bl}$	Masked	Demasked	$adr(D)$	MAP	nDCG
TF-IDF	0.234	0.411	41,816	8,867	0.788	0.211	0.386
BM25	0.234	0.411	41,816	8,867	0.788	0.212	0.386
DLH	0.226	0.403	41,816	8,867	0.788	0.204	0.38
DPH	**0.249**	**0.423**	41,816	8,867	0.788	**0.220**	**0.393**
InL2	0.238	0.413	41,816	8,867	0.788	0.216	0.389
MDL2	0.201	0.373	41,816	8,867	0.788	0.183	0.357

As expected, the results related to the resistance are the same across all IR models, as they clearly do not impact the possibility of demasking entities. From the IR effectiveness point of view, it is clear that performing a search on non-sanitized documents is more effective than on sanitized documents. Nevertheless, we note that the deterioration in performance is limited. It is also easy to observe that the DPH model slightly improves the other IR models, with MAP and nDCG values of 0.220 and 0.393 respectively.

Document Summarization and Masking. This set of experiments evaluates the combined adoption of document summarization and masking for sanitization. We test different summarization techniques considering the retrieval model (i.e., DPH) that showed the best results in the previous set of experiments.[12]

Results are reported in Table 2. From a macroscopic viewpoint, we observe that the joint adoption of summarization and masking proves effective for increasing resistance to demasking (the 0.877 value achieved by the model performing the worst for this task, *TextRank*, already improves the 0.788 value in Table 1). Comparatively evaluating the summarization models, it can be seen that *KLSummarizer* outperforms all other models regarding resistance to

---

[11] Recall that, in our experiments, the tokens to be masked are *named entities* in the document collection. Therefore, the number of masked tokens (i.e., Masked) in the texts indexed by the search engine is the same regardless of the employed IR model. Similarly, the number of demasked entities and, hence, demasked tokens (i.e., Demasked) is also the same regardless of the IR model.

[12] Recall that, in our experiments, the values we discuss are computed as the average of the scores computed considering each of the 8 summary lengths (ranging from 10% to 80% of the original document length).

Table 2. Evaluation metrics considering summarization and masking for sanitization (average over 8 summary lengths; IR model: DPH).

Model	$adr(D)_{as}$	MAP_{as}	$nDCG_{as}$
Luhn	0.884	0.194	0.357
KLSummarizer	**0.918**	0.174	0.339
Latent Semantic Analysis (LSA)	0.885	**0.205**	**0.374**
LexRank	0.891	0.195	0.357
TextRank	0.877	0.204	0.371
SBertSummarizer	0.899	0.184	0.351

demasking, but shows the poorest performance for retrieval effectiveness, for which the best model turns out to be LSA. The other model that achieves comparable results in terms of retrieval effectiveness is *TextRank*, which, however, as shown above, is the less effective in terms of resistance to demasking.

Document Summarization and Masking with Coreference Resolution and Query Expansion. This set of experiments evaluates the combined adoption of document summarization and masking for sanitization, CR, and QE. In particular, we show how the adoption of the best-performing retrieval and summarization models (DPH and LSA, respectively, as discussed above) can be impacted by performing different combinations (yes/no) of CR and QE. For space limitations, we report the results obtained using the *KLQueryExpansion* (KLQE) model only, which exhibited the best performance in comparative tests that we also performed with AxiomaticQE, RM3, and BolQE.[13]

Table 3. Evaluation metrics considering summarization and masking for sanitization, CR, and QE (average over 8 summary lengths; IR model: DPH; summarizer: LSA).

CR	QE	$adr(D)_{as}$	MAP_{as}	$nDCG_{as}$
No	No	0.885	0.205	0.374
Yes	No	**0.892**	0.199	0.366
No	KLQE	0.885	**0.222**	**0.416**
Yes	KLQE	**0.892**	0.214	0.405

Results are reported in Table 3. We can observe the diverse benefits of employing CR, within the sanitization phase, and QE, in the retrieval phase. Considering the adoption of CR only (second row in the table), resistance to demasking increases from 0.885 for basic DPH+LSA without CR nor QE (first

---

[13] https://pyterrier.readthedocs.io/en/latest/rewrite.html.

row in the table, which recalls the LSA result in Table 2) to 0.892. The same 0.892 value clearly is not impacted by the addition of KLQE (fourth row in the table), which does not impact the possibility of demasking a masked entity. IR effectiveness, on the other hand, shows a slight decrease from 0.205 (MAP_{as}) and 0.374 ($nDCG_{as}$) to 0.199 (MAP_{as}) and 0.366 ($nDCG_{as}$). Considering the adoption of KLQE only (third row in the table), IR effectiveness increases from 0.205 (MAP_{as}) and 0.374 ($nDCG_{as}$) for basic DPH+LSA without CR nor QE (first row in the table) to 0.222 (MAP_{as}) and 0.416 ($nDCG_{as}$). Resistance to demasking is clearly not impacted.

In summary, considering the joint adoption of both CR and KLQE in comparison to basic DPH+LSA, we can appreciate an increase in the performance related to both resistance to entity demasking and IR effectiveness, despite not achieving the best IR results achieved by adopting only QE. We note that this slight decrease is expected and not surprising, as CR inevitably causes alterations to the content that is to be indexed, impacting the evaluation of queries formulated to operate on the original unmodified contents.

5 Conclusions and Further Research

This study has provided insights into addressing data confidentiality concerns and enhancing document retrieval effectiveness in Data Marketplaces, specifically focusing on unstructured, textual documents. Our findings indicate that simple token masking alone is less effective at mitigating the risk of demasking compared to the combination of token masking with text summarization. However, this latter approach, while improving confidentiality, negatively impacts retrieval effectiveness. Our research demonstrates that a balanced approach can be achieved by incorporating Coreference Resolution during the masking process and employing Query Expansion during retrieval. This method successfully reduces the demasking risk while maintaining acceptable retrieval performance.

There is ample opportunity for future research to enhance our proposed solution. For example, more sophisticated summarization algorithms that inherently incorporate data confidentiality principles could be developed. Moreover, conducting comprehensive testing across various marketplace scenarios and datasets could validate the applicability and resilience of our approach.

Acknowledgments. This work was supported in part by the EC under grant GLACIATION (101070141), by the Italian MUR under PRIN 2022 project KURAMi (20225WTRFN, https://kurami.disco.unimib.it/) and by project SERICS (PE00000014) under the NRRP MUR program funded by the EU-NGEU.

Code and Data Availability. The code developed for this work and further information about the subset of data employed to test it can be accessed at: https://github.com/ikr3-lab/ControlledInformationDMs.

Disclosure of Interests. The authors have no competing interests to declare that are relevant to the content of this article.

References

1. Azcoitia, S.A., Laoutaris, N.: A survey of data marketplaces and their business models. ACM SIGMOD Rec. **51**(3), 18–29 (2022)
2. Batini, C., Scannapieco, M.: Data and information quality: Dimensions. Springer, Principles and Techniques (2018)
3. Buchanan, J., Kock, N.: Information overload: A decision making perspective. In: Proceedings of MCDM 2000, pp. 49–58 (2001)
4. Cao, X., Chen, Y., Liu, K.R.: Data trading with multiple owners, collectors, and users: an iterative auction mechanism. IEEE TSIPN **3**(2), 268–281 (2017)
5. Chang, Y., et al.: A survey on evaluation of large language models. ACM TIST **15**(3), 1–45 (2024)
6. De Capitani di Vimercati, S., Foresti, S., Livraga, G., Samarati, P.: Empowering owners with control in digital data markets. In: Proceedings of IEEE CLOUD 2019, pp. 321–328 (2019)
7. De Capitani di Vimercati, S., Foresti, S., Livraga, G., Samarati, P.: Toward owners' control in digital data markets. IEEE ISJ **15**(1), 1299–1306 (2020)
8. Fu, J., Wang, N., Cui, B., Bhargava, B.K.: A practical framework for secure document retrieval in encrypted cloud file systems. IEEE TPDS **33**(5), 1246–1261 (2021)
9. Fu, Z., Huang, F., Ren, K., Weng, J., Wang, C.: Privacy-preserving smart semantic search based on conceptual graphs over encrypted outsourced data. IEEE TIFS **12**(8), 1874–1884 (2017)
10. Hacigümüş, H., Iyer, B., Li, C., Mehrotra, S.: Executing SQL over encrypted data in the database-service-provider model. In: Proceedings of SIGMOD 2002, p. 216–227 (2002)
11. Lin, H., Ng, V.: Abstractive summarization: a survey of the state of the art. In: Proceedings of AAAI 2019, vol. 33, no. 01, pp. 9815–9822 (2019)
12. Moratanch, N., Chitrakala, S.: A survey on extractive text summarization. In: Proceedings of ICCCSP 2017, pp. 1–6 (2017)
13. Niu, C., Zheng, Z., Wu, F., Gao, X., Chen, G.: Trading data in good faith: integrating truthfulness and privacy preservation in data markets. In: Proceedings of ICDE 2017, pp. 223–226 (2017)
14. Pham, H., Woodworth, J., Amini Salehi, M.: Survey on secure search over encrypted data on the cloud. Concurrency Comput. Pract. Exp. **31**(17), e5284 (2019). e5284 cpe.5284
15. Popa, R., Redfield, C., Zeldovich, N., Balakrishnan, H.: CryptDB: protecting confidentiality with encrypted query processing. In: Proceedings of SOSP 2011, pp. 85–100 (2011)
16. Shibayama, N., Cao, R., Bai, J., Ma, W., Shinnou, H.: Evaluation of pretrained BERT model by using sentence clustering. In: Proceedings of PACLIC 2020, pp. 279–285 (2020)
17. Shree, A.R., Kiran, P.: Sensitivity context aware privacy preserving text document summarization. In: Proceedings of ICECA 2020, pp. 1517–1523 (2020)
18. Sukthanker, R., Poria, S., Cambria, E., Thirunavukarasu, R.: Anaphora and coreference resolution: a review. Inf. Fusion **59**, 139–162 (2020)
19. US Federal Committee on Statistical Methodology: Report on Statistical Disclosure Limitation Methodology – WP 22 (second version). USA, December 2005
20. Wu, Z., Shen, S., Lian, X., Su, X., Chen, E.: A dummy-based user privacy protection approach for text information retrieval. KNOSYS **195**, 105679 (2020)

The Impact of Web Search Result Quality on Decision-Making

Jan Heinrich Merker[1]((✉)), Lena Merker[2], and Alexander Bondarenko[1,3]

[1] Friedrich-Schiller-Universität Jena, Jena, Germany
{heinrich.merker,alexander.bondarenko}@uni-jena.de
[2] Martin-Luther-Universität Halle-Wittenberg, Halle, Germany
[3] Leipzig University, Leipzig, Germany

Abstract. People often search the Web for answers to comparative questions like "Is pasta healthier than pizza?" to inform everyday decisions. However, web search engines sometimes may return biased or low-quality results. Still, previous research has not considered the impact of varying search result quality, relevance, or stance on the users' decision-making process. To close this gap, we conducted a user study on quality, relevance, and stance assessments of 120 Google search results retrieved for eight comparative questions. We asked study participants about their decision and confidence before and after seeing the top-4 search results and which results influenced their decision. Our study showed that (1) high-quality search results are more likely to influence a user's decision, (2) topical relevance and search result quality have a similarly strong impact on decision-making, and (3) search results are more likely to influence decisions for factual comparative questions than for subjective questions.

1 Introduction

Decision-making is an integral part of everyday life when weighing pros and cons for simple questions like "Should I eat sandwiches or cereal for breakfast?" or more critical questions like "Is buying a house better than renting?" [3,23]. Nowadays, decisions are not only supported by prior knowledge and experience [26] but also by facts and arguments retrieved from the Web, e.g., when comparative questions are used as queries in web search [8]. While many studies have analyzed various kinds of web search biases and their impacts on the users [5,11,12,29,35,38], still only little is known about the impact of the web search result quality on the users' decisions. In this paper, we close this gap by conducting a systematic quality assessment of Google's results for comparative questions, followed by a study on the impact of the search result quality on the users' decisions.

To this end, we developed a set of evaluation criteria grounded in previous research to assess the quality, relevance, and stance of 120 documents retrieved by Google for 30 comparative questions. The individual documents' quality scores were combined to determine the average search result quality for each comparative question (Sect. 3). We further conducted a follow-up user study on decision-making with eight selected comparative questions with search results of varying quality (Sect. 4). In the study, we asked participants to decide on either of the comparison options (e.g., buying a house vs. renting) before and after seeing the search results, and to rate their decision confidence and the influence of individual search results on their decision. After collecting 554 responses from 442 participants, we enriched the user study data with the quality, relevance, and stance scores from the previous quality assessments and tested six hypotheses:

H1 Comparative questions on subjective topics lead to less confident decisions than questions on factual topics. (Intuition: Factual comparative questions (e.g., "Does cider or beer contain more calories?") are often "better" answered by search engines than subjective comparative questions (e.g., "Should I study philosophy or psychology?") [8]. Subjective questions are also more prone to cognitive biases.)

H2 Comparative questions with low-quality results lead to less confident decisions than questions with high-quality results. (Intuition: People seek to make the best decision based on the known information [26]. Accordingly, comparative questions with low-quality search results would be harder to answer, and high-quality results would be more likely to be used in the decision-making.)

H3 The higher a search result's quality, the more likely it influences the decision-making. (Intuition: Same as for Hypothesis H2).

H4 Users who are more confident in their decision before searching are less influenced by low-quality search results. (Intuition: Same as for Hypothesis H2).

H5 The quality of a search result has a higher impact on the decision-making process than its relevance. (Intuition: While relevance depends on the topic at hand, our search result quality criteria are topic-independent. We hypothesize a higher impact on decision-making than relevance.)

H6 Documents that take a stance towards one of the compared options have a higher impact on the decision. (Intuition: Relevant documents can take different stances towards the compared options, favoring either option [7]. We assume that documents that do not take a stance are less helpful in making a decision).

The significance tests indicate no significant difference between user confidence after seeing the search results for factual and subjective topics; thus, H1 cannot be confirmed. Similarly, we found no statistically significant evidence to confirm H2 that low-quality search results lead to less confident decisions than high-quality results. On the other hand, higher-quality results are still more likely to influence decisions; H3 is confirmed. We could also confirm H4 that more confident users in the decision before using web search are less influenced by low-quality search results. While our tests do not confirm H5 that the search result quality is more important than relevance in decision-making, combining both factors has a higher impact on the decision-making process than each factor alone. Finally, H6 is confirmed that search results that take a stance towards the compared options have a higher impact on the decision.

Our results entail several implications for web search. As quality and relevance are significantly correlated (high-quality results are also more often used to make decisions), it is important to consider document quality in document ranking. Since documents with a stronger stance have a higher impact on the users' decisions, the stance should also be considered a ranking signal. Moreover, our results show that high-quality documents are especially important to form decisions on high-stake subjective comparisons. Thus, search engines should potentially first identify whether a comparative question is subjective.

2 Related Work

How people decide on one or another option has been well studied by psychologists [3,23,26,33]. Decisions are made either intuitively or analytically [33], and can be influenced by prior knowledge or research made ad hoc [26]. Web search engines have become a common means for collecting facts, opinions, and arguments that guide decisions, with at least three percent of web search queries being comparative questions [8]. While factual questions (e.g., "Does cider or beer contain more calories?") can often be answered analytically based on facts, subjective comparative questions (e.g., "Should I study philosophy or psychology?") may require arguments that discuss the pros and cons of possible options [8]. With an increasing trend towards direct answers [28], web search engines became tools for rather intuitive or ready-to-use solutions than analytical decision-making.

This intuitive decision-making intensifies four types of cognitive biases [5] affect the decision-making: First, users are more likely to examine results that confirm their own prior beliefs, expectations, or hypotheses known as a *confirmation bias*, and adapt their search patterns accordingly [16,24,27,38–40]. Second, despite the diversity of viewpoints on the Web, search engines often favor results representing a particular point of view [12,35]. This *viewpoint bias* affects searchers' attitudes [11,31]. Third, users overestimate the trustworthiness of web search engines and their ranking models [6,13,16,22,34,37]. This *trust bias* is less pronounced for more experienced users [32]. Last, the *position bias* describes the tendency to prefer web pages placed at the top of the returned search results [2,17,27,28,32,34].

Furthermore, prior research on the impacts of search result quality is focused on *system-centered* evaluations [1,4,30]. So far, the impact of search result quality on the users' decision-making process has not been studied in detail. Our work contributes to a better understanding of the quality of search results for comparative questions and their impact on decision-making, by analyzing results retrieved by Google, and hence, is a first step to increasing the accountability of major search engines to their users' decisons [14,25].

3 Assessing Search Result Quality

To assess the search result quality for comparative questions, we (1) manually selected 30 questions from the 100 topics of the Touché shared task on comparative argument retrieval [9,10], (2) for each question, retrieved the top-4 results with Google, and (3) asked ten volunteer assessors to rate the quality of each search result following a set of predefined quality criteria.

Table 1. Quality, relevance, and stance criteria, their aspects, and answer options. The 'Score' column indicates points/multipliers for each choice and the criterion's weight in the aggregated quality. Agreement is Fleiss' κ; aspects without agreement are not used.

Aspect	Score	Aspect	Score	Aspect	Score
A *Content*	×4	**C** *Credibility*	×2	**D** *Up-to-dateness*	×1
A1 Completeness ($\kappa < 0.00$)		C1 Source ($\kappa = 0.52$)		D1 Date ($\kappa = 0.40$)	
A2 Scope ($\kappa = 0.24$)		news portal	+4	outdated	+2
scarce	+1	public institution	+4	up to date	+4
precise	+2	Q&A platform	±0	timeless	+4
appropriate	+3	encyclopedia	+2	D2 Updates ($\kappa = 0.15$)	
very detailed	+4	corporate website	+2	at least one update	+1
excessive	±0	blog	±0	no updates	±0
A3 Language ($\kappa = 0.32$)		C2 Author ($\kappa = 0.30$)			
objective / factual	+2	qualified author	+2	**E** *Relevance*	
entertaining	+1	unqualified author	+1	E1 Topical relevance ($\kappa = 0.19$)	
judgmental	±0	generated	±0	not relevant	
promotional	±0	unknown	+1	relevant	
		C3 Truthfulness ($\kappa = 0.29$)		highly relevant	
B *Usability*	×4	yes	×1		
B1 Media types ($\kappa < 0.00$)		no	×0	**F** *Stance*	
B2 Structure ($\kappa = 0.25$)		partially	×0.5	F1 Referral ($\kappa < 0.00$)	
unstructured	±0	unknown	×1	F2 Emphasis ($\kappa < 0.00$)	
roughly structured	+2	C4 Verifiability ($\kappa < 0.00$)		F3 Direction ($\kappa < 0.00$)	
well structured	+4			F4 Magnitude ($\kappa = 0.51$)	
very well structured	+6			strong	
				weak	
				no stance	
				neutral	

3.1 Data, Criteria, and Methodology

Data. Out of 100 Touché topics (each consisting of a comparative question, a description of the information need, and the relevance criterion) [9,10], we manually selected 30 topics that contain exactly two comparison options (we discarded, e.g., superlative questions like "What are the best dish detergents?").[1] Since the assessors were native German speakers, each topic's question and relevance criterion were translated into German. For each selected question, we retrieved the top-4 search results with Google, the most popular search engine in Germany,[2] excluding videos or PDFs, and using anonymous browsing to prevent personalization. In total, we collected 120 search results in German.

Criteria. Prior quality assessment frameworks for web documents (WebQual [20], 2QCV3Q [21], AIMQ [18], Touché [10]) do not directly apply to search results for comparative questions. Therefore, we developed a set of four search result quality criteria (content, usability, credibility, and up-to-dateness), which we complemented with relevance and stance. Each criterion is further narrowed to one or more aspects (Table 1): (1) *Content* quality is determined by a document's completeness, scope, and rhetoric style. High-quality documents cover the comparative information need comprehensively and provide reasoning supported

---

[1] Code and data available online: https://github.com/webis-de/CLEF-24.
[2] Retrieved on May 4–5, 2022. Archived results available online (see Footnote 1).

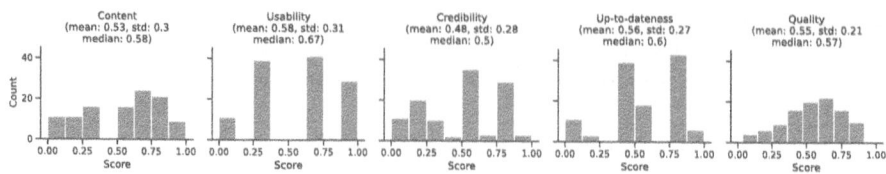

Fig. 1. Quality distributions for each quality criterion and aggregated quality.

by solid evidence [36]. (2) *Usability* hinges on the document structure and readability. Offering the same content in more than one media type (e.g., text, tables, figures) improves the accessibility of a document. High-quality documents are well-structured, do not contain disruptive content (e.g., advertisements), and are easy to read [19]. (3) *Credibility* is assessed by the document's source (e.g., newspaper or government), author's qualification, truthfulness, and verifiability [36]. Thus, credible documents come from reputable sources, are written by qualified authors, only contain truthful information, and provide references to their sources. (4) The *up-to-dateness* [19] describes whether a document is up-to-date (e.g., at most 40 days old [19]) and whether it has been updated at least once. Since a publication date was not always available, we considered a document up-to-date if it was not outdated or if it indicated that it was based on recent studies.

Relevance and stance were included to support the analysis of the hypohtheses H5 and H6 (Sect. 1) but were not used for measuring the quality of the search results. For the topical *relevance*, the topic narratives defined in the Touché shared task [9,10] were used and adapted to binary relevance labels (relevant or irrelevant). For the *stance*, we asked the assessors to judge which of the comparison options was mentioned in the document, which was discussed in more detail, and whether the document took a stance towards one of the options (e.g., "pro Pepsi" for the question of "Which is better, Pepsi or Cola?"). We also considered the stance magnitude, where direct recommendations indicate a strong stance and indirect supportive statements still indicate a weak stance.

Methodology. The 120 Google search results were assessed by ten volunteer assessors (German university students; 7 studied media science, 3 computer science). All assessors were provided with the codebook[1] and, for an initial pilot study, assessed the top-4 results of the same search query (topic 19, randomly selected) in random order. Table 1 shows the agreement (Fleiss' κ) for each evaluation aspect. The six aspects without agreement ($\kappa < 0.00$; i.e., A1, B1, C4, F1, F2, and F3) were removed from further analysis. In a follow-up video call, assessors discussed questions regarding the criteria and conflicting assessments. Afterwards, each assessor was given 12 search results for three queries to assess.

3.2 Evaluation

To analyze the quality of search results, we calculate quality scores for each quality criterion (content, usability, credibility, and up-to-dateness) based on their aspects, excluding four aspects without sufficient agreement. Scores for each criterion were calculated as the sum of answer points to its aspects (see the 'Score' column in Table 1). One exception is the truthfulness aspect, where the score is multiplied to account for the potential misinformation harmfulness. The resulting scores are normalized to a 0–1 range, 1 indicating a perfect score. If an aspect was not assessed (i.e., n/a was selected), we did not calculate a quality score for the corresponding criterion. Due to this filtering, 14 documents were excluded from the content quality score and seven documents from the credibility score. An aggregated score is then calculated as the weighted sum of the individual quality scores, where the weights (see the 'Score' column in Table 1) represent a media scientist's rated importance of the criteria. The weighted sum is again normalized to a 0–1 range. Documents that lack a score for at least one of the criteria are exempt from the aggregated score computation, leaving 103 documents.

The distributions of the quality scores are shown in Fig. 1. Usability tends to be the "easiest" criterion to fulfill (24% of the documents achieve a perfect usability score), whereas credibility is the "hardest" (median 0.5). Quality scores of all criteria vary largely and are not normally distributed, indicating a potential selection bias due to only selecting the top-4 results. The aggregated quality scores, however, are approximately normally distributed, with an average score of 0.55 and a median of 0.57. No correlation was found between the document ranks on the result page and their quality (Kendall's $\tau = 0.07$, $p = 0.37$, $\alpha = 0.05$). Additionally, we measured topical relevance and stance. Like quality, the topical relevance is not correlated to ranks ($\tau = -0.09$, $p = 0.29$), but relevance and quality have a significant positive rank correlation ($\tau = 0.21$, $p = 0.01$).

4 User Study

4.1 Data and Methodology

Data. To characterize and select queries for the user study, we compute the average quality score and standard deviation across all documents retrieved for each query. The 10 queries where at least one result's quality could not be assessed were excluded. From the remaining 18 topics, we first removed topics that would require extensive prior knowledge. Then, we manually selected eight topics that cover a wide range of the topic-wise average quality and standard deviations within the top-4 retrieved results by Google.[3] For example, topic 12 has a high quality and low standard deviation among the retrieved documents, topic 24 has a high std. deviation and high average quality, topic 22 has a consistently average-level quality, and topics 28 and 20 have deficient overall quality.

Methodology. After selecting the topics for the user study, we archived their top-4 search results and created a questionnaire for each topic.[1] Participants were asked to imagine the situation described in the topic and then reported

---
[3] Topics, results, and questionnaire: https://github.com/webis-de/CLEF-24.

Table 2. Contingency tables of the change of the users' decisions, the decision confidence after seeing search results, and change in confidence due to seeing the results, w.r.t. topic background or avg. search result quality for the topic. Significance marked bold (Pearson's χ^2 tests, $\alpha = 0.05$, Bonferroni correction). Changes to expected frequencies in grey font. Quality threshold: 0.57, confidence threshold: 5.

Predictor	Decision change			Final decision confid.			Decision confid. change			
	Equal	Changed	$\sum$	Low	High	$\sum$	Decr.	Equal	Incr.	$\sum$
Background	$\chi^2(1) = 0.45, p = 0.502$			$\chi^2(1) = 3.29, p = 0.070$			$\chi^2(2) = 18.76, \mathbf{p < 0.001}$			
factual	173 +4	101 -4	274	93 -11	181 +11	274	46 -1	98 -22	121 +23	265
subjective	168 -4	112 +4	280	117 +11	163 -11	280	50 +1	146 +22	78 -23	274
$\sum$	341	213	554	210	344	554	96	244	199	539
Quality	$\chi^2(1) = 5.59, p = 0.018$			$\chi^2(1) = 0.03, p = 0.859$			$\chi^2(2) = 4.81, p = 0.090$			
low quality	154 -14	119 +14	273	105 +2	168 -2	273	49 +1	132 +11	87 -12	268
high quality	187 +14	94 -14	281	105 -2	176 +2	281	47 -1	112 -11	112 +12	271
$\sum$	341	213	554	210	344	554	96	244	199	539

Table 3. Contingency tables of the self-assessed agreement with five statements about the decision-making w.r.t. topic background or avg. search result quality for the topic. Significance marked bold (Pearson's χ^2 tests, $\alpha = 0.05$, Bonferroni correction). Changes to expected frequencies in grey font. Quality threshold: 0.57.

Pred.	Conf. opinion			Better decis.			Did not help			Learned sth.			Contd. search		
	No	Yes	$\sum$	No	Yes	$\sum$	No	Yes	$\sum$	No	Yes	$\sum$	No	Yes	$\sum$
Backg.	$\chi^2(1)$=6.10 p=0.014			$\chi^2(1)$=**18.32** $\mathbf{p<0.001}$			$\chi^2(1)$=2.16 p=0.141			$\chi^2(1)$=3.49 p=0.062			$\chi^2(1)$=0.20 p=0.658		
factual	199 -13	75 +13	274	171 -23	103 +23	274	227 +7	47 -7	274	127 -11	147 +11	274	208 +3	66 +3	274
subj.	229 +13	51 -13	280	222 +23	58 -23	280	217 -7	63 +7	280	153 +11	127 -11	280	218 +3	62 -3	280
$\sum$	428	126	554	393	161	554	444	110	554	280	274	554	426	128	554
Qual.	$\chi^2(1)$=0.53 p=0.467			$\chi^2(1)$=**16.71** $\mathbf{p<0.001}$			$\chi^2(1)$=**13.57** $\mathbf{p<0.001}$			$\chi^2(1)$=**25.18** $\mathbf{p<0.001}$			$\chi^2(1)$=**10.96** $\mathbf{p<0.001}$		
low	215 +4	58 -4	273	216 +22	57 -22	273	201 -18	72 +18	273	168 +30	105 -30	273	193 -17	80 +17	273
high	213 -4	68 +4	281	177 -22	104 +22	281	243 +18	38 -18	281	112 -30	169 +30	281	233 +17	48 -17	281
$\sum$	428	126	554	393	161	554	444	110	554	280	274	554	426	128	554

whether they had prior knowledge of the topic. Before seeing the search results, they decided on one of the comparison options and indicated their confidence in their decision (1–6 rating scale). Then, after they were shown the top-4 search results (screenshot, title, and source), they were asked to decide again, to report their confidence, and to indicate which of the documents shown influenced their decision. We also asked the participants whether they agreed with five statements regarding the confirmation of the prior opinion, the helpfulness, the knowledge gained, and the necessity to do further research. Participants were allowed to skip reading documents they felt were irrelevant (as search engine users would normally do [15]) but reported which documents they read. The user study was conducted with 442 volunteer participants (German university students). They were given a link which randomly redirected to an online questionnaire corresponding to one of the eight topics. At the end of the questionnaire, the participants could volunteer to continue with another topic. A total of 554 submissions were received (1.25 submissions per participant on avg.; 69 per topic, min. 63, max. 80).

4.2 Evaluation

The majority of the participants (45%) did not change their decision after seeing the search results, while only 38% did (Table 2). The remaining 17% did not decide on either comparison option before or after seeing the results. Participants were already confident in their decisions before seeing the results (53% rated their confidence as 5/6 or 6/6) and further increased after seeing the results (64% rated 5/6 or 6/6). For 45% of the participants, their confidence did not change after seeing the results. For 37%, confidence increased, and for 18%, it decreased. Only 35% of the documents were reported to have influenced the users' decisions (Table 4) and only 29% of the participants reported that they could make a better decision based on the search results while 23% would continue their search (Table 3). Yet, half of the participants (49%) stated that they had learned something new about the topic and only 20% found the search results unhelpful. To verify each of the six hypotheses (Sect. 1), we perform significance tests (Pearson's χ^2, $\alpha = 0.05$, Bonferroni correction) on the contingency tables.

H1: *Comparative questions on subjective topics lead to less confident decisions than questions on factual topics.* No significant differences in the final users' confidence after seeing the search results were observed between factual and subjective topics (Table 2). However, the confidence increased significantly more for factual than subjective topics, resulting in higher final decision confidence. Participants reported they could make a better decision significantly more often for subjective than factual topics (Table 3). None of the remaining statements about the users' decision-making yielded significant differences between factual and subjective topics. Hence, we discard the hypothesis, even though factual topics lead to increased confidence more often than subjective topics.

H2: *Comparative questions with low-quality results lead to less confident decisions than questions with high-quality results.* To compare decision confidence w.r.t. search result quality, we first define low-quality topics as topics with an avg. document quality score below the median quality score of 0.57. High-quality topics have an avg. quality score of at least 0.57. Study participants changed their decision slightly more often for low-quality topics than for high-quality topics, and high-quality results led to a slightly increased decision confidence but none of the changes were significant (Table 2). Regarding the self-assessment of the users' decision-making, Table 3 shows that for high-quality topics, users more often reported that they could make a better decision and felt they had learned something new. For low-quality topics, users stated more often that the search results did not help and that they would continue the search. Due to the partially contradicting results for decision confidence and helpfulness in topics with different result quality, we discard the hypothesis.

H3: *The higher a search result's quality, the more likely it influences the decision-making.* For this hypothesis, we asked participants to report which documents had influenced their decision. Only documents that were at least par-

Table 4. Contingency tables of the influence of documents on the users' decisions w.r.t. result quality, relevance, quality and relevance, stance magnitude, initial confidence for low-quality documents. Significance marked bold (Pearson's χ^2 tests, $\alpha = 0.05$). Changes to expected frequencies in grey font. Quality threshold: 0.57, confidence threshold: 5.

Predictor	Document influence		
	No influence	Influence	Σ
Quality	$\chi^2(4) = 44.49, p < 0.001$		
low quality	648 +67	252 -67	900
high quality	492 -67	374 +67	866
Σ	1140	626	1766
Relevance	$\chi^2(1) = 41.77, p < 0.001$		
not relevant	649 +65	255 -65	904
relevant	491 -65	371 +65	862
Σ	1140	626	1766
Quality × relevance	$\chi^2(8) = 79.48, p < 0.001$		
not rel., low qual.	390 +65	113 -65	503
not rel., high qual.	259 ±0	142 ±0	401
rel., low qual.	258 +2	139 -2	397
rel., high qual.	233 -67	232 +67	465
Σ	1140	626	1766

Predictor	Document influence		
	No influence	Influence	Σ
Stance strength	$\chi^2(2) = 26.76, p < 0.001$		
no stance	273 -6	237 +6	510
weak stance	111 +26	45 -26	156
strong stance	75 -20	99 +20	174
Σ	459	381	840
Init. confid.	$\chi^2(1) = 4.51, p = 0.034$		
low confidence	302 -15	138 +15	440
high confidence	346 +15	114 -15	460
Σ	648	252	900
Ranking position	$\chi^2(3) = 10.62, p = 0.014$		
rank 1	286 -26	197 +26	483
rank 2	302 ±0	166 ±0	468
rank 3	281 +7	144 -7	425
rank 4	271 +19	119 -19	390
Σ	1140	626	1766

tially read were considered for this analysis. We again consider documents with a quality score of less than 0.57 as low quality and documents with a quality score of at least 0.57 as high quality. Table 4 shows that low-quality documents influence decisions significantly less often than high-quality documents. A document's rank also significantly affects a document's influence on the decision. However, a position bias can be ruled out as the ranks were not correlated to quality (Sect. 3.2). Hence, the hypothesis can still be confirmed.

H4: *Users who are more confident in their decision before searching are less influenced by low-quality search results.* We further analyze the influence of the users' prior decision confidence by filtering low-quality documents. Of the 900 low-quality documents that were at least partially read, 252 documents influenced the decision. In Table 4, we consider low-quality documents with below-median initial confidence and high initial confidence separately. The significance test reveals that low-quality documents influenced the decision significantly more often if the initial confidence was low, confirming the hypothesis.

H5: *The quality of a search result has a higher impact on the decision-making process than its relevance.* Even though a document's topical relevance and quality are conceptually different, our quality assessments revealed that both are highly correlated (Sect. 3). Compared to the significant influence of search result quality on decision-making, Table 4 also highlights a significant influence of topical relevance with only a slightly smaller effect than for result quality. Hence, the hypothesis cannot be confirmed. However, the tests also show that combining both factors has a higher impact on decision-making than the factors alone.

1: *Documents that take a stance towards one compared option have a higher impact on the decision.* Table 4 examines the impact of the stance magnitude in either direction. Documents with a strong stance (e.g., containing direct recommendations) influenced the decision significantly more often than those with a weak (e.g., indirect statements) or no stance, confirming the hypothesis.

4.3 Limitations

Our study results have several limitations. First, all participants were German university students, which might not represent the general population. Second, the study was conducted using a single search engine; thus results might not be generalizable to other search engines. Even though we used comparative questions from prior work claimed to represent real user information needs, for more robust findings, a larger study (more participants and questions) might be needed.

5 Conclusion

We evaluated the quality of web search results and the quality's impact on the decision-making for questions comparing two options. We derived guidelines to manually assess four quality criteria (content quality, usability, credibility, and up-to-dateness). The evaluation of the 120 assessed documents (top-4 results for 30 comparative questions) w.r.t. the search result quality, topical relevance, and stance showed substantial heterogeneity in the search result quality, significant correlation between relevance and quality, but no correlation of either quality or relevance with their ranks on Google's result page. Our quality assessments also highlighted that individual quality criteria on their own are not representative of a document's overall quality, motivating more systematic quality measurements for evaluation. Our criteria could serve as a starting point to design formal measures. Based on the quality assessments, we selected eight queries with varying result quality for a user study examining the search results' impact on user decisions. In the study, the participants were asked about their decisions and confidence before and after seeing the search results, which documents influenced their decision, and if they agreed with five statements about the decision-making process.

Our results showed that the quality of search results has a significant impact on being used in the decision-making process (H3) but not on the confidence of user decisions (H2). Quality can thus be considered an important factor in the search result ranking for comparative questions. As documents with a stronger stance also have a higher impact on the users' decisions (H6), we suggest that the stance magnitude should also be considered for ranking. Even though no significant difference was found between the confidence after seeing the search results of factual and subjective questions (H1), the topic background still significantly influences the change in decision confidence. Users gained confidence in their decisions significantly more often for factual than for subjective questions.

Improving search result quality, especially for subjective questions, could thus help users to make more confident decisions. The user study also showed that users with initially high decision confidence are less likely to be influenced by low-quality results (H4). Last, we observed a similarly pronounced impact of both quality and relevance on the decision-making process (H5) and that combining the two factors has a higher impact on decision-making than any factor alone. Because current evaluation merely considers relevance or quality on its own [10], combining both factors in future evaluations of comparative queries is worthwhile.

Acknowledgments. This work was partially supported by the DFG (German Research Foundation) through the project "ACQuA 2.0: Answering Comparative Questions with Arguments" (project 376430233) as part of the priority program "RATIO: Robust Argumentation Machines" (SPP 1999) and by the Stiftung für Innovation in der Hochschullehre under the "freiraum 2022" call (FRFMM-58/2022).

References

1. Abualsaud, M.: The effect of queries and search result quality on the rate of query abandonment in interactive information retrieval. In: CHIIR 2020 (2020)
2. Abualsaud, M., Smucker, M.D.: Exposure and order effects of misinformation on health search decisions. In: ROME@SIGIR 2019 (2019)
3. Ajzen, I.: The social psychology of decision making (1996)
4. Alanazi, A.O., Sanderson, M., Bao, Z., Kim, J.: The impact of ad quality and position on mobile SERPs. In: CHIIR 2020 (2020)
5. Azzopardi, L.: Cognitive biases in search: A review and reflection of cognitive biases in information retrieval. In: CHIIR 2021 (2021)
6. Bink, M., Schwarz, S., Draws, T., Elsweiler, D.: Investigating the influence of featured snippets on user attitudes. In: CHIIR 2023 (2023)
7. Bondarenko, A., Ajjour, Y., Dittmar, V., Homann, N., Braslavski, P., Hagen, M.: Towards understanding and answering comparative questions. In: WSDM 2022 (2022)
8. Bondarenko, A., et al.: Comparative web search questions. In: WSDM 2020 (2020)
9. Bondarenko, A., et al.: Overview of Touché 2020: Argument retrieval. In: Arampatzis, A., et al. (eds.) CLEF 2020. LNCS, vol. 12260, pp. 384–395. Springer, Cham (2020). https://doi.org/10.1007/978-3-030-58219-7_26
10. Bondarenko, A., et al.: Overview of Touché 2021: Argument retrieval. In: Candan, K.S., et al. (eds.) CLEF 2021. LNCS, vol. 12880, pp. 450–467. Springer, Cham (2021). https://doi.org/10.1007/978-3-030-85251-1_28
11. Draws, T., Tintarev, N., Gadiraju, U., Bozzon, A., Timmermans, B.: This is not what we ordered: Exploring why biased search result rankings affect user attitudes on debated topics. In: SIGIR 2021 (2021)
12. Gezici, G., Lipani, A., Saygin, Y., Yilmaz, E.: Evaluation metrics for measuring bias in search engine results. Inf. Retr. J. **24**(2) (2021)
13. Grimmelmann, J.: The Google dilemma. NYL Sch. L. Rev. **53**(2) (2008)
14. Gunning, D., Aha, D.W.: DARPA's explainable artificial intelligence (XAI) program. AI Mag. **40**(2) (2019)
15. Joachims, T.: Optimizing search engines using clickthrough data. In: KDD 2002 (2002)

16. Knobloch-Westerwick, S., Johnson, B.K., Westerwick, A.: Confirmation bias in online searches: Impacts of selective exposure before an election on political attitude strength and shifts. J. Comput. Mediat. Commun. **20**(2) (2015)
17. Lau, A.Y.S., Coiera, E.W.: Do people experience cognitive biases while searching for information? J. Am. Medical Informatics Assoc. **14**(5) (2007)
18. Lee, Y.W., Strong, D.M., Kahn, B.K., Wang, R.Y.: AIMQ: A methodology for information quality assessment. Inf. Manag. **40**(2) (2002)
19. Lewandowski, D., Höchstötter, N.: Web searching: A quality measurement perspective. In: Spink, A., Zimmer, M. (eds.) Web Search. Information Science and Knowledge Management, vol. 14, pp. 309–340. Springer, Heidelberg (2008). https://doi.org/10.1007/978-3-540-75829-7_16
20. Loiacono, E.T., Watson, R.T., Goodhue, D.L.: WebQual: A measure of website quality. In: AMA 2002 (2002)
21. Mich, L., Franch, M., Cilione, G.: The 2QCV3Q quality model for the analysis of web site requirements. J. Web Eng. **2**(1) (2003)
22. Narayanan, D., De Cremer, D.: 'Google told me so!' On the bent testimony of search engine algorithms. Philos. Technol. **35**(2) (2022)
23. Newell, B.R., Lagnado, D.A., Shanks, D.R.: Straight choices: The psychology of decision making (2022)
24. Nickerson, R.S.: Confirmation bias: A ubiquitous phenomenon in many guises. Rev. Gener. Psychol. **2**(2) (1998)
25. Nyholm, S.: Attributing agency to automated systems: Reflections on human-robot collaborations and responsibility-loci. Sci. Eng. Ethics **24**(4) (2018)
26. Peterson, M.: An introduction to decision theory (2017)
27. Pogacar, F.A., Ghenai, A., Smucker, M.D., Clarke, C.L.A.: The positive and negative influence of search results on people's decisions about the efficacy of medical treatments. In: ICTIR 2017 (2017)
28. Potthast, M., Hagen, M., Stein, B.: The dilemma of the direct answer. SIGIR Forum **54**(1) (2020)
29. Puschmann, C.: Beyond the bubble: Assessing the diversity of political search results. Digit. Journal. **7**(6) (2019)
30. Raiber, F., Kurland, O.: Using document-quality measures to predict web-search effectiveness. In: Serdyukov, P., et al. (eds.) ECIR 2013. LNCS, vol. 7814, pp. 134–145. Springer, Heidelberg (2013). https://doi.org/10.1007/978-3-642-36973-5_12
31. Rieder, B., Sire, G.: Conflicts of interest and incentives to bias: A microeconomic critique of Google's tangled position on the Web. New Media Soc. **16**(2) (2014)
32. Schultheiß, S., Lewandowski, D.: Misplaced trust? The relationship between trust, ability to identify commercially influenced results and search engine preference. J. Inf. Sci. **49**(3) (2023)
33. Tversky, A., Kahneman, D.: Judgment under uncertainty: Heuristics and biases. Science **185**(4157) (1974)
34. Unkel, J., Haas, A.: The effects of credibility cues on the selection of search engine results. J. Assoc. Inf. Sci. Technol. **68**(8) (2017)
35. Vaughan, L., Thelwall, M.: Search engine coverage bias: Evidence and possible causes. Inf. Process. Manag. **40**(4) (2004)
36. Wang, Y., Sarkar, S., Shah, C.: Juggling with information sources, task type, and information quality. In: CHIIR 2018 (2018)
37. Westerwick, A.: Effects of sponsorship, web site design, and Google ranking on the credibility of online information. J. Comput. Mediat. Commun. **18**(2) (2013)
38. White, R.: Beliefs and biases in web search. In: SIGIR 2013 (2013)

39. White, R.W.: Belief dynamics in web search. J. Assoc. Inf. Sci. Technol. **65**(11) (2014)
40. Yom-Tov, E., Dumais, S., Guo, Q.: Promoting civil discourse through search engine diversity. Soc. Sci. Comput. Rev. **32**(2) (2014)

Improving Laypeople Familiarity with Medical Terms by Informal Medical Entity Linking

Annisa Maulida Ningtyas[1,2(✉)], Alaa El-Ebshihy[1,3], Florina Piroi[1,3], and Allan Hanbury[1]

[1] Technische Universität Wien, Vienna, Austria
{alaa.el-ebshihy,florina.piroi,allan.hanbury}@tuwien.ac.at
[2] Universitas Gadjah Mada, Yogyakarta, Indonesia
annisamaulidaningtyas@ugm.ac.id
[3] Research Studios Austria, Data Science Studio, Vienna, Austria

Abstract. Social media is an important community engagement tool in the health domain for health promotion, patient education, and outreach. Considering its educational potential, we introduce an Informal Medical Entity Linking (EL) model with three components to enhance the laypeople comprehension of medical terminology by linking popularized medical phrases in social media posts to their specialized counterparts and to relevant Wikipedia articles. Medical experts assessed the accuracy and relevance of the EL model, finding that the Medical Concept Normalization (MCN) component of our model correctly classifies 89% of informal phrases. The second component, Entity Disambiguation (ED), effectively predicts relevant Wikipedia articles for identified popularized medical terms, and the third one, Learning to Rank (LTR), improves the identification of the most relevant Wikipedia articles based on popularized medical phrases and their specialized counterparts. These findings suggest our model can be a valuable tool for enhancing laypeople's comprehension of medical terminology by leveraging social media's educational potential.

Keywords: social media · medical entity linking · medical concept normalization · medical vocabulary

1 Introduction

Social media platforms, like online health forums, play a significant role in healthcare by facilitating community engagement, health promotion, patient education, and outreach. An increasing number of laypeople search for medical information on these platforms to understand specialized medical terminology, resulting in increased familiarity with medical terminology [7]. Health-related information on social media is often informally contributed by laypeople to express their health experience. Identifying medical concepts in social media posts and finding their specialized terminology is relevant to drug manufacturers who must collect and summarize product side effects. This task, known as *Medical Concept Normalization* (MCN), first detects popularized medical phrases in user texts then links them to medical concepts in knowledge bases like Systematized Nomenclature of Medicine - Clinical Terms (SNOMED-CT) [12]. While MCN has proven useful to medical organizations [12,19], its potential benefits for laypeople to improve their medical terminology comprehension have not yet been explored.

Fage-Butler et al. [7] emphasize that incorporating medical terminology in online patient-patient communication helps expand patient navigation skills in the medical environment [6]. Laypeople appreciate the consistent and appropriate use of medical terminology by professionals [9], though specialized terms are often difficult for those with limited medical domain knowledge. Considering the educational potential of social media, we introduce an *Informal Medical Entity Linking* (EL) model that enhances MCN. While MCN links popularized phrases to concepts in medical knowledge bases, where those concepts are specialized terms, our EL model specifically links popularized medical terms in social media posts to their corresponding specialized terms *and* Wikipedia articles that explain them. Our model accepts a user post, extracts popularized medical entities, and outputs their corresponding specialized medical entity and Wikipedia articles. The contributions of this study are: 1) the development of an Informal Medical EL model designed to support laypeople in learning specialized medical terminology in social media settings, 2) an expert assessment of our EL model's effectiveness, and 3) a re-ranking model to enhance EL performance. Figure 1 illustrates an overview of the proposed model.

2 Related Work

Prior research focuses on developing systems to help laypeople understand medical terminology in health-related documents, such as Electronic Health Records (EHRs), Personal Health Records (PHRs), and Scientific Publications [1,3,10,15,20]. These systems aim to empower patients by simplifying text through identifying medical terms and providing explanations. EHRs are highly challenging for laypeople to comprehend, and while translations improve understanding, this improvement is not statistically significant [20]. Zeng-Treitler et al. [20] proposed a three-step model to enhance the readability of EHRs by simplifying terminology. Similarly, Kandula et al. [10] developed a tool to address the semantic complexity of medical terms by replacing them with simpler synonyms or explanations. Another tool, NoteAid [3], uses supervised deep learning techniques to improve laypeople's understanding of medical terminology, with two main components: CodeMed, a lexical database, and MedLink, which connects medical terms to simpler explanations. Similarly, Alfano et al. [1] created SIMPLE, a web-based application that increases the accessibility of medical texts by identifying technical terms, mapping them to simpler equivalents, and providing easy-to-understand definitions. Unlike previous work [1,3,10,15,20] which employs text simplification techniques on EHR documents, our approach tackles the problem through domain entity linking from social media. Instead of simplifying specialized medical terms for laypeople, we introduce laypeople to specialized medical terms by linking these terms to the popularized medical phrases commonly used by laypeople and use Wikipedia articles of the specialized terms to provide explanations.

3 Model for Informal Medical Entity Linking

The Informal Medical Entity Linking (EL) we propose contains three distinct phases: (1) *Named Entity Recognition (NER)* identifies textual mentions in social media posts that represent popularized medical phrase referring to a disease or a drug; (2) *Medical Concept Normalization (MCN)* maps the identified mentions to corresponding concepts in SNOMED-CT; and (3) *Entity Disambiguation (ED)* links each of these SNOMED-CT concepts to corresponding entities in Wikipedia (Fig. 1). We will now detail each phase.

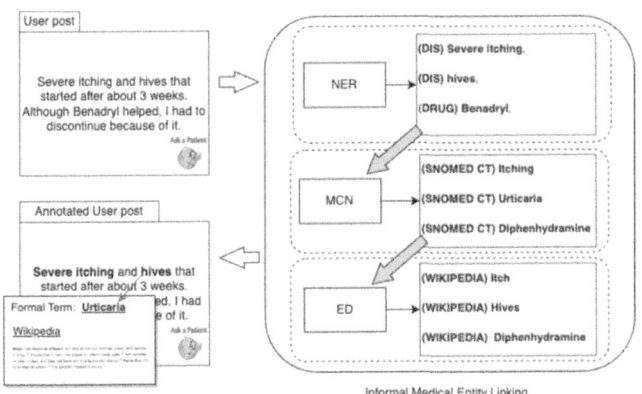

Fig. 1. Overview of the informal medical entity linking workflow.

We denote with V the vocabulary of tokens in a social media post[1]. A text sequence **t**, consists of sequentially ordered tokens from V, forming a sentence in the original text. We write it as $\mathbf{t} = (t_i)_{i=1}^{n}$, where n is the length of the token sequence. From a token sequence, **t** we can extract a set of phrases (set of consequent tokens, also called *spans*), $\mathbf{p(t)}$:

$$\mathbf{p(t)} = \{(p_j)_{j=1}^{m} | p_j \in \mathcal{L}(\mathbf{t})\}$$

where $\mathcal{L}(\mathbf{t})$ denotes the set of candidate spans over **t**.

For example, given a sample user text, "*Severe itching and hives that started after about 3 weeks. Although Benadryl helped, I had to discontinue because of*

---

[1] Tokens are extracted user text posts by the word tokenization.

it." we extract two sequences (as it contains two sentences), one of which is $\mathbf{t} =$ (Severe, itching, and, hives, that, ..., weeks). Then, $\mathcal{L}(\mathbf{t})$ includes possible candidate spans such as {(Severe, itching), (Severe, itching, and), (Severe, itching, and, hives), (hives), ...}.

NER Phase: We denote with $T = \{\mathbf{t}\}$ the set of all sentences, or token sequences, over V. The NER model takes T as input and returns a subset $P \subseteq \mathcal{L}(T)$, which is the list of popularized medical phrases. That is, $NER : T \to \mathcal{L}(T)$.

To identify the popularized medical phrases, we trained a NER model using a deep learning technique for sequence labeling. Building upon our previous work in [14], we use a BILSTM-CRF architecture, trained on the CADEC [11] and the MedRed [18] datasets[2] Using the example above, NER identifies the following set of popularized medical phrases, or *mentions*, $P = \{$(severe, itching), (hives), (benadryl)$\}$.

MCN Phase: In this phase, each identified popularized medical phrase $p \in P$, extracted in the NER phase, is mapped to a corresponding formal medical concept in SNOMED-CT. That is, denoting the set of formal medical concepts in SNOMED-CT with C, we have $MCN : P \to C$, and:

$$c_p = MCN(p) \quad \forall p \in P$$

where, c_p denotes the specialized medical term in SNOMED-CT mapped to the informal phrase p. For example, the phrase *(severe, itching)* is mapped to the SNOMED-CT code: 418290006, representing *Itching*.

We trained the MCN model as a multi-class classification task using a combination of CADEC, PsyTAR [21] and COMETA [2] datasets along with automatically labeled data generated by our approach in [13]. The training dataset was enriched by adding SNOMED-CT synonyms. The model architecture employed a gated recurrent unit (GRU)[3].

ED Phase: To disambiguate entities, we use the GENRE model [4], which takes an input text and generates a Wikipedia entity name one part at a time. It employs a Beam Search algorithm within a prefix tree structure, where each node represents tokens from the vocabulary, primarily Wikipedia titles [4]. This structure allows GENRE to propose potential Wikipedia entities based on the input sequence. It, then, connects proposed entities to actual Wikipedia entities by using a scoring and ranking mechanism, which evaluates the likelihood of each proposed entity name being a valid Wikipedia entity, considering the context provided by the input text. In this ED module, GENRE operates on two distinct inputs: popularized medical phrases, p, and their specialized medical term mapping, c_p, as follows:

1. **Tagging popularized medical phrases** (p): the original text, containing the popularized medical phrase p was identified, is tagged with placeholders

---

[2] https://github.com/maulidaannisa/data-augmentation.
[3] https://github.com/maulidaannisa/mcn_distant_supervision.

marking the start and end of the phrase. This tagged sentence is given to GENRE, which returns the top five Wikipedia entries fitting the tagged text and its phrase:

$$E_p = \text{GENRE}(\text{original text with [start]} + p + \text{[end] tagging})$$

where E_p is the set of top five Wikipedia entities given by GENRE.

2. **Tagging specialized medical terms** (c_p): the tagged text from the previous step is modified to include the specialized medical term (c_p) mapped to the popularized term (p). This sentence is given to GENRE, which returns the top five Wikipedia entries fitting the tagged text and the specialized medical term:

$$E_{c_p} = \text{GENRE}(\text{original text with [start]} + c_p + \text{[end] tagging})$$

where E_{c_p} is the set of Wikipedia entities given by GENRE.

We have two sets of Wikipedia entities: $E_p = \{e_{p1}, \ldots, e_{p5}\}$ for the informal phrases and $E_{c_p} = \{e_{c1}, \ldots, e_{c5}\}$ for the specialized medical terms, each ordered by relevance from GENRE, with e_1 being the most relevant entity for its respective input. Depending on the inputs to GENRE, the number of returned Wikipedia entries may range from five to ten unique entities. For instance, consider the sentence *"Severe itching and hives that started after about 3 weeks."* we have:

$$E_p = \text{GENRE}([\text{start}] \text{ Severe itching [end] and hives } \ldots \text{ 3 weeks})$$
$$= \{\text{Itch, Anorexia (symptom), Arthralgia, Allergic rhinitis, Erythema}\}$$

and for the specialized medical term, c_p, *itching*:

$$E_{c_p} = \text{GENRE}([\text{start}] \text{ Itching [end] and hives } \ldots \text{ 3 weeks})$$
$$= \{\text{Itch, Allergic rhinitis, Arthralgia, Herpes labialis, Infectious mononucleosis}\}$$

To refine entity selection, we introduce a learning-to-rank model *Rank*, which re-ranks the entities based on their relevance to informal and formal medical inputs. The re-ranking process for each medical concept c_p combines the entities E_p and E_{c_p} corresponding to the mapped specialized medical term c_p, ensuring that the most relevant entities for both informal and specialized medical terms are identified and prioritized effectively:

$$R_{c_p} = \text{Rank}(E_p \cup E_{c_p})$$

R_{c_p} is the ranked output for a given concept c_p, obtained by applying the *Rank* model to the union of the entity sets from both E_p and E_{c_p}. For our running example, we have:

$$R_{c=itching} = \{\text{Itch, Allergic rhinitis, Herpes labialis, Arthralgia, Erythema}$$
$$= \text{Anorexia (symptom), Infectious mononucleosis}\}$$

Finally, the informal medical entity linking model produces its end output as a structured set of tuples of three elements, each comprises: (1) a predicted popularized medical phrase from the text p, (2) its normalized counterpart c in SNOMED-CT, and (3) the most relevant Wikipedia entity e_1. For our example, where we have the informal phrase *(p = severe itching)*, the tuple is *(p = severe itching, c = itching, e_1 = "Itch")*.

4 Evaluation

We conducted an expert evaluation to assess the reliability of the information produced by the informal medical Entity Linking (EL) in helping laypeople understand medical terminology. Specifically, we focus on evaluating the outputs of the MCN and ED components of the EL model. We excluded evaluating the NER model since the failure analysis presented in [14] demonstrated that its predictions of popularized medical terms were sufficiently reliable.

4.1 Evaluation Data

We randomly selected 30 posts from the AskAPatient forum[4], where each post averages 3 sentences and 33 tokens. Using our informal medical EL model pipeline, we annotated these posts, extracting 225 terms that include predicted popularized medical phrases, their corresponding specialized medical terms, and Wikipedia article candidates.

4.2 Evaluation Design

We involved three experts with different medical background: a general practitioner, a midwife, and a medical coder[5]. To ensure a comprehensive evaluation, each expert assessed all 225 annotated terms from 30 posts. They used a dedicated interface to complete two tasks for each annotated term in the social media posts (Fig. 2):

1. **Verify the correctness of the MCN model output** (the box marked with 1): Indicate whether the displayed specialized term shown in "Formal term"row is a correct mapping to the popularized term by selecting "Yes" or "No" accordingly.
2. **Select the appropriate Wikipedia article** (marked with 2): Mark the most relevant Wikipedia article from a list of the top 5 ranked articles, given by the GENRE model, for both popularized and specialized medical terms. When no Wikipedia articles was considered relevant, the experts selected the option "0".

---

[4] Ask a Patient - Drug Ratings and Patient Reviews https://www.askapatient.com/.
[5] Due to limited resources, we collaborated with these experts who agreed to voluntarily participate in the evaluation tasks.

Fig. 2. Expert Evaluation Platform

5 Results and Discussion

We use the outcome from the expert annotation experiment to assess the output quality of our MCN model and of the GENRE model used in the ED phase.

5.1 Expert Agreement

We computed the *Inter-Annotator Agreement* (IAA) to observe differences in how the experts annotated the data. Table 1 shows the average Cohen's Kappa score along with the standard deviation, minimum, and maximum scores of the pair-wise agreement. This IAA We find that the agreement between experts in verifying the correctness of the MCN output can be seen as *fair*. This finding emphasizes the challenging nature of verifying the accuracy of predicted specialized medical terms starting from popularized terms. For example, while the MCN model classified 'feeling useless' to the 'feeling hopeless (SCUI:307077003)' SNOMED-CT term, the broad nature of 'feeling useless' makes it challenging for experts to confirm its correctness, since it can also be mapped to 'Depression mood (SCUI: 366979004)'.

Table 1. The pairwise agreement between experts

Description	Cohen's Kappa Score		
	Avg. Pairwise	Min	Max
MCN Output Correctness	0.37 ± 0.07	0.32	0.46
Wikipedia Selection (specialized term)	0.51 ± 0.01	0.50	0.52
Wikipedia Selection (popularized term)	0.53 ± 0.01	0.52	0.55

For the second task, selecting suitable Wikipedia articles to explain specific medical terms, showed *moderate* agreement among experts for both types of input queries provided to the GENRE model. The IAA scores were higher for the second task than the first, indicating that selecting appropriate Wikipedia articles for medical terms is more straightforward than verifying the accuracy of informal-to-specialized term mappings.

5.2 Model Performance Analysis Based on the Annotation

We use the collected annotated data (Sect. 4.2) as correct reference data to analyze the quality and performance of our proposed MCN and the GENRE-based ED models.

MCN Model Accuracy: For each popularized term assessed, we determined the correct specialized term mapping by a majority vote from the experts' annotations. In case of a tie, we treated the popularized term as having no correct specialized term mapping. Based on this the accuracy of the MCN model prediction is calculated:

$$\text{accuracy} = \frac{\sum_{i=1}^{N} \text{correct\_mappings}_i}{N} \times 100$$

where N is total number of popularized phrases evaluated, i is the index for each popularized phrase, ranging from 1 to N, and $correct\_mappings_i$ is 1 if the i-th phrase is correctly mapped ("yes"), 0 if not ("no").

We find that the MCN model accurately classified 89.3% (201 out of 225 terms) of the popularized medical terms into their specialized terms. Upon analysis, the cases that were incorrectly predicted, according to the expert annotators, were difficult to decide on by the experts themselves, too. For example, *"poor bowel movement"* is a very broad term that can mean many different bowel problems. It is not clear in the user post if it means *constipation* or *diarrhea*[6]. The predicted specialized term is *Infectious diarrheal disease (SCUI:19213003)* which has the more specific meaning of diarrheal disease caused by viruses[7]. While it correctly identifies the topic ("bowel problems"), it inaccurately narrows down the broad popularized term to a specific type of diarrheal disease. A clearer failure of the MCN module, the terms *"want to cry all the time"* can be interpreted as a constant feeling of sadness, however the model classified as *Hypersomnia (SCUI:77692006)* (a person feel excessively tired during the day)[8]. Based on these findings, we can argue that the MCN model has a capability to accurately classify popularized terms into counterparts, although it faces challenges with very colloquial expressions.

---

[6] https://medlineplus.gov/bowelmovement.html.
[7] https://www.cdc.gov/disasters/disease/infectevac.html.
[8] https://www.ninds.nih.gov/health-information/disorders/hypersomnia.

ED Model Evaluation. For both popularized and specialized medical terms, the expert evaluators had to review the top five Wikipedia articles, as returned by GENRE on the two types of sentences. They had to select the article that, in their opinion, best explained the corresponding medical concept. At the end of the evaluation, for each popularized medical term we selected one Wikipedia article by majority voting. In cases of a tie, we assign the *0* value, indicating that there is no relevant Wikipedia article for the respective popularized medical term. When an expert chose 'No' to mark the incorrect MCN model prediction (Task 1), we set the Wikipedia selection automatically to *0*. This assumption is based on the idea that, when the specialized term is incorrect, it becomes challenging to find a suitable Wikipedia article that aligns with both the popularized and specialized term, affecting the overall ranking quality of the MAP score. For this expert evaluation we measured the ED-GENRE model performance by computing *P@1* (Precision at 1) and *MAP* (Mean Average Precision). The model, when using specialized medical terms as inputs, achieves a P@1 score of 0.66, which is slightly higher than the score of 0.67 for inputs using popularized medical phrases. However, the model scores higher in Mean Average Precision (MAP) at 0.73 when using informal phrases, compared to 0.70 with specialized terms (see Table 2).

Table 2. ED-GENRE model performance based on the experts evaluation

Query Input	P@1	MAP
specialized medical terms	0.67	0.70
popularized medical terms	0.66	0.73

We are interested in having the relevant candidate from Wikipedia articles at the top of the list. The P@1 shows that 66% of the candidates in the top-ranked. To improve this, we used the information collected from the annotation to train a re-ranking algorithm, referred to as Learning-To-Rank (LTR) from the output of GENRE-ED module which is presented in the following section.

6 Learning-to-Rank for Reranking the ED Module

We formulate the re-ranking problem as the LTR task. To develop a LTR model for re-ranking Wikipedia candidate articles, we follow two key steps: (1) determining the relevance score between the query (the input to the GENRE-ED module, including both popularized and specialized terms, and the user posts), and documents (union set of Wikipedia articles output by the GENRE-ED module), and (2) performing feature engineering. Once these two steps are completed, we apply algorithms from different LTR categories using the RankLib library[9] to train and evaluate the re-ranking model.

---
[9] https://sourceforge.net/p/lemur/wiki/RankLib.

6.1 Relevance Score Between Medical Terms and Wikipedia

To build the LTR model, we filtered the expert-annotated data to only include popularized terms that were correctly mapped to formal medical concepts based on majority votes of the experts in Task 1. We then used these 201 correctly mapped terms to train the LTR model for re-ranking Wikipedia articles retrieved in Task 2 (Sect. 4.2), with a 75/25 split for training and testing data respectively. The relevance of each Wikipedia article was determined by counting the number of times it was selected by the experts for each query (including both popularized and specialized terms). This count served as the relevance score for each article. If a Wikipedia article appeared in both the popularized and specialized term sets, we used the higher count from either set as the final relevance score for that article. The detailed process shown in Fig. 3.

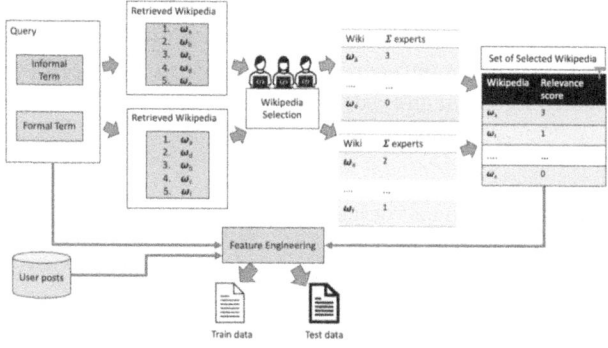

Fig. 3. A process to calculate the relevance score and prepare the dataset for training the Learning-to-Rank (LTR) model

6.2 Feature Engineering

The effectiveness of an LTR model depends heavily on feature engineering. We incorporate query-level and interaction-level features adapted from the Microsoft LTR dataset [8]. Query-level features are derived directly from input queries of the GENRE-ED model, namely popularized terms, specialized terms, and user posts.

1. **Covered Query Term Number**: Count of query terms found in Wikipedia titles.
2. **Covered Query Term Ratio**: Proportion of query terms found in Wikipedia titles, calculated as the number of covered query terms divided by the total number of terms in the query.
3. **Number of Characters in Queries**: Total character count in query terms.
4. **Query IDF**: Importance of a query term relative to its frequency across Wikipedia titles, calculated as the reciprocal of the number of titles containing the term.

5. **Query TF-IDF**: A set of features combining term count and IDF for each query term.

Interaction-level features are derived from the correlation between queries and Wikipedia articles. We use the first paragraph of each article for feature extraction. The features at this level include:

1. **Term Frequency (TF)**: A set of features quantifying query term count in corresponding Wikipedia introduction paragraphs.
2. **BM-25**: Relevance score measuring Wikipedia article suitability for a query [17].
3. **Semantic Similarity**: Measure of how closely related queries and Wikipedia introduction paragraph are in terms of their meaning by calculating their cosine similarity using sentence transformers. We used all-MiniLM-L6-v2 and PubMedBERT [16].

6.3 Results and Discussion

We evaluate various Learning to Rank (LTR) models using MAP and P@1 (see Table 3). The evaluation is based on the test data used to assess the LTR model (Fig. 3), with the baseline derived from the rank of correctly predicted MCN in the test data. The Coordinate Ascent algorithms demonstrate the best overall performance across both metrics. Upon further analysis, it is observed that all listed LTR algorithms have higher MAP and P@1 scores when retrieving Wikipedia articles using both popularized and specialized medical terminology than the original results obtained with ED-GENRE. Specifically, the original Wikipedia retrievals had MAP and P@1 scores of 0.83 and 0.85 for popularized terms, and 0.80 and 0.79 for specialized medical terms, which empirically demonstrates the effectiveness of LTR algorithms for the task at hand.

Table 3. Performance of Learning-To-Rank Algorithms

Algorithm	$P@1$	MAP
MART	0.92	0.91
RankBoost	0.90	0.90
AdaRank	**0.94**	0.92
Coordinate Ascent	**0.94**	**0.93**
LambdaMART	0.88	0.90
RandomForests	0.86	0.89
Baseline (popularized terms)	0.83	0.85
Baseline (specialized medical terms)	0.80	0.79

We investigate the importance of the features in the case of the Coordinate Ascent algorithm using the forward feature selection [5] process to determine which of the contributed features have the highest impact on the prediction result (i.e., MAP and P@1), and the feature with the best metrics is selected and appended to the list.

This iterative process is continued until the coordinate ascent no longer shows any significant improvements. Figure 4 shows the list of features that have improved the performance of the Coordinate Ascent model. The three most important features were based on the semantic similarity features, a statistical feature (i.e. the ratio of covered words to sentences), and lexical features.

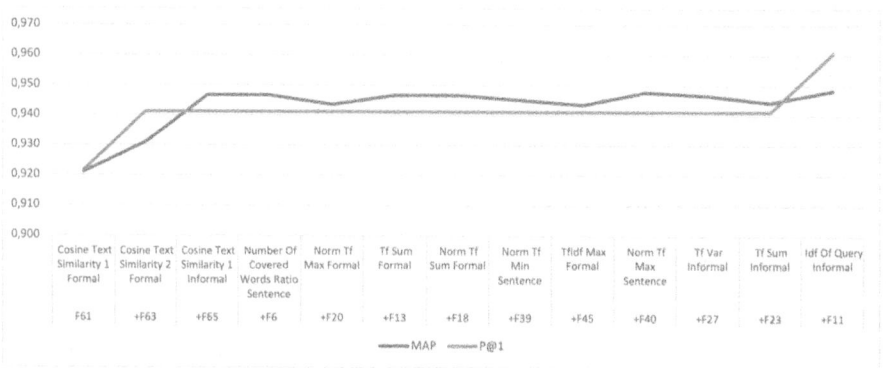

Fig. 4. Forward feature selection to understand the feature importance of Coordinate Ascent algorithm. The x-axis shows the sequence of feature until maximum performance.

7 Conclusion

We introduce an informal medical entity linking model to support laypeople in understanding medical terminology, comprising three phases: 1) NER identifies popularized medical phrases, 2) MCN maps each popularized phrase to a specialized term, and 3) ED selects the most relevant Wikipedia entity for each specialized term. Expert evaluation showed fair IAA for MCN and moderate for ED. The MCN model correctly classified 89% of popularized phrases. In the ED phase, the GENRE model effectively predicted relevant Wikipedia articles, which is further enhanced by the LTR model.

While our expert evaluation showed promising results, we identified limitations and areas for improvement. The MCN model struggled with highly colloquial expressions. Additionally, we aim to include annotator feedback on missing Wikipedia pages and expanding LTR datasets to enhance model generalization across diverse queries. Additionally, we aim to explore neural-reranker models as an alternative to improve article re-ranking compared to traditional methods.

References

1. Alfano, M., Lenzitti, B., Lo Bosco, G., Muriana, C., Piazza, T., Vizzini, G.: Design, development and validation of a system for automatic help to medical text understanding. Int. J. Med. Informatics **138**, 104109 (2020)
2. Basaldella, M., Liu, F., Shareghi, E., Collier, N.: COMETA: a corpus for medical entity linking in the social media. In: Proceedings of the 2020 Conference on EMNLP, pp. 3122–3137. ACL, Online (2020)
3. Chen, J., et al.: A natural language processing system that links medical terms in electronic health record notes to lay definitions: system development using physician reviews (2018)
4. De Cao, N., Izacard, G., Riedel, S., Petroni, F.: Autoregressive entity retrieval. In: 9th International Conference on Learning Representations, ICLR 2021, Virtual Event, Austria, 3–7 May 2021. OpenReview.net (2021)
5. Dhal, P., Azad, C.: A comprehensive survey on feature selection in the various fields of machine learning. Appl. Intell. 1–39 (2022)
6. Fage-Butler, A.M., Jensen, M.N.: The interpersonal dimension of online patient forums: how patients manage informational and relational aspects in response to posted questions. HERMES-J. Lang. Commun. Bus. **51**, 21–38 (2013)
7. Fage-Butler, A.M., Nisbeth Jensen, M.: Medical terminology in online patient-patient communication: evidence of high health literacy? Health Expect. **19**(3), 643–653 (2016)
8. Han, X., Lei, S.: Feature selection and model comparison on microsoft learning-to-rank data sets. CoRR abs/1803.05127 (2018)
9. Harrison, R., MacFarlane, A., Murray, E., Wallace, P.: Patients' perceptions of joint teleconsultations: a qualitative evaluation. Health Expect. **9**(1) (2006)
10. Kandula, S., Curtis, D., Zeng-Treitler, Q.: A semantic and syntactic text simplification tool for health content. In: AMIA Annual Symposium Proceedings, vol. 2010, p. 366. AMIA (2010)
11. Karimi, S., Metke-Jimenez, A., Kemp, M., Wang, C.: Cadec: a corpus of adverse drug event annotations. J. Biomed. Inform. **55**, 73–81 (2015)
12. Miftahutdinov, Z., Tutubalina, E.: Deep neural models for medical concept normalization in user-generated texts. In: Proceedings of the 57th Annual Meeting of the ACL: Student Research Workshop, Florence, Italy, pp. 393–399. ACL (2019)
13. Ningtyas, A.M., El-Ebshihy, A., Herwanto, G.B., Piroi, F., Hanbury, A.: Leveraging wikipedia knowledge for distant supervision in medical concept normalization. In: Barrón-Cedeño, A., et al. (eds.) CLEF 2022. LNCS, vol. 13390, pp. 33–47. Springer, Cham (2022). https://doi.org/10.1007/978-3-031-13643-6_3
14. Ningtyas, A.M., Hanbury, A., Piroi, F., Andersson, L.: Data augmentation for layperson's medical entity linking task. In: Proceedings of the 13th Annual Meeting of the Forum for Information Retrieval Evaluation, pp. 99–106. ACM, New York (2022)
15. Polepalli Ramesh, B., Houston, T., Brandt, C., Fang, H., Yu, H.: Improving patients' electronic health record comprehension with noteaid. In: MEDINFO 2013, pp. 714–718. IOS Press (2013)
16. Reimers, N., Gurevych, I.: Sentence-BERT: sentence embeddings using Siamese BERT-networks. In: Proceedings of the 2019 Conference on EMNLP. ACL (2019)
17. Robertson, S., Zaragoza, H., Taylor, M.: Simple BM25 extension to multiple weighted fields. In: Proceedings of the Thirteenth ACM International CIKM, pp. 42–49. ACM, New York (2004)

18. Scepanovic, S., Martin-Lopez, E., Quercia, D., Baykaner, K.: Extracting medical entities from social media. In: Proceedings of the ACM Conference on Health, Inference, and Learning, CHIL 2020, pp. 170–181. ACM, New York (2020)
19. Tutubalina, E., Miftahutdinov, Z., Nikolenko, S., Malykh, V.: Medical concept normalization in social media posts with recurrent neural networks. J. Biomed. Inform. **84**, 93–102 (2018)
20. Zeng-Treitler, Q., Goryachev, S., Kim, H., Keselman, A., Rosendale, D.: Making texts in electronic health records comprehensible to consumers: a prototype translator. In: AMIA Annual Symposium Proceedings, vol. 2007, p. 846. AMIA (2007)
21. Zolnoori, M., et al.: The psytar dataset: from patients generated narratives to a corpus of adverse drug events and effectiveness of psychiatric medications. Data Brief **24**, 103838 (2019)

Mapping the Media Landscape: Predicting Factual Reporting and Political Bias Through Web Interactions

Dairazalia Sánchez-Cortés[1(✉)], Sergio Burdisso[1], Esaú Villatoro-Tello[1], and Petr Motlicek[1,2]

[1] Idiap Research Institute, Martigny, Switzerland
{dairazalia.sanchez-cortes,sergio.burdisso,esau.villatoro,
petr.motlicek}@idiap.ch
[2] Brno University of Technology, Brno, Czech Republic

Abstract. Bias assessment of news sources is paramount for professionals, organizations, and researchers who rely on truthful evidence for information gathering and reporting. While certain bias indicators are discernible from content analysis, descriptors like political bias and fake news pose greater challenges. In this paper, we propose an extension to a recently presented news media reliability estimation method that focuses on modeling outlets and their longitudinal web interactions. Concretely, we assess the classification performance of four reinforcement learning strategies on a large news media hyperlink graph. Our experiments, targeting two challenging bias descriptors, factual reporting and political bias, showed a significant performance improvement at the source media level. Additionally, we validate our methods on the CLEF 2023 *CheckThat!* Lab challenge, outperforming the reported results in both, F1-score and the official MAE metric. Furthermore, we contribute by releasing the largest annotated dataset of news source media, categorized with factual reporting and political bias labels. Our findings suggest that profiling news media sources based on their hyperlink interactions over time is feasible, offering a bird's-eye view of evolving media landscapes.

Keywords: news media profiling · media bias descriptors · factual reporting · political bias

1 Introduction

Given its open and distributed nature, the World Wide Web (WWW) has become the main information source worldwide, democratizing content creation and making it easy for everybody to share and spread information online. On the bright side, this phenomenon enables a faster dissemination of information compared to what was possible with traditional newspapers, radio, and TV. On the downside, at the moment of removing the "gate-keeper" role from traditional

---

© The Author(s), under exclusive license to Springer Nature Switzerland AG 2024
L. Goeuriot et al. (Eds.): CLEF 2024, LNCS 14958, pp. 127–138, 2024.
https://doi.org/10.1007/978-3-031-71736-9_7

media, it opens the door for additional problems, e.g., the spread of misinformation, at breaking-news speed, that can potentially mislead the users and even impact their behavior [3,25].

Thus, while the goal of this democratic channel is to provide users with the necessary tools to acquire greater knowledge about a topic, the reality is that in the way this knowledge (i.e., news) is presented and reported is not necessarily always impartial [2,18], and there is a growing concern regarding the biases of different media outlets when reporting specific events [14]. For example, in polarizing topics like politics, many of the news can be biased towards one political perspective or the other, i.e., *political bias*, which may influence citizens' voting decisions and preferences of undecided individuals [13].

To mitigate the impact of misinformation and to favor critical assessment for the newsreaders, independent bias assessment services like MBFC[1] and allsides[2] perform information verification. The review process is performed manually by professionals at the event or article level, clearly this is a challenging schema to maintain on the long term given the fast-speed proliferation of both news media websites and news articles. Automation comes handy to perform certain fact-checking tasks, like gathering information (e.g. articles with similar topics, metadata on the media-publisher, etc.); for the more complex parts of the verification analysis, advances in AI continues pushing the boundaries in order to provide valuable tools (for example, search and retrieval, summarization, transformers, LM and LLMs). While the latest LLMs performance on several tasks is remarkable, they are still prone to carry unauthenticated information [17].

While many existing tools are being adopted to support verification tasks at the article level (with and without human supervision), there are very few advances to fully automate news media profiling at the source level (other than popularity). Previous research has shown evidence that some news bias descriptors can be inferred by just inspecting the outlet website metadata [10,16]. Other approaches have addressed source reliability, factuality of reporting or political bias, by assembling information from multiple external and social media sources, metadata and/or content-based features [3–5,7,9,21,22]. Unfortunately, methodologies relying on social media metadata can not longer be reproduced at scale given the current access restrictions.

A recent research shifting from the social media and text-based approach, is presented in [8]. Burdisso *et al.*, proposed a highly performing and robust graph-based methodology to score news media reliability. Their method considers the longitudinal interactions on the web to learn a reliability value from their source neighbors. Based on the research evidence that neighboring properties can be spread among news media outlets, we extend their work and we propose to address the following research question: to what extent it is possible to profile news media outlets (i.e., different properties) based solely on their interaction with other media sources? To address this question, in this paper we focus on two challenging media bias descriptors: factuality of reporting and political bias.

---

[1] https://mediabiasfactcheck.com.
[2] www.allsides.com.

We choose to extend Burdisso et al. methodology given that it is both language and content independent (political, religious, racial, etc.), it can be applied at a larger scale and their 17k English news outlets dataset is publicly available.

Our main contributions are as follows: (a) we show that it is possible to predict/estimate bias descriptors, i.e., political bias and factual reporting; of the source media based on their interactions with other sources (outperforming the baseline); (b) we validate the robustness of our approach on the publicly available dataset from the CLEF CheckThat! challenge, specifically collected to classify political bias with currently active news outlets, and we established a new SOTA result; (c) we release the biggest dataset at the source media level with standard political bias and factual reporting labels[3].

2 Related Work

The bias in news media is a pervasive and ubiquitous problem [14,15,25]. The need for applied research on news media descriptors has increased since 2000 due to the generalized adoption of social media platforms, and the proliferation of tools that facilitate both websites and news-content creation [6,12]. Bias in news media has a wide descriptors spectrum [14,15,27], for example Racial Bias refers to preferences of coverage or not of events related to minorities or group of individuals [24]. Gender Bias refers to the inclination towards one gender over another, resulting in unequal treatment, coverage and perception [1,23]. Political Bias, refers to partial representation of political issues or tendency to favor a particular political ideology.

A significantly large NLP community has reported advances on news media bias at the article level (i.e., based on content), also referred as bias at the event-label or a short-term bias on a selected event [14]. However, in this paper, we contribute towards the news media source profiling (i.e., at the source level). We focus our research work on the following two long-term bias descriptors:

Factual Reporting. Recent task challenges, particularly the *CheckThat!* Lab challenge at CLEF 2023, have addressed the factuality of reporting based on three classes (*High*, *Mixed* and *Low*) at the article level [21]. Submitted models range from traditional supervised models (such as SVMs, Random Forest, gradient Boost) to Deep Learning-based ones [19,21]. Due to the challenging nature to perform factuality assessment, graph-based models emerged to address the problem disclosing better performance when combined with text-based approaches [3,4,11,22]. Fairbanks et al., [11] proposed a structural model based on the metadata from the article's news web links. Their findings revealed that credibility, a descriptor in close relation with factual reporting, is harder to determine from merely the content. Baly et al. [3] analyzed the factual reporting focusing on the source media. Their approach used text-based features from articles content and metadata including Wikipedia pages, Twitter, URL-related features (domain, orthography, char n-grams), and Web traffic (Alexa service). Also targeting the factuality at the media level, Panayotov et al., [22] proposed

---

[3] https://github.com/idiap/Factual-Reporting-and-Political-Bias-Web-Interactions.

to model the factuality of reporting using graph neural network and similarity between news media based on their audience overlap. Although the latest models revealed significant improvements at the media level, the methods in [3,22] rely on the Alexa website ranking and web traffic information, which is now discontinued.

More recent approaches are focusing on state-of-the-art LMs and LLMs, from adversarial training, ensemble of models based on RoBERTa or GPTs [20,26]. Li *et al.*, heuristics on adversarial training revealed the importance of semantics in the title and the summary of the news captured at the beginning and end of the article. Their best performing political inference results from a majority voting from four implemented models from which, two are RoBERTa-based. Tran *et al.*, examined the impact of imbalanced training data between *High*, *Mixed* and *Low* factual reporting. The authors introduced a RoBERTa-based back-translation framework that significantly surpassed the baseline performance. Their approach ranked among the top three performers at the *CheckThat!* Lab challenge in 2023. To the best of our knowledge, the state-of-the-art methodology in media profiling, outperforming ensembles of content-based and external data was recently introduced in [8]. Burdisso *et al.*, propose an hyperlink-interactions graph to infer News source reliability degree (a continuous value) based on reinforcement learning techniques. In addition to the standing performance, authors contribute with the largest reported dataset in source media profiling with 17k English-speaking news outlets.

Political Bias. In the recent years, the inference of political bias at the outlet level has been approached by applying SVMs, CatBoost and applied oversampling techniques, mostly enhancing content-features from articles [2,4,9]. Baly *et al.*, [3,4] proposed a framework based on SVMs reporting significant results when complementing content-based data with Wikipedia and social media metadata. Recently, Azizov *et al.*, [2] proposed a majority voting ensemble of CatBoost models and TF-IDF, showing better performance than LM-frameworks at the *CheckThat!* lab challenge at CLEF 2023 [9] given a benchmark dataset with three political classes (*Left*, *Center*, *Right*). In Panayotov *et al.* [22], the political bias was modeled using a graph neural network augmented with audience/social media data. Graph-based approaches showed evidence that metadata capturing information other than the article content improved classification of political stance. Given the still open challenge to accurately infer political bias at the news source level, more recent approaches are exploring the pertinence of using LMs [26,27]. Tran *et al.* [26], analyzed and addressed the three-class (*Left*, *Center*, *Right*) imbalance by translating to Spanish and back to English the classes with less articles. Then, they fine-tuned RoBERTa English-large, and performed a majority voting at the article-level to infer the news source political leaning, showing a significant performance above the baseline. Wessel *et al.*, [27] proposed a framework using transformers to infer 9 bias descriptors. For the case of political bias, the original bias annotation provided at the outlet level is transformed into two classes bias and not-bias. Despite the 2 million political news articles used in this work, they were exclusively gathered from the top 11 most popular US media outlets. Authors concluded that cognitive and political bias at the content-level are the most challenging bias descriptors to detect, in contrast with for example gender or racial bias.

Although some approaches show significant improvement over majority baselines, the robustness and scalability of the models is not sufficient to consider the factual reporting and political bias problem solved. Contrary to previous research that depends on content, audience feedback, and/or metadata, in this paper we extend a very recent work that models the problem in a scalable fashion relaying on network interactions among the news sources [8]. Following sections describe the proposed methodology and obtained results.

3 Methodology and Strategies

In order to validate our research question and based on the evidence presented by Burdisso et al. that longitudinal interactions can spread the news media reliability degree among their neighbors [8], we extend their work to address factual reporting and political bias.

The introduced approach consists of first building a news media graph from the WWW and then applying different reinforcement learning strategies to infer the reliability values. More precisely, constructing a weighted directed graph $G = \langle S, E, w \rangle$ where there is an edge $(s, s') \in E$ if source s contains articles (hyper) linked to s' and where the weight $w(s, s') \in [0, 1]$ is the proportion of total hyperlinks in s linked to s'.[4] In this work, we hypothesize that the political bias and factual reporting of sources s can be estimated from the sources it interacts with, by inheriting their properties.

Following the original work in [8], we model the estimation as a Markov Decision Process (MDP) $\langle S, A, P, r \rangle$ such that: (1) The set of states S are all the news outlets websites—i.e. $S = S$; (2) The set of actions A contains only one element, the *"move to a different news media website"* action; (3) The probability P of moving from the origin s to s' will be given by the proportion of hyperlinks in s connecting to s'—i.e. we have $P(s, s') = w(s, s')$; and (4) The reward r of moving to another news source (s') is determined only by the origin source(s), and it will be positive or negative depending on the known property—e.g. $r(s) = 1$ if we know for this s we have *Right* or *High*, for political bias or factual reporting, respectively; $r(s) = -1$ if s is *Left* or *Low*, for political bias or factual reporting, respectively; $r(s) = 0$ otherwise. Finally, the property (political bias or factual reporting level) value for all news sources s in the graph will be estimated by a function $\rho(s)$ following 4 different strategies:

- **F**-*property*: The property value is proportional to the *expected* perceived reward given by the following Bellman equation where π is the unique policy

---

[4] Note that this simple hyperlink-based representation is also implicitly capturing content-based references to and from other sources.

(i.e. the probability of taking action $a \in A$ in state s) and $\gamma \in [0, 1)$ the discount factor:[5]

$$\rho(s) = \sum_{s' \in \mathbb{S}} P^{\pi}(s, s')[r(s') + \gamma \rho(s')] \qquad (1)$$

That is, under this strategy, the value of source s will be inherent from the sources it connects in the **Future**.

- **P**-*property*: The property value is interpreted a proportion of the accumulated perceived reward, i.e., the value is inherited by the sources that lead to it in the **Past**. The value is thus, giving by the following the reverse Bellman equation:

$$\rho(s) = r(s) + \gamma \sum_{s' \in \mathbb{S}} P^{\pi}(s', s) \rho(s') \qquad (2)$$

- **FP**-*property*: This strategy combines the previous two strategies by considering **Future** and **Past** information. A source s increases its positive value $\rho(s)$ as more positive sources link to it ($\rho_{\mathbf{P}}^+(s)$), while losing value as it links to more negative sources ($\rho_{\mathbf{F}}^-(s)$).[6] Thus, $\rho(s)$ is simply defined as:

$$\rho(s) = \rho_{\mathbf{F}}^-(s) + \rho_{\mathbf{P}}^+(s) \qquad (3)$$

- **I**-*property*: Investment Strategy (invest and collect credits) consisting of two iterative steps, repeated n times: (1) all sources invest their property value to the neighboring sources proportionally to the strength of their links ($w(s, s')$) following Eq. 4, (2) sources collect the credits back proportionally to the investment and update its own property value following Eq. 5.

$$totalcredits(s) = \sum_{s' \in S} w(s', s) \cdot \rho(s') \qquad (4)$$

$$\rho(s) = \rho(s) + \sum_{s' \in S} w(s, s') \cdot credits_s(s') \qquad (5)$$

where credits are distributed among investors s', in proportion to their contribution to s, i.e., $credits_{s'}(s) = w_{s'}(s) \cdot totalcredits(s)$.

3.1 Datasets

There are several attempts to unify existing datasets to assess Bias in news media. Recently, a unified bias dataset was presented including several Bias descriptors [27], nevertheless, the collection of articles, sentences, comments, etc., are on one hand targeting rather short-term bias (text-based), and on the other hand large part of the data do not have URLs to existing news media

---

[5] The discount factor controls the distance of looking back/forward; $\gamma \approx 0$ focuses mostly on present reward $r(s)$, while $\gamma \approx 1$ considers all history/future to compute $\rho(s)$.
[6] Algorithm 1 and Algorithm 2 in [8] detail how these updates are applied.

Table 1. Label distribution on both datasets.

Dataset	Political Bias			Factual Rep.			Total
	Left	Center	Right	Low	Mixed	High	
MBFC (ours)	2078	763	1079	408	1391	2121	3920
CLEF *CheckThat!*	272	359	392	-	-	-	1023

sources. Recently, [8] released the largest dataset with URLs annotated with reliability labels constructed by collecting and consolidating annotations from different sources. In this work, we follow a similar process as described by the authors in [8] to build our own dataset with political bias and factual reporting annotation, which we refer to as "MBFC".

MBFC. Following the methodology described in [8], we crawled 3920 news media URL domains from the *Media Bias/Fact Check* (MBFC)[4] service including annotated bias descriptors that are further transformed and normalized into political and factual reporting labels as follows: for Political bias, the final normalized categories are *Left, Center, Right*; for the case of Factual Reporting, labels include *High* (which aggregated high and very high), *Mixed* and *Low* (which aggregates low and very low).

CLEF *CheckThat!* Additionally, in order to compare results with previously published approaches, we use the dataset released for the CLEF 2023 *CheckThat!* lab which focused on political bias identification. This dataset contains a total of 1023 news media URL domains with political bias labels crawled from *allsides*[5], a website that gathers news articles with balanced representation of the different political perspectives. The data is officially divided into fixed train, dev, and test set splits containing 817 *(Left-216, Center-296 and Right-305)*, 104 *(Left-31, Center-34 and Right-39)*, and 102 *(Left-25, Center-29 and Right-48)* news sources, respectively. More details about the data and the labeling process can be found in [9]. Table 1 summarizes the label distribution and size of both introduced datasets.

4 Experiments and Results

In this work we used the graph G built in [8] consisting of 17K news sources obtained after processing 100M news articles from Common Crawl News. Following [3,8] we report 5-fold cross-validation evaluation results on our MBFC datasets, whereas for CLEF's *CheckThat!* we report results on the official test set. In order to estimate the factual score of reporting from the graph, we first convert the factuality/bias ground truth labels from the training set into rewards as follows: $r(s) = 1$ if the media label is *High/Right*, $r(s) = -1$ if *Low/Left*, and $r(s) = 0$ otherwise. Then, at inference time, sources s are classified with the label *Right/High* if $\rho(s) > 0$ and *Left/Low* otherwise. Even though one limitation of the proposed strategies is that they are essentially binaries, in order to

Table 2. 5-fold cross-validation average results for Political Bias and Factual Reporting classification. The best-performing values are <u>underlined</u>, while the 2nd-best results appear in **bold** font.

Task	Strategy	F_1 score			Accuracy
		Macro avg.	High/Righ	Low/Left	
Factual Rep.	Majority	38.94 ± 0.04	87.88 ± 0.09	0.00 ± 0.00	83.84 ± 0.16
	Random	36.44 ± 0.88	65.69 ± 1.64	7.18 ± 1.45	49.93 ± 1.69
	F-Factuality	57.60 ± 4.38	95.00 ± 0.97	20.19 ± 7.86	90.60 ± 1.76
	P-Factuality	**85.13** ± 2.73	**98.70** ± 0.35	**71.55** ± 5.15	**97.52** ± 0.66
	FP-Factuality	71.35 ± 2.33	96.76 ± 0.65	45.93 ± 4.09	93.89 ± 1.19
	I-Factuality	<u>**87.99**</u> ± 4.60	<u>**99.02**</u> ± 0.43	<u>**76.96**</u> ± 8.79	<u>**98.12**</u> ± 0.81
Political Bias	Majority	38.04 ± 0.07	0.00 ± 0.00	76.08 ± 0.14	65.40 ± 0.18
	Random	45.42 ± 1.84	30.29 ± 2.86	60.55 ± 1.75	49.65 ± 1.73
	F-Political	60.42 ± 3.74	41.56 ± 6.27	79.29 ± 1.42	69.44 ± 2.30
	P-Political	**74.08** ± 2.31	**65.80** ± 3.23	**82.36** ± 1.39	**76.73** ± 1.95
	FP-Political	64.90 ± 3.15	52.47 ± 4.82	77.33 ± 1.94	69.34 ± 2.55
	I-Political	<u>**77.77**</u> ± 2.45	<u>**70.97**</u> ± 3.39	<u>**84.56**</u> ± 1.54	<u>**79.85**</u> ± 2.12

compare results in *CheckThat!* three-label classification task, we use the official dev set to find an ϵ value to classify sources s as follow: *Left/Low* if $\rho(s) < -\epsilon$; *Right/High* if $\rho(s) > \epsilon$; *Center/Mixed* otherwise. More precisely, we selected the hyper-parameters $\epsilon = 3e-3$, $\gamma = 0.15$ (Eq. 1, 2, 3), and $n = 2$ (Eq. 5) after performing a grid search maximizing the Macro avg. F1 score with $\epsilon \in [1e-3, 1e-1]$ (1e−3 increments), $\gamma \in [0.05, 0.95]$ (0.05 increments), $n \in [1, 10]$, respectively.

4.1 Factuality of Reporting

Table 2 shows the results from the 5-fold cross-validation for Factual Reporting. The baseline for comparison includes Random and Majority class classification. The **F**-Factuality strategy performed at 57.60 F1-score overall, for the individual classes *High* Factual reporting performance is 95.00 F1-score and 20.19 for *Low* Factual Reporting. For all cases there is significant improvement with respect to the baselines. For **P**-Factuality F1-score performance is 85.13, and 98.7 and 71.55 for the *High* and *Low* classes. The significantly high performance reveals that indeed the graph with past reward strategy captures close interacting networks on both sides, *High* score and *Low* score of factual reporting. The strategy **FP**-Factuality performs at 71.35 F1-score, although it outperforms **F**-Factuality and the baselines, it remains behind **P**-Factuality. Finally, the **I**-Factuality strategy outperforms all the other strategies up to 87.99 F1-score, 76.96 for class Low and 99.02 for the class High. The results show that for the case of **I**-Factuality (the invest and collect strategy), the gathered information from the hyperlinks and its neighbors can accurately capture the level of factuality, significantly better for the class Low.

Table 3. Results on CLEF's *CheckThat!* dataset on Political Bias of news media. MAE: Mean Absolute Error. The smaller MAE value translates into better predictions.

Team	MAE(↓)	F$_1$ score(↑)	Accuracy(↑)
Baseline [9]	0.902	-	-
Awakened	0.765	-	-
Accenture [26]	0.549	0.625	0.627
Frank [2]	0.320	0.727	0.725
F-Political	0.333	0.632	0.667
P-Political	**0.238**	**0.760**	**0.762**
FP-Political	0.309	0.670	0.690
I-Political	**0.214**	**0.784**	**0.786**

I-Factuality accurately identifies almost all sources with *Low* Factual Reporting, which is indeed a key contribution of this paper. We assume that high performance of the reward value might be due to capturing unintentionally the lifespan of a news media domain, which has been reported as a high contributor in the identification of disinformative websites [16]. Both strategies **P**-Factuality and **I**-factuality are highly performing on F1-score and Accuracy, similarly to findings on Reliabilty of news media in [8], disclosing an accurate profiling of Bias given only their network interactions overtime.

4.2 Political Bias

Results on MBFC. Table 2 shows the 5-fold cross-validation results for F1-score and Accuracy, we included two baselines Random, and Majority class for comparison. For the political leaning the **F**-Political performs at 60.42 F1-score, and 79.29 F1-score for Right at the class level, showing a modest improvement over the baseline (76.08). For **P**-Political the overall F1-score performance is 74.08, with 65.8 for the class *Left* and 82.36 for the class *Right*. For the combined **FP**-Political the F1-score of 64.90 outperforms the **F**-Political but does not improve the **P**-Political performance, for both the overall and the class level, which indicates that past information contributes more to the predictions. The best performing strategy is **I**-Political performing at 77.77 F1-score and, 70.97 and 84.56 for the classes *Left* and *Right* respectively. At the class level, our results on political bias show significantly better performance on the class *Right*. Figure 1 shows part of the graph for the news media source www.newrepublic.com, where the values are estimated with **I**-political. The size of the node is proportional to their political bias, as newrepublic predominantly engages with *Left*-wing sources, its final value leaned significantly towards the *Left* (red).

Results on the CLEF *CheckThat!* Table 3 shows the F1-score performance and the official scoring metric MAE (Mean Absolute Error) for the Labs at CLEF 2023. The political labels were coded as ordinal values (*Left*-0, *Center*-1, *Right*-2), a smaller MAE value translates into better predictions from the

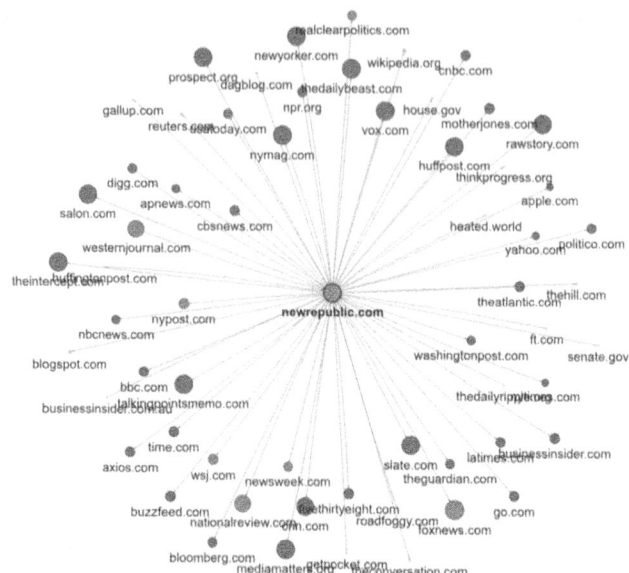

Fig. 1. Example showing how `newrepublic.com` relates with neighboring news sources. *Left* and *Right* wing sources are colored red and blue respectively, in addition, size of the node reflects the degree of the bias (learned by our **I**-political strategy). We can see that since `newrepublic.com` interacts mostly with *Left*-wing sources, its final bias degree ended up being considerable *Left*-wing. (Color figure online)

proposed models. The baseline with MAE of 0.902 uses an SVM classification model based on N-Grams. The top performed participating model [2] achieved a MAE of 0.320, outperforming the baseline and the other participating models. However, our proposed strategies (**P**-Political and **I**-Political) outperform the best-performing participating model in all reported metrics the top (MAE, F1-score and Accuracy). The MAE top performance (smaller MAE) indicates that the miss-predictions are less severe (from *Center* to the extremes or vice-versa), otherwise inferences will result on a higher penalization if predicting completely opposite extremes *Left* ↔ *Right*.

5 Conclusions

This research extends the methodology proposed in [8] by addressing long-term news media profiling, contrasting with approaches focused solely on short-term bias. Our experiments on two challenging bias descriptors-factual reporting and political bias-utilize four reinforcement learning strategies for classification performance evaluation. We provide compelling evidence supporting the longitudinal view of news media and their web interactions as a robust and scalable proxy for profiling, particularly regarding political bias and factual reporting. Concretely, performed experiments show that the proposed approach allows superior performance in estimating outlet media bias descriptors compared to baseline methods. Furthermore, we present promising results from comparisons with other participating models submitted to the CLEF 2023 *CheckThat!* lab, designed for inferring political bias in currently active news outlets. Our approach surpasses top results in both F1-score and the official MAE performance measure, establishing a new SOTA result for this particular task. Finally, as an additional contribution, we release the largest dataset at the source media level, annotated with standard political bias and factual reporting labels.

As part of future efforts, we aim to investigate the dynamics of political bias changes over time within news media, such as shifts from center to extreme positions. Additionally, we plan to explore the integration of other bias descriptors, such as press freedom, in multi-task bias identification.

Acknowledgments. This work was supported by CRiTERIA, EU project funded under the Horizon 2020 program, grant agreement number 101021866.

Disclosure of Interests. The authors have no competing interests.

References

1. Asr, F.T., et al.: The gender gap tracker: using natural language processing to measure gender bias in media. PLoS ONE **16**(1), e0245533 (2021)
2. Azizov, D., Liang, S., Nakov, P.: Frank at checkthat! 2023: detecting the political bias of news articles and news media. Working Notes of CLEF (2023)
3. Baly, R., Karadzhov, G., Alexandrov, D., Glass, J., Nakov, P.: Predicting factuality of reporting and bias of news media sources. In: Proceedings of the Conference on Empirical Methods in Natural Language Processing, pp. 3528–3539 (2018)
4. Baly, R., et al.: What was written vs. who read it: news media profiling using text analysis and social media context. In: Proceedings of the Association for Computational Linguistics (2020). https://doi.org/10.18653/v1/2020.acl-main.308
5. Baly, R., Karadzhov, G., Saleh, A., Glass, J., Nakov, P.: Multi-task ordinal regression for jointly predicting the trustworthiness and the leading political ideology of news media. In: Proceedings of the North American Chapter of the Association for Computational Linguistics (2019). https://doi.org/10.18653/v1/N19-1216
6. Billard, T.J., Moran, R.E.: Designing trust: design style, political ideology, and trust in "fake" news websites. Digit. Journal. **11**(3), 519–546 (2023)
7. Bozhanova, K., Dinkov, Y., Koychev, I., Castaldo, M., Venturini, T., Nakov, P.: Predicting the factuality of reporting of news media using observations about user attention in their youtube channels. In: Proceedings of the International Conference on Recent Advances in Natural Language Processing, pp. 182–189 (2021)
8. Burdisso, S., Sanchez-Cortes, D., Villatoro-Tello, E., Motlicek, P.: Reliability estimation of news media sources: birds of a feather flock together. In: Proceedings of the North American Chapter of the Association for Computational Linguistics (2024). https://aclanthology.org/2024.naacl-long.383
9. Da San Martino, G., Alam, F., Hasanain, M., Nandi, R.N., Azizov, D., Nakov, P.: Overview of the clef-2023 checkthat! lab task 3 on political bias of news articles and news media. Working Notes of CLEF (2023)

10. Esteves, D., Reddy, A.J., Chawla, P., Lehmann, J.: Belittling the source: trustworthiness indicators to obfuscate fake news on the web. In: Proceedings of the First Workshop on Fact Extraction and VERification (FEVER), pp. 50–59 (2018)
11. Fairbanks, J., Fitch, N., Knauf, N., Briscoe, E.: Credibility assessment in the news: do we need to read. In: Proceedings of the MIS2 Workshop Held in Conjunction with 11th International Conference on Web Search and Data Mining, pp. 799–800. ACM (2018)
12. Fang, X., Che, S., Mao, M., Zhang, H., Zhao, M., Zhao, X.: Bias of AI-generated content: an examination of news produced by large language models. Sci. Rep. **14**(1), 1–20 (2024)
13. Gezici, G.: Quantifying political bias in news articles (2022)
14. Hamborg, F.: Media bias analysis. In: Hamborg, F. (ed.) Revealing Media Bias in News Articles, pp. 11–53. Springer, Cham (2023). https://doi.org/10.1007/978-3-031-17693-7_2
15. Hanimann, A., Heimann, A., Hellmueller, L., Trilling, D.: Believing in credibility measures: reviewing credibility measures in media research from 1951 to 2018. Int. J. Commun. **17**, 214–235 (2023)
16. Hounsel, A., Holland, J., Kaiser, B., Borgolte, K., Feamster, N., Mayer, J.: Identifying disinformation websites using infrastructure features. In: USENIX Workshop on Free and Open Communications on the Internet (FOCI) (2020)
17. Kamalloo, E., Dziri, N., Clarke, C.L., Rafiei, D.: Evaluating open-domain question answering in the era of large language models. In: Proceedings of the North American Chapter of the Association for Computational Linguistics (2023). https://doi.org/10.18653/v1/2023.acl-long.307
18. Kulshrestha, J., et al.: Search bias quantification: investigating political bias in social media and web search. Inf. Retrieval J. **22**, 188–227 (2019)
19. Leburu-Dingalo, T., Thuma, E., Motlogelwa, N., Mudongo, M., Mosweunyane, G.: UBCS at checkthat! 2023: stylometric features in detecting factuality of reporting of news media. Working Notes of CLEF (2023)
20. Li, C., Xue, R., Lin, C., Fan, W., Han, X.: Cucplus at checkthat! 2023: text combination and regularized adversarial training for news media factuality evaluation. Working Notes of CLEF (2023)
21. Nakov, P., et al.: Overview of the clef-2023 checkthat! lab task 4 on factuality of reporting of news media. Working Notes of CLEF (2023)
22. Panayotov, P., Shukla, U., Sencar, H.T., Nabeel, M., Nakov, P.: Greener: graph neural networks for news media profiling. In: Proceedings of the 2022 Conference on Empirical Methods in Natural Language Processing, pp. 7470–7480 (2022)
23. Van der Pas, D.J., Aaldering, L.: Gender differences in political media coverage: a meta-analysis. J. Commun. **70**(1), 114–143 (2020)
24. Pope, D.G., Price, J., Wolfers, J.: Awareness reduces racial bias. Manage. Sci. **64**(11), 4988–4995 (2018)
25. Strömbäck, J., et al.: News media trust and its impact on media use: toward a framework for future research. Ann. Int. Commun. Assoc. **44**(2), 139–156 (2020). https://doi.org/10.1080/23808985.2020.1755338
26. Tran, S., Rodrigues, P., Strauss, B., Williams, E.: Accenture at checkthat! 2023: learning to detect factuality levels of news sources. Working Notes of CLEF (2023)
27. Wessel, M., Horych, T., Ruas, T., Aizawa, A., Gipp, B., Spinde, T.: Introducing mbib-the first media bias identification benchmark task and dataset collection. In: Proceedings of the International ACM SIGIR Conference on Research and Development in Information Retrieval, pp. 2765–2774 (2023)

Under-Sampling Strategies for Better Transformer-Based Classifications Models

Marcin Sawiński[✉][iD], Krzysztof Węcel[iD], and Ewelina Księżniak[iD]

Poznań University of Economics and Business, 61-875 Poznań, Poland
marcin.sawinski@ue.poznan.pl

Abstract. This paper presents findings from the Check-That! Lab Task 1B-English submission at CLEF 2023. The research developed a method for evaluating the check-worthiness of short English texts. The first iteration focused on identifying optimal model architectures and adaptation techniques, while the second iteration involved curating the dataset for improved results. The study included fine-tuning several GPT and BERT models, applying zero-shot, few-shot, and Chain-of-Thought prompting strategies, and utilizing dataset sampling techniques informed by quality and training dynamics metrics.

Team achieved first place in the competition by fine-tuning the OpenAI GPT-3 curie model. Findings suggest that fine-tuned BERT models can perform comparably to GPT models, but dataset curation was pivotal in obtaining superior results across various model architectures.

Keywords: check-worthiness · GPT · BERT · DeBERTa · dataset curation · class imbalance · under-sampling · model training

1 Introduction

Fact-checking is the rigorous process of verifying information to ensure its accuracy and authenticity. It involves critically examining claims, statements, or data presented in various forms of media. Check-worthiness estimation is the process of identifying which statements should be fact-checked based on their potential impact and likelihood of being false. This study investigates the potential for automatically evaluating the check-worthiness of claims in unimodal (text-only) English content, based on experiments conducted within the CheckThat! Lab, Task 1B-English at CLEF 2023 [16]. The task is defined as a binary classification problem with two labels: check-worthy or not.

The original study [16], published as part of the CheckThat! Lab at CLEF 2023, aimed to compare the performance of twelve GPT and BERT models and corresponding model adaptation techniques trained on two versions of the dataset. This paper extends the research with experiments and analysis of fifteen dataset variants created using four under-sampling techniques, tested by training two DeBERTa V3 models. Additionally, two novel under-sampling techniques are introduced in this paper.

The original version of the chapter has been revised. The author names are now corrected. A correction to this chapter can be found at
https://doi.org/10.1007/978-3-031-71736-9_19

2 Related Work

The research examines the problem of optimizing training data selection. Due to annotation ambiguity, training datasets can contain biased or wrongly labeled examples [2].

We drew inspiration from the Data Maps introduced by Swayamdipta et al. [19], which partition dataset examples into three regions: easy-to-learn, ambiguous, and hard-to-learn. These regions are defined based on two training dynamics measures: confidence and variability, calculated from the sequence of logits produced for each example during evaluation steps throughout model training. Easy-to-learn examples exhibit low variability and high model confidence, whereas hard-to-learn examples are characterized by high variability and low confidence. Ambiguous examples have high variability and medium confidence. The hypothesis is that examples from each region play different roles in learning and generalization. Therefore, selecting data based on these regions could improve model predictions. The paper presents experiments on the WinoGrande, SNLI, MultiNLI, and QNLI datasets. Two baselines were established: one by training the model on the entire dataset and the other by training on a randomly sampled one-third of the examples. The study presents the results of models trained on ambiguous and hard-to-learn examples, as well as those trained on samples selected using other metrics such as correctness, forgetting, AL-uncertainty, AL-greedyK, and AFLite. The test sets include both in-distribution and out-of-distribution data. The findings indicate that using ambiguous examples for training improves accuracy in comparison to the baseline for out-of-distribution samples, with no significant change observed for in-distribution test sets. Training on hard-to-learn examples also enhanced classification results, although the improvement was less pronounced compared to training on easy-to-learn examples. Additionally, the study noted that hard-to-learn examples might indicate annotation errors.

Sar-Shalom et al. [15] proposed a method for computing training dynamics measures based on separate training processes rather than between epochs, achieving the best results by training on ambiguous examples. Ince et al. [9] focused on compositional generalization tasks, using training dynamics to select specific subsets of hard-to-learn samples, which consistently yielded superior generalization performance. Shi et al. [17] addressed commonsense question-answering, finding that training models with ambiguous and hard-to-learn data led to the most significant improvements in baseline performance. However, this trend is not always consistent. Snijders et al. [18] emphasized the importance of not only the training dataset's selection but also the test set's difficulty level. They tested training on various combinations of easy (E), medium (M), hard (H), and impossible (I) samples. The results revealed that incorporating more difficult examples in the training set reduced performance on easier test sets but enhanced performance on more challenging ones.

3 Methodology

The study was conducted iteratively, exploring two areas for possible improvement of prediction results: i) selection and fine-tuning of prediction models, ii) preparation of the training dataset. The initial motivation was to train many variants of models and choose the best one, but it became apparent that the preparation of the training dataset was also important. The study involved executing multiple training and evaluation runs using various combinations of models and dataset preparation methods. Each run used disjoint datasets to perform: i) training, ii) evaluation with loss or F1 score metric over the positive class, iii) testing using the F1 score metric over the positive class.

In the first iteration, described in a previous paper [16], a selection of models and model adaptation techniques were tested against an unmodified dataset provided by the lab organizers. The dataset was provided with three splits: *train*, *dev*, and *dev_test*, which were used for training, evaluation, and testing, respectively. Following best practices, we performed fine-tuning of the GPT-3 and BERT family of models, which included a search of the hyper-parameter space to find the combination of parameters best suited for the dataset. Additionally, prompting strategies—zero-shot learning and few-shot learning—were tested with GPT-4 models. The result of the first iteration was an internal ranking of models with their corresponding best configurations (hyper-parameters, prompts, and other external components setup).

The second iteration extended our previous research by evaluating the performance of new variants of the training dataset. The best models from the first iteration were retrained and re-evaluated to test the impact of dataset changes on the final score. The best-performing combination of the training dataset and model was then selected to make predictions on the final test dataset, which was submitted for the Check-That! Lab Task 1B-English competition at CLEF 2023. Due to compute resource constraints, we were not able to execute and evaluate all training runs before the Check-That! Lab submission deadline. Nevertheless, we observed an interesting phenomenon: in many cases, reducing the size of the training dataset resulted in better models.

This paper is a follow-up to the previous study [16], wherein we analyze the impact of various under-sampling methods on model training results measured by the F1 score for the positive class. We specifically focus on DeBERTa v3 models. The following research questions were formulated:

- **RQ1.** Does a DeBERTa v3 large model outperform the base model?
- **RQ2.** Does changing the class ratio in the training dataset improve model training?
- **RQ3.** Are any specific balancing methods better than the others?

Random initialization of neural network parameters causes the model to converge to different states. Therefore, we executed training runs multiple times with different random seeds. For the second iteration, we planned the execution of 120 training runs: 60 with each model variant (DeBERTa v3 large and base) on 15 variants of datasets, with each run repeated 4 times with different random seed initializations for the models. We used ANOVA (Analysis of Variance) to answer our research questions, i.e., to determine whether the various under-sampling methods considering different factors have a significant impact on model training quality.

4 Models

In 2022, the majority of teams used BERT models [12], and only one team [1] used GPT models. Their solution won first place in the subtask 1B - 'Verifiable Factual Claims Detection' and third place in the subtask 1A - 'Check-Worthiness Estimation' in English. Given the development of GPT models, we decided to compare the potential of large GPT models against BERT models in detecting check-worthy claims.

In the first iteration, we selected eleven BERT models and four GPT models to start our experiments. Meaningful results were reported for nine models: ○ DistilBERT base uncased [14] ○ DeBERTa v3 base [8] ○ RoBERTa base [11] ○ ALBERT large v2 [10] ○ ELECTRA base [6] ○ GPT-neo 125M [4] ○ GPT-3 curie[1] ○ GPT-3 davinci ○ GPT-4.

OpenAI models were chosen for experiments for two reasons: they showed an advantage in multiple performance tests and they were cost-effective both in terms of fine-tuning and inference. At the time of conducting the experiments, the most advanced language model available for fine-tuning was GPT-3, and we used the *curie* and *davinci* variants for fine-tuning. The GPT-4 model was used for zero-shot and few-shot learning.

In the second iteration, we selected only two models for experiments: DeBERTa v3 base and DeBERTa v3 large [7].

4.1 Model Adaptation Techniques

Fine-Tuning with Hyper-Parameter Search. We fine-tuned all BERT models for a binary classification task with the F1 score over the positive class as the optimization goal. The search of the hyper-parameter space included testing the AdamW and AdaFactor optimizers, batch sizes of 8 and 16, and two floating-point precision options: FP16 and FP32, along with several learning rates. We also tested layer-wise learning rate decay since lower layers primarily process general language data, whereas upper layers are more involved in task-specific functions like classification.

Fine-Tuning GPT-3. Fine-tuning of GPT-3 models was performed using two OpenAI GPT-3 variants: *davinci* and *curie*. The same hyper-parameter values were used for all fine-tuning experiments: batch size of 8, a learning rate multiplier of 0.1, and a prompt loss weight of 0.01. Training lasted for four epochs.

Ensemble Models. To leverage classifications from multiple BERT models, we trained a LightGBM ensemble model combining outputs from our fine-tuned models with other state-of-the-art models. The following variables were included:

---

[1] https://platform.openai.com/docs/models.

- Predictions and probabilities from the fine-tuned models: RoBERTa, RoBERTa with layer-wise learning rate decay, XLM-RoBERTa, DeBERTa v3 base, DistilBERT, ALBERT, ELECTRA, YOSO, and GPT-neo.
- Emotion probabilities from the BERTemo model.[2]
- Sentiment probabilities calculated with ReBERTa for sentiment analysis.[3]
- Logits returned by the ELECTRA discriminator.[4]

Prompting. The study explored *zero-shot learning* [13], *few-shot learning* [5], and simple *Chain-of-Thought* [20] strategies for GPT models. In the *zero-shot learning* experiment, to avoid the ambiguity of the term *check-worthiness*, the concept was briefly explained within the prompt (a claim must be factual, verifiable, and potentially harmful) [16]. In the *few-shot learning* experiment, we created three prompt templates filled with four positive and four negative examples. The examples were dynamically selected from the training dataset based on the highest cosine similarity of claim embeddings, which were generated using the all-mpnet-base-v2 model.[5] In the *Chain-of-Thought* experiment, the check-worthiness estimation was decomposed into intermediate steps, using only four examples (two labeled *Yes* and two labeled *No*). The examples were selected manually and used as the assistant prompts.

5 Dataset Analysis and Curation

The dataset for the experiments consisted of 23,533 statements extracted from U.S. general election presidential debates, annotated by human coders and originally published in 2015, known as the ClaimBuster dataset [3]. The dataset was split into *train*, *dev*, and *dev_test* with 16,876, 5,625, and 1,032 examples in each split, respectively. The dataset was imbalanced, with only a 24% share of the positive class in the *train* and *dev* splits, and 23% in the *dev_test* split.

5.1 Curating the Dataset for the First Iteration of Experiments

A comparison of the CheckThat! Lab 2023 dataset with the original ClaimBuster dataset revealed that the *train* and *dev* splits were generated from examples with *crowd-sourced* labels, whereas the *dev_test* split was identical to the ClaimBuster dataset, referred to as *ground-truth*. The *ground-truth* dataset was labeled by three experts and was used to screen spammers and low-quality participants in the *crowd-sourced* part of the dataset.

This distinction between *ground-truth* and *crowd-sourced* labels surfaced during the evaluation of the initial results of baseline models. The models trained on the *train* dataset achieved, on average, an F1 score 0.1 higher when tested

---

[2] https://huggingface.co/bhadresh-savani/bert-base-uncased-emotion.
[3] https://huggingface.co/cardiffnlp/twitter-roberta-base-sentiment-latest.
[4] https://huggingface.co/google/electra-large-discriminator.
[5] https://huggingface.co/sentence-transformers/all-mpnet-base-v2.

on *dev_test* (i.e., *ground-truth*) than on *dev* (i.e., *crowd-sourced*). The difference could be attributed to the composition of the split (e.g., fewer borderline examples in *dev_test*) or the quality of labels (e.g., higher consistency in *dev_test*), with the latter being more probable as it correlates with the annotation process (i.e., experts vs. crowd-sourced labels).

The authors of the ClaimBuster dataset introduced screening criteria to exclude low-quality labels and published three filtered datasets with class ratios of 1:2, 1:2.5, and 1:3[6]. The 2:1 dataset was used for curating the training dataset in the first iteration of the study. We prepared two datasets with 8,706 examples each. The first dataset (called *curated*) consisted of high-quality labels (the examples included in the ClaimBuster 2xNCS.json file), and the second dataset (called *raw*) was an equal-sized subset with randomly selected examples. The datasets were split into 1,000 validation examples and 7,706 training examples. No new examples or features were added to the dataset, and the entire training and validation process was executed using subsets of the dataset provided by the CheckThat! 2023 organizers. The only information derived from ClaimBuster was the list of examples with higher expected quality labels that were used to filter examples.

5.2 Dataset Variants for the Second Iteration of Experiments

For the second iteration of experiments, we planned to further analyze the impact of changing class ratios and shares of quality examples on model prediction results. For this research, we decided not to investigate over-sampling techniques. Instead, we tested random under-sampling and proposed novel approaches to under-sampling using supportive measures. For clarity, each method tested in the experiments was assigned a unique code: *RUS*, *QUS*, *DUS*, and *HUS*, as explained below.

Random Under-Sampling (*RUS*). Five variants of the training dataset were created by applying random under-sampling to bring class ratios down to 1:1.5, 1:1.2, and 1:1, using three random seed values. Dataset variants created with this method were assigned a unique code: *RUS*.

Quality Based Under-Sampling (*QUS*)). From ClaimBuster files, we've derived a general annotation quality feature. All examples in the dataset that were contained in any of the three ClaimBuster files (2xNCS.json, 2.5xNCS.json, and 3xNCS.json) were marked as high quality, while those not included were marked as low quality. The annotation quality feature was used for under-sampling: three *QUS* datasets contain all positive examples (minority class) and only high-quality negative examples (majority class).

Under-Sampling Based on Training Dynamics Metrics (*DUS* and *HUS*). Six dataset variants was created using training dynamics metrics [19]. We first executed a training run on the complete dataset and collected logits after each of five training epochs. The logits were later used to calculate *variability*

---

[6] https://zenodo.org/record/3836810.

and *confidence* measures, which allow us to map dataset examples into one of three categories: *easy-to-learn*, *hard-to-learn*, and *ambiguous* (see Fig. 1). Under-sampling was performed by removing the most *easy-to-learn* and *hard-to-learn* examples from the majority class to ensure that the most *ambiguous* examples were not excluded following results reported in other studies [9]. Under-sampling with Training Dynamics metrics was performed in two ways:

i) By symmetrically removing the most *easy-to-learn* and *hard-to-learn* examples (*DUS*). All majority class examples were sorted in descending order by their ℓ_2 distance from the reference point (*variability, confidence*) = (0.5, 0.5) and removed until the desired class count was reached.

ii) By first removing all *hard-to-learn* examples (defined as examples having an ℓ_2 distance from (*variability, confidence*) = (0.5, 0.5) greater than 0.35 while having a confidence < 0.5) and then removing *easy-to-learn* examples sorted by descending distance from (*variability, confidence*) = (0.5, 0.5) until the desired class count was reached.

See visualization of the thresholds for *DUS* and *HUS* in Fig. 1.

Class Count Ratios. The most balanced high-quality subset made available by ClaimBuster has a 1:2 ratio of positive to negative example counts. To test if further reduction of the majority class count can yield a positive effect, we created additional datasets with 1:1.5, 1:1.2, and 1:1 ratios.

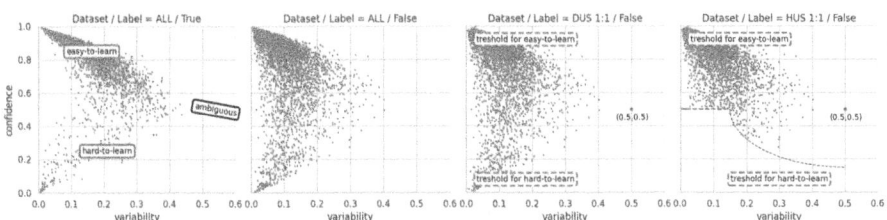

Fig. 1. Data map for three train datasets, based on a DeBERTa V3 base classifier plotted separately for the positive class (shared across all datasets) and the negative class (for *ALL* examples, under-sampled symmetrically - *DUS*, and under-sampled by removing *hard-to-learn* examples first - *HUS*)

6 Experiment Results

6.1 First Iteration Results: Finding Best Model

We fine-tuned BERT and GPT models for the first iteration using the techniques described in Sect. 4.1. The best results for training BERT-family models without data curation were obtained with DeBERTa v3 base fine-tuned (0.894), RoBERTa v3 base fine-tuned (0.862), and RoBERTa base fine-tuned with layer-wise decay (0.860). By applying the ensemble technique, we achieved an F1 score of 0.854.

For GPT models without dataset curation, the best results were obtained with GPT-3 curie (0.826) and GPT-Neo 125M (0.80). Applying prompting techniques to GPT-4 produced the worst results (GPT-4 few-shot learning: 0.788, GPT-4 zero-shot learning: 0.778, GPT-4 Chain-of-Thought: 0.722). The specific results of adapting models are described in [16].

Application of data curation (see Sect. 5) led to the improvement of the results of fine-tuned GPT-3 models. Dataset curation improved the F1 score of GPT-3 curie to 0.898 and davinci to 0.876. GPT-3 curie fine-tuned on the curated dataset achieved the best score, and the result was submitted for the competition [16]. Additionally, we trained RoBERTa v3 and DeBERTa v3 on a curated dataset, achieving F1 scores of 0.896 and 0.818, respectively.

The best results were obtained by fine-tuning the GPT model, but the difference between the performance of GPT models and BERT-family models was small. It is worth noting that dataset curation significantly improved the results for fine-tuned GPT models; for some BERT models, the improvement was less pronounced or even led to worse results.

6.2 Second Iteration Results: Finding Best Training Dataset

In this part of the experiment, we investigated the impact of various dataset under-sampling methods on two selected models: DeBERTa V3 base and large. We executed all planned 120 training runs (15 dataset variants, 2 models, and 4 random seed initializations). Analysis of the results (see Fig. 2) implies that improvements from balancing the dataset might be more pronounced than improvements from changing the model. Moreover, going beyond random under-sampling (RUS) with methods informed by data quality can improve both the maximum and average score achieved, as well as reduce the variance of the results, which may lead to more predictable performance of the model in actual use.

Acknowledging the fact that results achieved by the same model using the same dataset may vary greatly just by changing the random seed, we avoid a simple ranking approach.

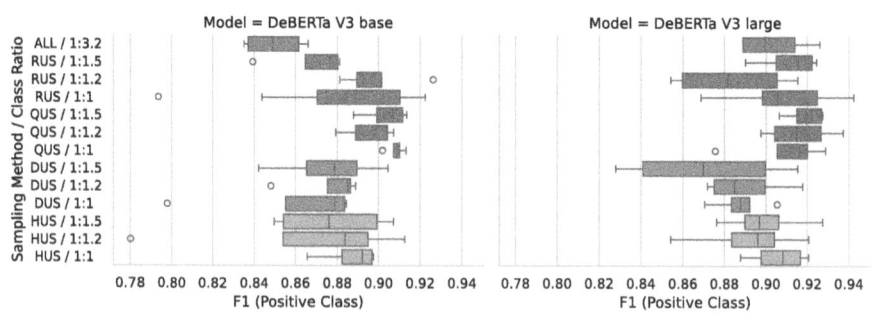

Fig. 2. Box plots of F1 Score (Positive Class) achieved by multiple training runs by Model, Sampling Method, and Class Ratio. Variance of results within each group comes from random seed initialization for the model trainer and the random seed used for random under-sampling of the dataset.

6.3 Significance of Differences Between Various Model Training Factors

We applied an analysis of variance (ANOVA) to compare several unknown population means concerning three factors: the Model used in training, the Class Ratio, and the Sampling Method. Our target variable is the F1 (Positive Class) score, calculated on predictions for the dev_test dataset.

Before we can use the ANOVA hypothesis test, we need to check several assumptions. First, data are randomly selected from a populations and randomly assigned to each of the treatment groups. Second, values in each sampled group should follow a normal distribution. Third, variances within groups are equal, i.e., practically the ratio of the largest to the smallest standard deviation should be smaller than 2.

To check normality we used the Shapiro-Wilk and a Q-Q plot tests. For Shapiro-Wilk test the following hypotheses were formulated:

- H_0 (null hypothesis): the population is distributed normally,
- H_1 (alternative hypothesis): the distribution differs from normal distribution.

P-values calculated for the Class Ratio groups are: group '1:1' – 0.0002, group '1:1.2' – 0.0026, group '1:1.5' – 0.0439, group 'ALL' – 0.6947. In three out of four groups, with p-values less than $\alpha = 0.05$, and we should reject hypothesis H_0 – that groups are distributed normally.

P-values calculated with Shapiro-Wilk statistics for the factor Sampling Method are: group 'ALL' – 0.6947. group 'RUS' – 0.0456, group 'QUS' – 0.7598, group 'DUS' – 0.0307, group 'HUS' – 0.0015. In three out of five groups, with p-values less than $\alpha = 0.05$, we should reject hypothesis H_0 – that groups are distributed normally.

P-values calculated with Shapiro-Wilk statistics for the factor Model are: group 'DeBERTa V3 base' – 0.0001, group 'DeBERTa V3 large' – 0.0284. In both groups, with p-values less than $\alpha = 0.05$, we should reject hypothesis H_0 – that groups are distributed normally. We cannot confirm normal distribution with Shapiro-Wilk statistics, however the ANOVA procedure with fixed factors and equal sample sizes works well even when the assumption of normality is violated, unless one or more of the distributions are highly skewed or the variances are very different. We can resort to a Q-Q plot test, which shows roughly straight lines, hence the assumption about normal distribution or at least absence of highly skewed distribution is confirmed (see Fig. 3).

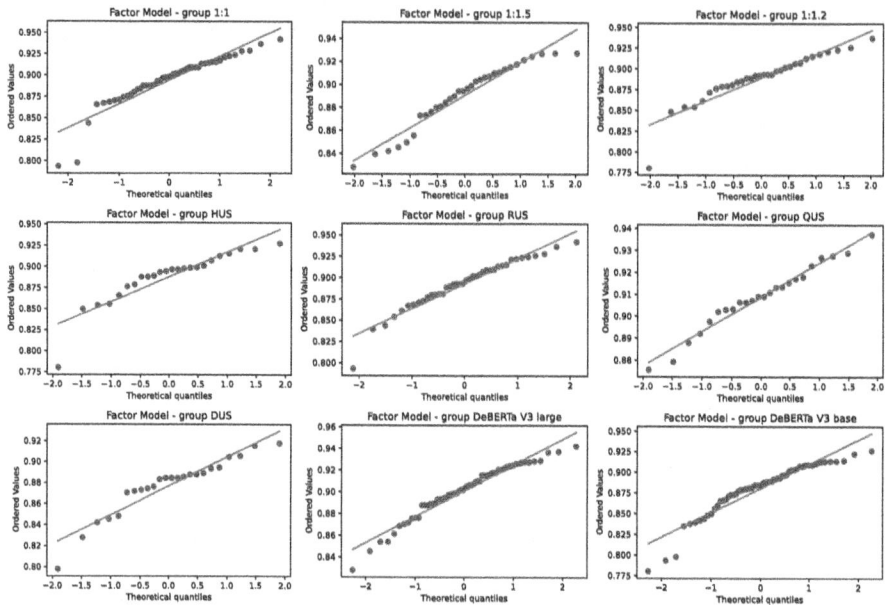

Fig. 3. Q-Q plot for groups within analyzed factors

Standard deviations for `Class Ratio` are: $\sigma_{1:1} = 0.029$, $\sigma_{1:1.2} = 0.029$, $\sigma_{1:1.5} = 0.028$, $\sigma_{ALL} = 0.033$. The `Class Ratio` of the largest to the smallest sample standard deviation is 1.16 for all groups and only 1.04 when skipping group *ALL*.

Standard deviations for `Sampling Method` are: $\sigma_{ALL} = 0.033$, $\sigma_{RUS} = 0.029$, $\sigma_{QUS} = 0.015$, $\sigma_{DUS} = 0.028$, $\sigma_{HUS} = 0.031$. The ratio of the largest to the smallest sample standard deviation is 2.16 for all groups and only 2.05 when skipping group *ALL*.

Standard deviations for `model` are: $\sigma_{DeBERTaV3base} = 0.03$, $\sigma_{DeBERTaV3large} = 0.024$. The ratio of the largest to the smallest sample standard deviation is 1.28.

Even though not all the ratios of the largest to the smallest variances were less than 2, we cannot assume that the variances are significantly different, with the maximum value being 2.16. Therefore, we decided to apply a one-way ANOVA to study single factors and a two-way ANOVA to consider interactions between factors.

For one-way ANOVA the following hypotheses were formulated

- H_0 (null hypothesis): the means of all populations are equal,
- H_1 (alternative hypothesis): at least one population mean differs from the rest.

Results for F-statistics and p-values presented in Table 1 reveal that there is a difference in the performance of the models considering the Sampling Method and Model, but not for Class Ratio.

Table 1. Results of one-way ANOVA

Factor	F statistic	p-value	conclusion
Class Ratio	0.9	0.4606	null hypothesis accepted
Sampling Method	4.8	0.0013	null hypothesis rejected
Model	16.8	0.0001	null hypothesis rejected

In the case of two-way ANOVA we check how two factors affect a response variable and whether there is an interaction between the two factors. The following hypotheses are verified:

- H_0 (null hypothesis): the effect of one factor does not depend on the effect of the other factor, i.e., there is no interaction between factors,
- H_1 (alternative hypothesis): there is an interaction between factors.

Results are presented in Table 2. The p-values for single factors– Model, Sampling Method–are less than 0.05 which implies that analyzed factors have a statistically significant effect on model training. Class Ratio is not statistically significant. The p-values for all interaction effects are greater than 0.05, so we failed to reject the null hypotheses. We do not have enough evidence that there are significant interaction effects between analyzed factors.

Table 2. Results of two-way ANOVA for interaction between the factors

Factors	d.f.	sum sq	mean sq	F statistic	p-value
C(Model)	1	0.0126	0.0126	18.4689	0.0000
C(Class Ratio)	3	0.0022	0.0007	1.0826	0.3601
C(Sampling Method)	4	0.0128	0.0032	4.7147	0.0016
C(Model):C(Class Ratio)	3	0.0028	0.0009	1.3951	0.2487
C(Model):C(Sampling Method)	4	0.0007	0.0002	0.2669	0.8986
C(Sampling Method):C(Class Ratio)	12	0.0067	0.0006	0.8160	0.6335
Residual	100	0.0681	0.0007	-	-

7 Conclusions and Future Work

The study proved that selection of under-sampling method is statistically significant. Balancing the dataset greatly improves F1 scores, and random under-sampling (RUS) appears to be an effective starting point. The single highest result (F1 0.942) with DeBERTa V3 large was achieved when training with a randomly under-sampled dataset with a class ratio of 1:1. On average, random under-sampling helped increase the F1 score by 0.016 (0.893 vs. 0.877). Even more improvements were observed with Quality-based Under-Sampling (QUS), where the average F1 score was higher by 0.032 (0.909 vs. 0.877). Application of QUS also halved the standard deviation compared to other methods.

Under-sampling based on Training Dynamics metrics did not surpass the results of random under-sampling (RUS). Symmetrical removal of the most easy-to-learn and hard-to-learn examples (DUS) performed comparably to training on the uncurated dataset, while removal of hard-to-learn examples first (HUS) delivered an average 0.011 better F1 score. The primary conclusion of the study is that while random under-sampling offers significant improvements, it can be surpassed by under-sampling informed by quality metrics. Given such features are not usually available, alternative proxies need to be identified. We investigated whether proposed approach derived from Training Dynamics could serve as viable alternatives but were unable to provide evidence supporting their efficacy for either of the proposed methods (DUS and HUS).

Model selection is statistically significant, and DeBERTa V3 large apparently outperforms the base model in most cases. Although individual runs may yield better results for the base model, both maximum and mean F1 scores (Positive Class) tend to be higher with the large model, on average by 0.02 (0.901 vs. 0.881).

Selection of specific class ratio (1.1 vs 1.2 vs 1.1.5) for under-sampling does not have a statistically significant effect on results.

Future research may include the creation of other formulas based on Training Dynamics metrics that could improve results beyond random under-sampling. Additionally, other methods for identifying incorrect labels may be tried out as a way to inform the curating or balancing of a training dataset.

References

1. Agrestia, S., Hashemianb, A., Carmanc, M.: PoliMi-FlatEarthers at CheckThat! 2022: GPT-3 applied to claim detection. In: Working Notes of CLEF 2022 - Conference and Labs of the Evaluation Forum, CLEF 2022, Bologna, Italy (2022)
2. Anand, A., et al.: Don't blame the data, blame the model: understanding noise and bias when learning from subjective annotations. arXiv preprint arXiv:2403.04085 (2024)
3. Arslan, F., Hassan, N., Li, C., Tremayne, M.: A benchmark dataset of check-worthy factual claims. In: 14th International AAAI Conference on Web and Social Media. AAAI (2020)
4. Black, S., Gao, L., Wang, P., Leahy, C., Biderman, S.: GPT-Neo: Large scale autoregressive language modeling with meshtensorflow (2021). https://doi.org/10.5281/zenodo.5551208
5. Brown, T.B., et al.: Language models are few-shot learners (2020)
6. Clark, K., Luong, M.T., Le, Q.V., Manning, C.D.: ELECTRA: Pre-training Text Encoders as Discriminators Rather Than Generators (2020)

7. He, P., Gao, J., Chen, W.: Debertav3: improving deberta using electra-style pre-training with gradient-disentangled embedding sharing (2021)
8. He, P., Liu, X., Gao, J., Chen, W.: DeBERTa: Decoding-enhanced BERT with Disentangled Attention (2021)
9. Ince, O.B., Zeraati, T., Yagcioglu, S., Yaghoobzadeh, Y., Erdem, E., Erdem, A.: Harnessing dataset cartography for improved compositional generalization in transformers. arXiv preprint arXiv:2310.12118 (2023)
10. Lan, Z., Chen, M., Goodman, S., Gimpel, K., Sharma, P., Soricut, R.: ALBERT: A Lite BERT for Self-supervised Learning of Language Representations (2020)
11. Liu, Y., et al.: RoBERTa: A Robustly Optimized BERT Pretraining Approach (2019)
12. Nakov, P., et al.: Overview of the clef-2022 checkthat! lab task 1 on identifying relevant claims in tweets. In: CLEF 2022, Bologna, Italy (2022)
13. Ouyang, L., et al.: Training language models to follow instructions with human feedback (2022)
14. Sanh, V., Debut, L., Chaumond, J., Wolf, T.: DistilBERT, a distilled version of BERT: smaller, faster, cheaper and lighter (2020)
15. Sar-Shalom, A., Schwartz, R.: Curating datasets for better performance with example training dynamics. In: Findings of the Association for Computational Linguistics: ACL 2023, pp. 10597–10608 (2023)
16. Sawiński, M., et al.: Openfact at checkthat! 2023: head-to-head GPT vs. BERT - a comparative study of transformers language models for the detection of check-worthy claims. In: Working Notes of CLEF 2023 - Conference and Labs of the Evaluation Forum. CLEF 2023, Thessaloniki, Greece (2023)
17. Shi, H., et al.: Qadynamics: training dynamics-driven synthetic QA diagnostic for zero-shot commonsense question answering. arXiv preprint arXiv:2310.11303 (2023)
18. Snijders, A., Kiela, D., Margatina, K.: Investigating multi-source active learning for natural language inference. arXiv preprint arXiv:2302.06976 (2023)
19. Swayamdipta, S., et al.: Dataset cartography: mapping and diagnosing datasets with training dynamics. arXiv preprint arXiv:2009.10795 (2020)
20. Wei, J., et al.: Chain-of-thought prompting elicits reasoning in large language models (2023)

Classification of Social Media Hateful Screenshots Inciting Violence and Discrimination

Davide Buscaldi[1(✉)], Paolo Rosso[2,3], Berta Chulvi[4,5], and Ting Wang[6]

[1] LIPN, Université Sorbonne Paris Nord, Villetaneuse, France
buscaldi@lipn.fr
[2] PRHLT, Universitat Politécnica de Valéncia, Valencia, Spain
prosso@upv.es
[3] valgrAI - Valencian Graduate School and Research Network of Artificial Intelligence, Valencia, Spain
[4] Universitat de València, Valencia, Spain
[5] Symanto Research, Nuremberg, Germany
berta.chulvi@symanto.es
[6] IMT Atlantique, Nantes, France

Abstract. In this paper, we address the challenge of categorizing hateful content on social media through the analysis of online screenshots. Such screenshots may contain only text, only images with a caption, or images with embedded text. OCR-based techniques may help only in the first case, while for the other two cases it would be necessary to leverage visual language models to classify the type of content and its source. We leverage various techniques both from OCR'd text and large or visual language models to classify the type of content and its source. The results show that the task is a difficult one, although our experiments shed some light on the possible effective solutions for this task.

Keywords: Hate Speech · Online screenshots · Incitement to violence and discrimination · Large Language Models

1 Introduction

Fighting the spread of hate speech in digital communication is a central concern for the United Nations[1] and for the Council of Europe[2]. The problem is not a small one, and some authors speak about a Hate Speech Epidemic that leads to political radicalization and deteriorates intergroup relations [1]. Given the enormous amount of user-generated content, the task of automatic identification and, if possible, counteracting the spread of hate speech on social media, is becoming a fundamental aspect of the fight against violence and intolerance.

---

[1] https://www.un.org/en/hate-speech.
[2] https://www.coe.int/en/web/no-hate-campaign/coe-work-on-hate-speech.

Detecting and managing toxic content on microblogging platforms poses a significant challenge, often necessitating human supervision. One effective approach involves capturing instances of harmful content and cataloguing them based on various criteria such as the content type, its origin, or any other factor considered important for studying this type of content.

Most of the work in hate speech detection in social media has been done on texts and also on memes (for a recent review see [4]) but not considering screenshots. With the increasing accuracy of AI techniques for detecting hate speech in textual content, common speech masking strategies may use screenshots that convert text, or text plus an image, into a single image.

In our research, we explore various models aimed at automating the classification of detrimental online content. By employing state-of-the-art Large Language Models (LLMs) and Visual Language Models (VLMs), we seek to enhance the efficiency and accuracy of identifying and categorizing harmful material, thereby contributing to the development of more robust content moderation systems for microblogging platforms.

2 Dataset

In our experiments, we used 3,180 screenshots of hate speech messages shared on five social media: 918 on Facebook, 732 on Twitter, 651 on Instagram, 355 on TikTok, and 524 on Youtube. The messages were classified by a team of experts that monitor hate speech in social media for OBERAXE (Observatorio Español del Racismo y la Xenofobia)[3]. The messages were spread between July 2022 and May 2023. The screenshots' content could be only text (2,457) or text with images (723). The experts who work for OBERAXE classified these screenshots into two principal categories: (1) messages that only **incite violence** (610) and (2) messages that only **incite discrimination** (1,130) and (3) messages that incite both (1,440). These two categories are consistent with the ones used by [5], which distinguish between "calls for violence" and "assaults on human dignity". They are also supported by the two-class specification given by UNESCO [3] for hate speech: (a) "expressions that advocate incitement to harm (particularly, discrimination, hostility or violence) based upon the target being identified with a certain social or demographic group" and (b) "expressions that foster a climate of prejudice and intolerance on the assumption that this may fuel targeted discrimination, hostility, and violent acts".

An example of the textual content of a screenshot classified as promoting violence is *"Tie a block of concrete to his foot and let him swim back to his fucking country"* because it gives ideas about how to kill migrants that arrive by boat. An example for the discrimination category is *"Fucking Chinese and their fucking virus"*.

This dataset does not contain a negative category (that is, there is no "no hateful content" category), but it can be used for other classification tasks,

---

[3] https://www.inclusion.gob.es/oberaxe/es/index.htm

for instance distinguishing between inciting violence, inciting discrimination, or both. It can also be used for the classification of other information such as the source of content. This may be important to study trends and to identify which platform is preferred for hate dissemination campaigns.

3 Models

We applied various models, considering text and image dimensions together or individually. The details of the models are as follows:

Tf.idf + logistic regression classifier: this is a baseline model, in which the input is the OCR'd text extracted from the image and the textual features are represented as Bag-of-Words, with minimum frequency 2. The classifier is a standard Logistic Regression classifier. This model has also been useful to identify the keywords (i.e., the features) that are most relevant for the "violence" and "discrimination" categories.

BERT (Bidirectional Encoder Representations from Transformers) [2] and **RoBERTa (Robustly optimized BERT approach)** are both state-of-the-art transformer-based models developed by Google and Facebook respectively. BERT employs a transformer architecture that learns contextual word representations by pre-training on large corpora using masked language modeling and next-sentence prediction tasks. RoBERTa builds upon the success of BERT by optimizing several training hyperparameters and removing the next sentence prediction task, resulting in improved performance and robustness. In our experiments, we fine-tuned RoBERTa on the training split of the corpus.

CLIP (Contrastive Language-Image Pretraining) [8] is a deep learning model developed by OpenAI in 2021, in which the embeddings for images and text share the same space. This feature is obtained by pre-training an image and a text encoder to predict which images were paired with which texts in a very large dataset. The model can then be used to perform classification, or to extract captions from images. In our experiments, we used CLIP to obtain a caption of the image and append it to the original text. Then RoBERTa is fine-tuned on the full training texts.

The **Swin transformer** [7] is a hierarchical visual transformer model. It builds hierarchical feature maps by merging image patches in deeper layers. Image patches are made of window partitions. Each window partition is shifted between consecutive self-attention layers, providing connections among them that significantly enhance modeling power. The advantages of the Swin transformer are exploited in the **Donut** [6] model to create an OCR-free document understanding model, that can reply to questions on a scanned document without the need to convert it into text. In our experiments, Donut is used as an encoder to produce an image embedding. The embedding is obtained as a pooled vector of size 768 from the last hidden states. The vectors are then used as input to: an XGBoost[4] model with softmax objective function and AUC loss, with 20 training epochs and depth max set to 4, or to a fully-connected neural network

---

[4] https://github.com/dmlc/xgboost.

(FCNN) with one hidden layer of 256 units, trained for 100 epochs with Adam optimizer and learning rate of 0.001. When the Swin Transformer is used alone, it is fine-tuned for 5 epochs on the training data. It didn't show clear convergence which may explain its negative results in comparison with Donut used as encoder.

4 Results

In this section we present the results we obtained on two sub-tasks: hate speech type prediction and source prediction. All experiments have been run on the same 80:20 split, with 2544 instances in the training set and 636 instances in the test set.

4.1 Hate Speech Type Prediction

In this subtask, our objective is to predict two types of content: violent and discriminative. It's important to clarify that our goal is not to discern whether a message contains hate speech; rather, we already acknowledge its presence and aim to categorize it into one of two types: violent or discriminative. The results for this sub-task are shown in Table 1.

Observing the results, it's evident that all models exhibit comparable performance in the "violence" category, except for the RoBERTa model, which achieves a 5% improvement over the others. In contrast, the performance in the "discrimination" category shows more variability. Once more, RoBERTa demonstrates superior performance, particularly when coupled with captions generated via CLIP. This is mainly because in the test split, there are more pictures in the discrimination category, while for the violence category most images show only textual information, whereas CLIP can't produce anything useful.

We extracted the most relevant textual features for the Tf.idf based classifier, and we observed that out of the top 20 most relevant words, 14 are the same for both categories. This explains why it could be difficult to distinguish between these two types of content.

Table 1. Results predicting the type of content: Acc for accuracy and F1 (macro average).

Model	Violence		Discrimination	
	Acc	F1	Acc	F1
Tf.Idf + LogReg classifier	0.808	0.446	0.628	0.386
RoBERTa	0.850	0.563	0.762	0.639
RoBERTa + CLIP	0.804	0.531	0.767	0.640
Donut embeddings + XGB	0.805	0.531	0.594	0.522
Donut embeddings + FCNN	0.806	0.462	0.613	0.405
Swin Transformer	0.808	0.446	0.629	0.386

4.2 Source Prediction

In this section we present the results on source prediction. The task consists in, given the text and the capture of the toxic comment, predicting from which source the comment has been generated. The possible sources are: Facebook, Twitter, Instagram, TikTok and Youtube. The results are shown in Table 2.

It can be noted in Fig. 1 that the text-only model cannot classify the sources. The BERT fine-tuned model is even worse as it classifies everything as originating

Table 2. Results predicting the source of content

Model	Accuracy	F1 (macro)
Tf.Idf, text only	0.358	0.135
BERT fine-tuned, text only	0.295	0.091
Donut embeddings + XGB	0.558	0.542
Donut embeddings + FCNN	0.577	0.566
Swin Transformer	0.295	0.091

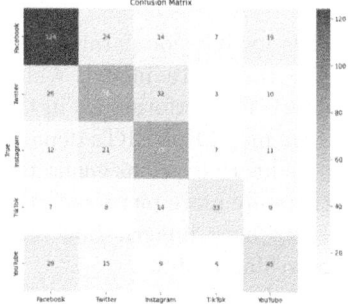

(a) Donut + XGB head

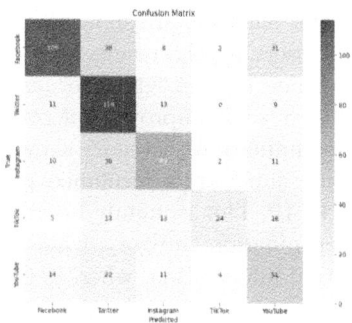

(b) Donut + FCNN head

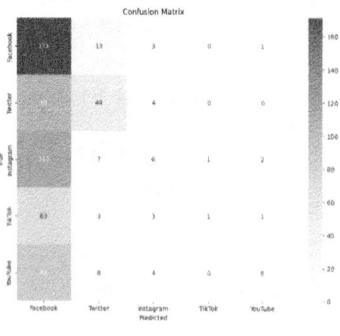

(c) tf.idf text-only

Fig. 1. Confusion matrices obtained with the text-only tf.idf text-only method, Donut embeddings with XGB head and Donut embeddings with a FCNN head.

from Facebook (we omit its confusion matrix for this reason). On the other hand, we can see that the Donut embeddings allow to discriminate quite well among the sources, indicating that the Donut model is able to capture the visual style of each platform. The only major confusion seems to be between Twitter and Instagram comments, particularly when the XGB head is used.

5 Conclusions

We studied the problem of hateful content categorization from screen captures. While text-based representations prove effective in discerning instances of discrimination and violence, visual features significantly aid in identifying content under the "discrimination" category, mainly because more images are used in this kind of message. However, predicting the source solely based on textual features appears unfeasible. Leveraging the Donut hybrid visual-language model for encoding proved its effectiveness in detecting the source of the content, albeit with occasional confusion, particularly between comments originating from Twitter and Instagram.

Acknowledgements. This work was carried out in the framework of the FAKEnHATE-PdC research project on FAKE news and HATE speech (Grant PDC2022-133118-I00), funded by MCIN/AEI/10.13039/501100011033 and by European Union NextGenerationEU/PRTR. We would like to thank the Spanish Observatory of Racism and Xenophobia (OBERAXE) of the Government of Spain for their collaboration in providing the data for this research. Berta Chulvi's work was carried out during her period as a researcher at the PRHLT of the Universitat Politècnica de València.

References

1. Bilewicz, M., Soral, W.: Hate speech epidemic. the dynamic effects of derogatory language on intergroup relations and political radicalization. Political Psychol. **41**(S1), 3–33 (2020). https://doi.org/10.1111/pops.12670. https://onlinelibrary.wiley.com/doi/abs/10.1111/pops.12670
2. Devlin, J., Chang, M.W., Lee, K., Toutanova, K.: BERT: Pre-training of deep bidirectional transformers for language understanding. In: Burstein, J., Doran, C., Solorio, T. (eds.) Proceedings of the 2019 Conference of the North American Chapter of the Association for Computational Linguistics: Human Language Technologies, Volume 1 (Long and Short Papers), Minneapolis, Minnesota, pp. 4171–4186. Association for Computational Linguistics (2019). https://doi.org/10.18653/v1/N19-1423. https://aclanthology.org/N19-1423
3. Gagliardone, I., Gal, D., Alves, T., Martinez, G.: Countering online hate speech. Unesco (2015)
4. Gandhi, A., et al.: Hate speech detection: a comprehensive review of recent works. Expert Syst. e13562 (2024). https://doi.org/10.1111/exsy.13562
5. Kennedy, B., et al.: Introducing the gab hate corpus: defining and applying hate-based rhetoric to social media posts at scale. Lang. Resour. Eval. (1), 7–108 (2022). https://doi.org/10.1007/s10579-021-09569-x

6. Kim, G., et al.: Donut: document understanding transformer without OCR. arXiv preprint arXiv:2111.15664, vol. 7, p. 15 (2021)
7. Liu, Z., et al.: Swin transformer: hierarchical vision transformer using shifted windows. In: Proceedings of the IEEE/CVF International Conference on Computer Vision, pp. 10012–10022 (2021)
8. Radford, A., et al.: Learning transferable visual models from natural language supervision. In: International Conference on Machine Learning, pp. 8748–8763. PMLR (2021)

SessionPrint: Accelerating kNN via Locality-Sensitive Hashing for Session-Based News Recommendation

Mozhgan Karimi[✉]

Antwerp University, Antwerp, Belgium
mozhgan.karimi2@uantwerpen.be

Abstract. Traditional kNN methods, while proven to be accurate in session-based scenarios such as news recommendation, suffer from computational inefficiencies, especially when dealing with large datasets typical of real-world applications. This can lead to high costs for computing infrastructure as well as slow response times, during online recommendation generation. We propose an approach, called SessionPrint, that employs locality-sensitive hashing to reduce the time it takes to find neighboring sessions. Furthermore, we devise a multi-stage variant of our approach as well as a version that utilizes a final precision pass so as to drill down to the most fitting set of neighboring sessions in the most efficient way possible.

We evaluate the performance of our approach in terms of both accuracy and efficiency on four real-world news datasets of varying sizes. The results confirm that SessionPrint not only reduces the time to generate recommendations but also maintains high accuracy compared to a traditional session-based kNN implementation, providing a scalable solution for real-world applications where rapid response times are crucial.

Keywords: Session-based Recommender Systems · News Recommendation · Locality-Sensitive Hashing · Performance Evaluation

1 Introduction and Related Work

Recommender systems, especially news recommender systems [9], have increasingly become a focus of academic research. However, evaluations of these systems mostly concentrate on accuracy metrics, while often neglecting other important factors such as computational efficiency and memory consumption. Challenges that feature real-world benchmarks, e.g., CLEF/NewsREEL [11], underscore that while aligning recommendations with user preferences is crucial, system responsiveness can be equally important.

In the news domain, where users demand timely content and are often not logged in, many algorithms are impractical due to their reliance on comprehensive user profiles or need for extensive (overnight) training. Thus, session-based algorithms are often the only viable option, i.e., only the clicks from the current user's most recent session are used to predict what to recommend. K-nearest neighbor (kNN) approaches stand out in the literature for their accuracy in addressing this task, often even surpassing modern deep-learning approaches [8,12]. However, as traditional kNN implementations do not build a condensed model during training, similarity calculations have to be executed on demand, delaying recommendation list generation. As a remedy, locality-sensitive hashing (LSH) [6] has been successfully applied to more traditional *user-user* kNN recommendation scenarios. Several performance analyses [1,5,10] have shown a high reduction in computation time, while keeping accuracy losses manageable.

A few approaches have also been proposed that apply LSH to *session-based* recommendation, e.g., to incorporate multi-modal features into a neural-network model [3], to facilitate content-based recommendation [2], or to calculate word-vector centroids in web query recommendation [4]. These works primarily focus on using LSH to reduce input dimensionality of additional content or metadata. However, none of these session-oriented approaches aim to primarily enhance the efficiency of traditional methods, such as click-session-based kNN, and as a result, there are no comparative evaluations regarding runtime performance.

In this paper, we tailor an LSH-enhanced kNN approach called *GoldFinger* to the session-based recommendation scenario. We propose several extensions to the original scheme to optimize the trade-off between runtime and accuracy. As a key contribution of this work, we evaluate our approaches' viability in a session-based context by conducting an in-depth "replay" evaluation on four real-world news recommendation datasets with respect to both accuracy and runtime.

2 Proposed Approach

2.1 Problem Definition

In traditional kNN approaches applied to session-based recommendation, the primary objective is to compare the current user's session with historical sessions of other users with respect to a predefined similarity measure. Once the k most similar sessions have been determined, then recommendation candidates are retrieved from these neighboring sessions and ranked based on each neighboring session's similarity score. Commonly used similarity measures include Jaccard similarity, Cosine similarity, or more complex measures designed to capture temporal session dynamics, such as V-SkNN [7].

Although such approaches are often effective in session-based recommendation scenarios [8,12], they can become impractical in environments with large item selections or extensive user bases. Even when using efficient data structures, a simple Jaccard comparison of two sessions has a least linear runtime complexity with respect to the number of items. Efficiency can be improved by employing heuristics like comparing only the N most recent sessions, while still retaining effectiveness in domains like news recommendation where fresh items are preferred [7]. Nevertheless, even when such heuristics are applied, session-based kNN approaches are among the slowest baselines in terms of computation time per recommendation retrieval. In practice, this can result in higher costs for computation infrastructure and a delay for users browsing a web page.

2.2 Approximate kNN via Fingerprinting of Sessions

To address the aforementioned efficiency shortcomings, we employ locality-sensitive hashing (LSH) to reduce session comparisons to a constant time complexity, with only a marginal increase in memory requirements. To this end, we tailor the GoldFinger [5] hashing approach to the context of session-based news recommendation. Together with further extensions and variants, we collectively call our approach *SessionPrint*. Formally, in line with how GoldFinger calculates fingerprints of whole user profiles [5], we generate a fingerprint of each session s as a bit array of size b such that $B(s) = (\beta_x)_{x \in [0..b-1]}$, where the value of each bit β_x within $B(s)$ is calculated as

$$\beta_x = \begin{cases} 1, & \text{if } i \in s : h(i) = x \\ 0, & \text{otherwise.} \end{cases} \quad (1)$$

Here, $h(i)$ is a uniform hashing function that assigns each item i a value between 0 and b. Subsequently, an approximate Jaccard similarity between the two sessions can be calculated based on their fingerprints $B(s_1)$ and $B(s_2)$ as

$$sim(s_1, s_2) = \frac{||B(s_1) \text{ AND } B(s_2)||_1}{||B(s_1)||_1 + ||B(s_2)||_1 - ||B(s_1) \text{ AND } B(s_2)||_1} \quad (2)$$

where AND denotes the bitwise *and* operation of two bit arrays and $||\cdot||_1$ denotes the number of true bits within a bit array (see the original publication [5] for a proof of the approximative power of the similarity formula).

In practice, we can pre-compute each session's bit array hash $B(s)$. Thus, when generating recommendations for a session s, only the bit array $B(s)$ for the current session has to be calculated on-the-fly. Unlike traditional kNN comparisons, where the computation time grows with the number of items in the item space, $sim(s_1, s_2)$ is computationally bound by the number of bits b in the fingerprint, which can be chosen at will. Thus, a tradeoff can be achieved between efficiency (with smaller b) and effectiveness (with larger b). Furthermore, the additional space requirements are practically negligible when choosing a small hash size, such as 128 bits. Combined, this allows us to efficiently compare the current session with a large number of other sessions, and not just a small recent subset.

2.3 Multi-stage Fingerprinting and Final Precision Pass

In addition to tailoring GoldFinger to a session-based recommendation context, we propose a strategy to potentially improve its effectiveness while impacting its runtime only marginally. To this end, we extend the original idea by introducing a *multi-stage session fingerprinting* scheme. Formally, we define a session's extended fingerprint as

$$B_{ext}(s) = \left(B^b(s),\ B^{b\cdot 2^1}(s),\ B^{b\cdot 2^2}(s),\ \ldots,\ B^{b\cdot 2^{(l-1)}}(s)\right) \tag{3}$$

where l is the number of stages and $B^n(s)$ is the fingerprint bit array $B(s)$ with bit size n, calculated via Eq. 1.

This allows for successive comparisons, e.g., first via a 128-bit hash, then 256 bits, etc., to whittle down the number of potential neighbors, stage by stage. Each time, the remaining number of neighbors shrinks, and thus, more precise comparisons using a larger hash size only have to be executed on a relatively small number of sessions, resulting in a negligible impact on overall runtime.

Lastly, we suggest incorporating an (optional) *final precision pass* to be applied after single- or multi-stage fingerprinting. In either case, after the number of potential neighbors has been reduced to a manageable size via hash comparisons, we propose to incorporate another final pass of a standard similarity measure, such as Jaccard, Cosine, or V-SkNN. By doing so, we can aim to reduce noise introduced through preceding stages and obtain more precise similarity scores to rank candidate items derived from the final set of neighbor sessions.

3 Experimental Evaluation

3.1 Methodology

We conduct an evaluation on four real-world datasets: the 2017 CLEF/News-REEL challenge dataset with $1.1\,m$ clicks (publisher 418), the 2017 Outbrain challenge dataset with $1.1\,m$ clicks (publisher 43), the SmartMedia Adressa dataset with $18\,m$ clicks, and the "small" 2024 RecSys challenge dataset with $3.3\,m$ clicks. We use the StreamingRec framework [7] to perform a realistic "replay" evaluation on these datasets, in which recommendation models can be improved on-the-fly. However, given that kNN approaches do not maintain a traditional model, the only consideration to make our approach streaming-compatible is the need to compute fingerprints of new sessions, whenever they become available.

We compare our proposed approach to a precise kNN implementation. As a similarity measure, we use the V-SkNN [7] approach and we employ recency-based session pre-filtering to keep the runtime manageable. Recency filter thresholds and the value for k are pre-selected per dataset via a validation set. In addition, we include a session-based item co-occurrence algorithm, known for its execution speed and reasonable accuracy compared to V-SkNN.

We contrast these baselines against various variants of our proposed Session-Print approach. For the hashing function, we apply a simple modulo operation on the item ID, expressed as $h(i) = \text{ID}(i) \bmod b$, which yields a sufficiently uniform distribution. Based on preliminary experiments on validation sets, we employ $b = 128$, as it results in a good trade-off between accuracy and runtime. To evaluate the accuracy of our approach, we use both the F1 measure and the Mean Reciprocal Rank (MRR) metric. To assess computational efficiency, we measure the average time required to generate a recommendation list.

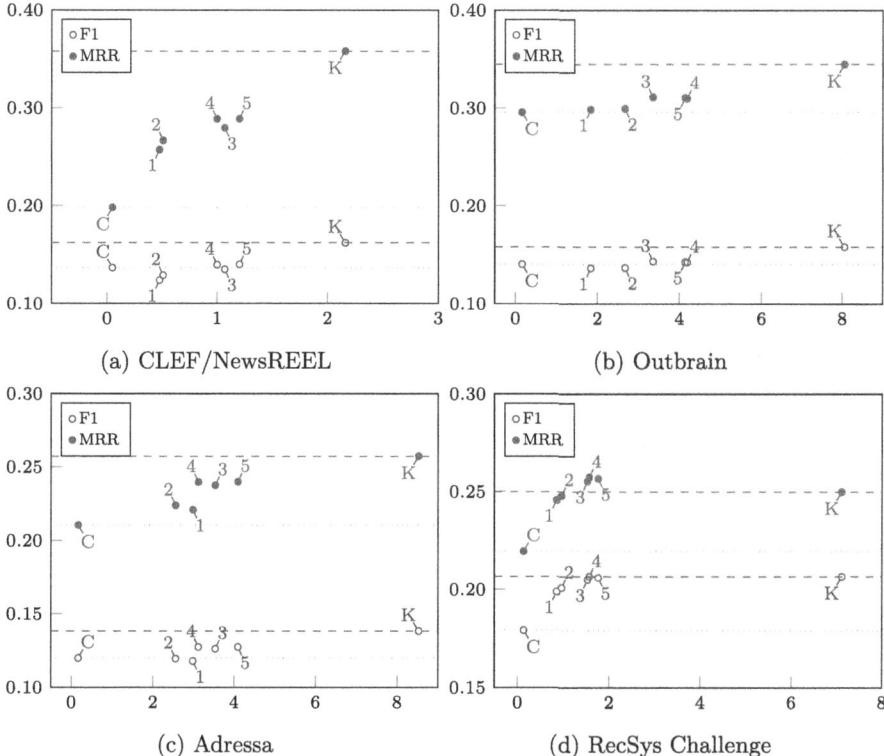

Fig. 1. Results of the experimental evaluation. X-axis: average recommendation time in ms. Points marked C and K are the co-occurrence and kNN baselines. Numbered points show SessionPrint: (1) basic hashing (2) with recency post-filtering (RF); (3) with final V-SkNN precision pass (PP); (4) with RF and PP; (5) multi-stage hashing with RF and PP ($l = 2$).

3.2 Results

Figure 1 presents the results of our experimental evaluation. Most notably, all reported variants of SessionPrint exhibit drastically reduced runtimes compared to the kNN baseline. For instance, the simplest hashing strategy (1) achieves runtime reductions of at least 64%, while preserving more than 71% of the original accuracy. For the RecSys dataset, SessionPrint even retains 96% accuracy and reduces runtime by 88%, indicating that it is best suited for larger datasets.

Higher accuracy retention can be achieved by applying more sophisticated variants, at the cost of slightly longer runtimes. Interestingly, variants 4 and 5 even outperform the kNN baseline in terms of MRR on the RecSys dataset, while being 78% faster. This is likely because the kNN baseline needs to employ recency *pre*-filtering of sessions (see Sect. 2.1), whereas SessionPrint can utilize all sessions in the dataset due to the low computational cost of hash comparisons.

Across all datasets, incorporating a final precision pass consistently improves accuracy with minimal runtime overhead. In contrast, multi-stage fingerprinting does not consistently outperform single-stage fingerprinting. Even though the multi-stage approach (5) exhibited the best accuracy on the CLEF dataset among all tested variants, the difference to its single-stage counterpart (4) was not statistically significant. Overall, the most consistent variant seems to be 4. Depending on the dataset, it was able to reduce the runtime compared to the baseline by 48% to 78%, and for all of the datasets, it significantly outperformed the co-occurrence baseline in both F1 and MRR.

4 Conclusions

Our empirical analysis demonstrates that locality-sensitive hashing can be applied to session-based kNN approaches to achieve significant reductions in terms of average runtimes for generating recommendation lists, while accuracy compared to the baseline is retained to a high degree. A variant of our approach that first applies hashing to identify an initial set of neighbors followed by a V-SkNN precision pass exhibits the most reliable results, consistently outperforming a co-occurrence baseline and reducing runtime compared to a traditional kNN approach by 48% to 78%. In future work, we plan to further investigate the effect of hash size on the trade-off between runtime and accuracy.

References

1. Aytekin, A.M., Aytekin, T.: Real-time recommendation with locality sensitive hashing. J. Intell. Inf. Syst. **53**(1), 1–26 (2019)
2. Beleveslis, D.: Heuristic approach for content based recommendation system based on feature weighting and LSH. Ph.D. thesis, Int. Hellenic University (2020)
3. Dąbrowski, J., et al.: An efficient manifold density estimator for all recommendation systems. In: Mantoro, T., Lee, M., Ayu, M.A., Wong, K.W., Hidayanto, A.N. (eds.) ICONIP 2021. LNCS, vol. 13111, pp. 323–337. Springer, Cham (2021). https://doi.org/10.1007/978-3-030-92273-3_27
4. Fernández-Tobías, I., Blanco, R.: Memory-based recommendations of entities for web search users. In: CIKM 2016, pp. 35–44 (2016)
5. Guerraoui, R., Kermarrec, A.M., Niot, G., Ruas, O., Taïani, F.: GoldFinger: fast & approximate Jaccard for efficient KNN graph constructions. IEEE Trans. Knowl. Data Eng. **35**(11), 11461–11475 (2022)
6. Indyk, P., Motwani, R.: Approximate nearest neighbors: towards removing the curse of dimensionality. In: STOC 1998, pp. 604–613 (1998)
7. Jugovac, M., Jannach, D., Karimi, M.: StreamingRec: a framework for benchmarking stream-based news recommenders. In: RecSys 2018 (2018)
8. Karimi, M., Cule, B., Goethals, B.: On-the-fly news recommendation using sequential patterns. In: INRA @ RecSys 2019 (2019)
9. Karimi, M., Jannach, D., Jugovac, M.: News recommender systems-survey and roads ahead. Inf. Process. Manag. **54**(6), 1203–1227 (2018)
10. Liu, C.L., Wu, X.W.: Fast recommendation on latent collaborative relations. Knowl.-Based Syst. **109**, 25–34 (2016)

11. Lommatzsch, A., et al.: CLEF 2017 NewsREEL overview: a stream-based recommender task for evaluation and education. In: CLEF 2017 (2017)
12. Shehzad, F., Jannach, D.: Performance comparison of session-based recommendation algorithms based on GNNs. In: Goharian, N., et al. (eds.) ECIR 2024. LNCS, vol. 14611, pp. 115–131. Springer, Cham (2024). https://doi.org/10.1007/978-3-031-56066-8_12

Who Will Evaluate the Evaluators? Exploring the Gen-IR User Simulation Space

Johannes Kiesel[1]( ), Marcel Gohsen[1], Nailia Mirzakhmedova[1], Matthias Hagen[2], and Benno Stein[1]

[1] Bauhaus-Universität Weimar, Bauhausstr. 9a, 99423 Weimar, Germany
{johannes.kiesel,marcel.gohsen,nailia.mirzakhmedova, benno.stein}@uni-weimar.de
[2] Friedrich-Schiller-Universität Jena, Ernst-Abbe-Platz 2, 07743 Jena, Germany
matthias.hagen@uni-jena.de

Abstract. The reliable and repeatable evaluation of interactive, conversational, or generative IR systems is an ongoing research topic in the field of retrieval evaluation. One proposed solution is to fully automate evaluation through simulated user behavior and automated relevance judgments. Still, simulation frameworks were technically quite complex and have not been widely adopted. Recently, however, easy access to large language models has drastically lowered the hurdles for both user behavior simulation and automated judgments. We therefore argue that it is high time to investigate how simulation-based evaluation setups should be evaluated themselves. In this position paper, we present GenIRSim, a flexible and easy-to-use simulation and evaluation framework for generative IR, and we explore GenIRSim's parameter space to identify open research questions on evaluating simulation-based evaluation setups.

Keywords: Conversational search · Generative IR · User simulation

1 Introduction

Generative retrieval systems (Gen-IR) typically return generated texts instead of existing documents [6] and often allow users to follow up on the responses in chat-like interfaces (e.g., You.com). Such conversational systems have been the focus of, for example, the TREC CAsT [10] and iKAT [2] tracks. Systems participating in these tracks are asked to continue a given fixed interaction sequence between a user and another system for one next step. This setup enables standard Cranfield-style evaluation but at the cost of neglecting that different system responses may lead to different plausible user interactions in an ongoing conversation. An often propagated alternative is to evaluate a retrieval system against simulated user interactions [3]. Human relevance judgments for specific interaction sequences then may not be reusable for other sequences, but automated judgments could be a way out as they correlate with human ones [5,12].

Still, despite the IR community's interest in such fully automated evaluation, it had remained more of a theoretical idea. Reasons for this could be that setting up and running user simulations was perceived as quite complex and that it was not clear how the quality of the simulated behavior can and should be evaluated. In this position paper, we aim to support a more practically-oriented discussion. We contribute the flexible and easy to deploy simulation and evaluation framework GenIRSim (Section 2), and we explore GenIRSim's parameter space to highlight open research questions on evaluating simulation-based evaluation setups (Sect. 3). Our collected questions show that user simulation offers many new research opportunities on the evaluation of retrieval systems.

2 Automating Interactive Gen-IR System Evaluation

To showcase how Gen-IR systems can be easily evaluated in a fully automated way in a shared task or in a research project, we present the new open source framework GenIRSim.[1] The framework just requires the specification of a set of topics—just like in Cranfield-style evaluation—, and of user model configurations. Simulations are then run and evaluated by the framework for each combination of user model, topic, and to-be-tested retrieval system. Instead of human judgments, automatically aggregated evaluation scores can be used to compare the retrieval systems. GenIRSim's main features are:

Command Line and Web Interface. GenIRSim can be used with the same configuration files both in a web browser (to refine, test, and demonstrate configurations, cf. Fig. 1) and from the command line (for batch system evaluation and continuous development). Both interfaces produce the same output.

File-Based Configuration and Quick Deployment. Every aspect of a simulation and evaluation can be configured through configuration files in JSON format (example excerpt shown in the 'Configuration' part of Fig. 1). For starters, GenIRSim's README file describes how to create a Docker-based setup in just a few minutes to evaluate a basic Gen-IR system consisting of an open language model and an Elasticsearch index of Wikipedia. Different Gen-IR configurations can be tested by simply changing the Elasticsearch query or the result synthesis prompt in the configuration file. In principle, GenIRSim can used without GPUs, but GPU usage can drastically reduce run time.

Interlinked Simulation, Search, and Evaluation. The search or simulation outputs can be enriched and then easily used for evaluation in a flexible manner. For example, our default user simulator prompts the language model to generate (as JSON-formatted output) both the user utterance and an abstract description of what a user would expect a good response to contain. The user utterance and/or expectation can then be inserted into the prompt for evaluation showcased in the configuration excerpt in Fig. 1 for the `expectation` that is inserted via a template parameter `{{variables.userTurn...}}` into the prompt to determine the `ExpectationMatch` of a response.

---

[1] https://github.com/webis-de/GenIRSim.

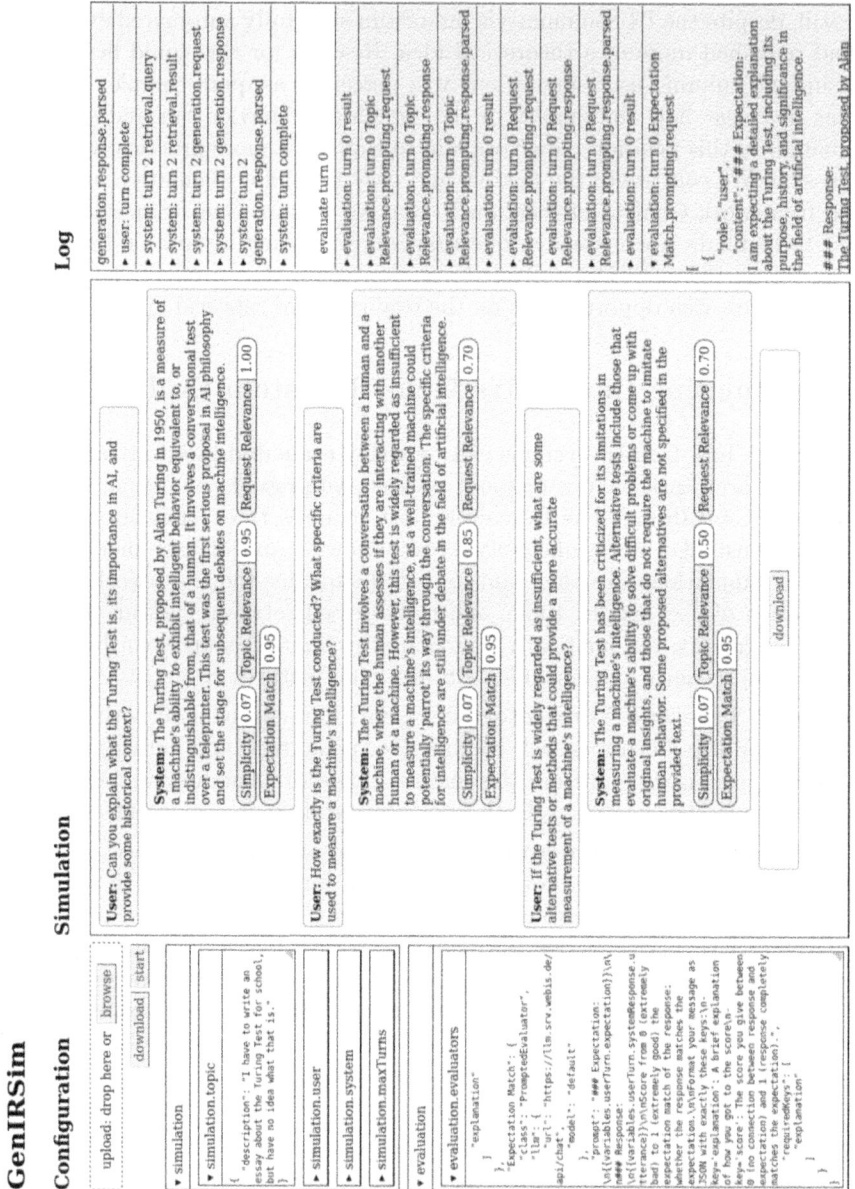

Fig. 1. Screenshot of GenIRSim's three-part web interface after a simulation and evaluation run: (1) the 'Configuration' part allows to load, inspect, edit, and download the configuration; (2) the 'Simulation' part shows the created user–system interactions, including automated judgments in badges for the system's responses; and (3) the 'Log' part shows messages including those exchanged with the language model and the search servers. In the screenshot, the configuration part shows the settings for the Expectation Match evaluator, including a prompt with template variables {{...}}, and the log part shows the start of the prompt for evaluating the model's response to the first turn.

Flexibility and Extensibility. Like SimIIR 2.0 [15], GenIRSim simulates user behavior to evaluate retrieval systems. However, while SimIIR 2.0 focuses on models for traditional list-based result pages, GenIRSim allows any user models that provide utterances and follow-up utterances on a topic / information need. Furthermore, the current Elasticsearch-based setup can easily be replaced by more sophisticated frameworks for creating conversational Gen-IR systems, like Macaw [14] or DECAF [1], as long as they have an API to interact with.

3 Exploring the Gen-IR Simulation and Evaluation Space

GenIRSim is designed for flexibility, making only few assumptions about the user simulation, the Gen-IR system, and the evaluation component. In particular, the user simulation and the Gen-IR system are just required to generate utterances when provided with either an utterance from the other party or with a topic at the start of the simulation, and the evaluation just needs to return a numeric score for each system turn based on the simulation and meta-information. However, this flexible design raises several questions about the best way to simulate and evaluate interactions—and what "best" means in this context. In the following, we outline such questions and highlight potential areas for future research.

User Information and Knowledge. TREC iKAT [2] used a personal text knowledge base to represents a user's personal information and knowledge as a list of short statements (e.g., "I am vegetarian," "I like Lord of the Rings," or "I know everything about rocket science"). Integrating such statements into a language model prompt for user simulation is easy and is done for our default user. However, if the statements are interconnected, knowledge graphs might offer a better presentation [3]. If so, how should the simulator employ relations from such graphs? How can relations be pre-filtered in case the graphs are large and detailed, as opposed to the short abstract statements above?

User Selection. How diverse should the simulated users be in terms of cultural, economical, and social background? Which age groups should be represented? What about minorities? Is it problematic if language models represent stereotypical users? Should user groups be selected based on abstract attributes or even be sampled along certain dimensions (e.g.,'curious', 'naive/asking simple questions', 'extroverted: 4 out of 5')? How can it be ensured that the language models faithfully simulate such users [7]?

Multilingualism. Many state-of-the-art large language models are actually multilingual, which opens up the possibility of multilingual retrieval experiments. For example, the open Llama3 model generates sound French answers when prompted "Why is the sky blue? Answer in French." This raises the question: can we simulate users that interact with a system in languages other than English, even if the indexed dataset contains only English documents?

User Model Updates. One way to model the past conversation is to just fill the language model's context window with the chat history. Alternatively, a user state in form of a TREC iKAT statement list or in form of a knowledge graph can be updated over the course of a simulation; for example, by extracting and incorporating structured knowledge from messages as RDF triples [4]. Can one also incorporate meta-information [8] in this way? Should models also forget [3]?

Evaluation Aspects. As for a system response's quality, there are many different interpretations of relevance (with respect to the topic, the current query, the expectations behind the current query, etc.) and also many other proposed measures. For example, Sakai [11] proposed 21 measures related to correctness, ethical behavior, personalization, and user satisfaction, while Gienapp et al. [6] integrated 10 measures of response utility in their suggested evaluation model for ad hoc Gen-IR. In pilot experiments, we found that prompting language models within GenIRSim provides for a quick way to implement different measures. But for which measures are language models reliable? Should language models be used for evaluation at all, given that they are black boxes [11]? Moreover, some measures are not applicable for some turns or tasks [11]. For example, measuring "correctness" is not that applicable if the user asks for counterfactual reasoning. How can measures be selected and weighted?

"Thought" Processes. To the best of our knowledge, our Expectation Match measure showcased in the example in Fig. 1 is the first to utilize meta-information from a user's "thought" process for system response evaluation. Traditionally, system responses are judged by trained human assessors and not the actual users. While human assessors have no access to a user's thoughts, chain-of-thought prompting [13] can be used to access the "thoughts" of simulated users. Can this approach also be used to quantify information scent [9]? For example, a language model could be prompted to first "think" about different available actions (e.g., different next utterances), evaluate them internally based on expected gained information, and then choose the action with the highest expected gain. Moreover, can we use expectations to measure serendipity of results? Serendipitous results would have a low Expectation Match, but a high match to a user's interest. In our view, the use of a simulated user's "thoughts" in evaluation is an especially interesting avenue for research as in reality the human users of a retrieval system also re ideal candidates to judge whether the system performed well in the user's sessions or in specific turns.

4 Conclusion

In this paper, we have argued that simulation-based evaluation systems are now easy to set up, showcased by our new GenIRSim framework, and that it is time to investigate how simulation-based evaluation systems themselves should be evaluated. In this regard, we have identified open research questions in six directions. However, we do not necessarily see the open questions as obstacles to using simulation techniques in IR today. Instead, the questions should rather be seen as an inspiration and as opportunities for future IR evaluation research.

Acknowledgements. This work was partially supported by the European Commission under grant agreement GA 101070014 (https://openwebsearch.eu) and by the Thüringer Ministerium für Wirtschaft, Wissenschaft und Digitale Gesellschaft (TMWWDG) under grant agreement 5575/10-5 (MetaReal).

References

1. Alessio, M., Faggioli, G., Ferro, N.: DECAF: a modular and extensible conversational search framework. In: Proceedings of SIGIR 2023, pp. 3075–3085 (2023). https://doi.org/10.1145/3539618.3591913
2. Aliannejadi, M., Abbasiantaeb, Z., Chatterjee, S., Dalton, J., Azzopardi, L.: TREC iKAT 2023: the interactive knowledge assistance track overview. In: Proceedings of TREC 2023 (2023). https://trec.nist.gov/pubs/trec32/papers/Overview_ikat.pdf
3. Balog, K.: Conversational AI from an information retrieval perspective: remaining challenges and a case for user simulation. In: Proceedings of DESIRES 2021, pp. 80–90 (2021). https://ceur-ws.org/Vol-2950/paper-03.pdf
4. Carta, S., Giuliani, A., Piano, L., Podda, A.S., Pompianu, L., Tiddia, S.G.: Iterative zero-shot llm prompting for knowledge graph construction. arXiv 2307.01128 (2023). https://doi.org/10.48550/ARXIV.2307.01128
5. Faggioli, G., et al.: Perspectives on large language models for relevance judgment. In: Proceedings of ICTIR 2023, pp. 39–50 (2023). https://doi.org/10.1145/3578337.3605136
6. Gienapp, L., et al.: Evaluating generative Ad Hoc information retrieval. In: Proceedings of SIGIR 2024 (2024)
7. Kiesel, J., Gohsen, M., Mirzakhmedova, N., Hagen, M., Stein, B.: Simulating follow-up questions in conversational search. In: Proceedings of ECIR 2024, pp. 382–398 (2024). https://doi.org/10.1007/978-3-031-56060-6_25
8. Kiesel, J., Meyer, L., Potthast, M., Stein, B.: Meta-information in conversational search. Trans. Inf. Sys. (TOIS) **39**(4) (2021). https://doi.org/10.1145/3468868
9. Maxwell, D., Azzopardi, L.: Information scent, searching and stopping - modelling SERP level stopping behaviour. In: Proceedings of ECIR 2018, pp. 210–222 (2018). https://doi.org/10.1007/978-3-319-76941-7_16
10. Owoicho, P., Dalton, J., Aliannejadi, M., Azzopardi, L., Trippas, J.R., Vakulenko, S.: TREC CAsT 2022: going beyond user ask and system retrieve with initiative and response generation. In: Proceedings of TREC 2022 (2022). https://trec.nist.gov/pubs/trec31/papers/Overview_cast.pdf
11. Sakai, T.: SWAN: a generic framework for auditing textual conversational systems. arXiv 2305.08290 (2023). https://doi.org/10.48550/ARXIV.2305.08290
12. Thomas, P., Spielman, S., Craswell, N., Mitra, B.: Large language models can accurately predict searcher preferences. arXiv 2309.10621 (2023). https://doi.org/10.48550/ARXIV.2309.10621
13. Wei, J., et al.: Chain-of-thought prompting elicits reasoning in large language models. In: Proceedings of NeurIPS 2022 (2022). http://papers.nips.cc/paper_files/paper/2022/hash/9d5609613524ecf4f15af0f7b31abca4-Abstract-Conference.html
14. Zamani, H., Craswell, N.: Macaw: an extensible conversational information seeking platform. In: Proceedings of SIGIR 2020, pp. 2193–2196 (2020). https://doi.org/10.1145/3397271.3401415
15. Zerhoudi, S., et al.: The SimIIR 2.0 framework: user types, Markov model-based interaction simulation, and advanced query generation. In: Proceedings of CIKM 2022, pp. 4661–4666 (2022). https://doi.org/10.1145/3511808.3557711

De-noising Document Classification Benchmarks via Prompt-Based Rank Pruning: A Case Study

Matti Wiegmann[1]([✉])[iD], Benno Stein[1][iD], and Martin Potthast[2][iD]

[1] Bauhaus-Universität Weimar, Weimar, Germany
matti.wiegmann@uni-weimar.de
[2] University of Kassel, hessian.AI, and ScaDS.AI, Kassel, Germany

Abstract. Model selection is based on effectiveness experiments, which in turn are based on benchmark datasets. Benchmarks for "complex" classification tasks, such as tasks with a high subjectivity, are prone to label noise in their (manual) annotations. For such tasks, experiments on a given benchmark may therefore not reflect the actual effectiveness of a model. To address this issue, we propose a three-step de-noising strategy: Given labeled documents from a complex classification task, use large language models to estimate "how strong the signal within a document is in the direction of its class label", rank all documents according to their estimated signal strengths, and omit documents below a certain threshold. We evaluate this strategy in a case study on the assignment of trigger warnings to long fan fiction texts. Our analysis reveals that the documents retained in the benchmark contain a higher proportion of reliable labels, and that model effectiveness assessments are more meaningful and models become easier to distinguish (Code and Data: https://github.com/webis-de/CLEF-24).

1 Introduction

There are text classification tasks for which providing a sufficient amount of labeled data is difficult. The difficulty may be due to the subjectivity of the task (Is this text a product *description* or a product *advertisement*?), a high number of classes (Which of the 188 cognitive biases occur in this text?), a missing dichotomy since only one class can be characterized (Does this text has an enticing writing style?), the need for expert knowledge (Is argument A more convincing than argument B?), or a combination of these characteristics. For such tasks, LLMs have shown great performance, even in zero-shot settings.

But, just as powerful as LLMs are in this respect, they are obviously not a panacea: Time, cost, and latency are among their main limiting factors, especially for classification tasks that require ad hoc decisions and high throughput. Consider, for example, the generation of a search engine result page (SERP) on which documents containing product advertising, undesirable prejudices, or sarcasm are to be filtered out. The practical and efficient approaches, instead, fine-tune neural networks based on dense document representations, such as

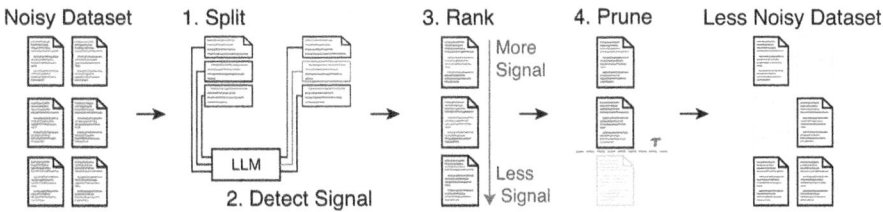

Fig. 1. Overview of the proposed method of pruning documents with a label depending on how strong the signal for this label is according to an LLM classifier.

BERT or RoBERTa [8]. Their limiting factor, however, is the knowledge acquisition bottleneck, i.e. the lack or the quality of labeled data. This lack of labeled data is often countered by collecting data from weakly-supervised sources. One example of this is the extraction of trigger warnings from online blogs, where authors signal if their work contains harmful content.

However, weakly-supervised data acquisition leads to noisy data due to errors or inconsistencies in the distant knowledge source. The use of noisy data to benchmark classification models (which is the focus of this paper) is problematic: model performances may be underestimated, model differences may be smaller or vanish, or, in the worst case, leaderboard rankings change. Or the other way around: reducing label noise in benchmark data increases model scores and may increase the performance difference between models, which makes it easier to assess which model is actually better and by how much.

This is where our contribution comes in: The paper in hand proposes prompt-based text classification to reduce label noise, especially false positives, in difficult document classification benchmarks (i.e. test datasets) (cf. Fig. 1). We use the LLMs to detect how much signal is present in each document to justify the label assigned to it, and we remove the documents with the weakest signal (Sect. 3). We evaluate our method using three common models (XGBoost, RoBERTa, Longformer) on a multi-label trigger detection dataset [16] (as used in a joint task on CLEF 2023 [15]), which provides some organic information about label reliability (Sect. 4).

Our results (Sect. 5) show that our method increases the ratio of noisy to reliable documents in the benchmark from 1:1 up to 1:6, that models tested on de-noised data score up to 0.15 F_1 higher than when tested on "noisy" documents, and that models may scores the same on noisy data but significantly different on de-noised dataset.

2 Related Work

Although current (pre-trained) deep learning models are somewhat robust to label noise given sufficient training data [13,18], reducing label noise is still essential when training non-neural models [3,9] or with limited training data. Most related work focuses on training data de-noising neural classifiers [6,17], especially with semi-supervised methods like adapting the loss function [11,14],

Table 1. Number and length of eligible source and sampled evaluation documents.

Warning	Source Data		Sample used in this Work			Length	
	Unknown	Reliable	Unknown	Flipped	Reliable	Mean	Std
Death	124,958	1,579	600	200	200	3,351	2,717
Violence	119,684	1,736	600	200	200	4,021	2,853
Homophobia	22,688	558	600	200	200	4,125	2,809
Self-harm	23,029	1,343	600	200	200	3,478	2,688

by over-parameterization [7], or by rank pruning [10] via predicted probabilities. Some related works also use weak supervision methods to estimate label reliability [5,12] from (multiple) external sources. For our work, we adapt the rank pruning idea but use an external source (an LLM) instead of a semi-supervised signal. However, the most notable difference of our work is that we do not focus on de-noising the training data to improve the model but the test data to improve the benchmark reliability, which is why we study organic noise instead of only injecting synthetic noise like the related work (e.g., on TREC question-type and AG-News datasets [4]).

3 Finding and Pruning Noisy Documents

Our label de-noising procedure assumes the following: First, the input dataset contains a set of documents, and each document has one or more labels from a finite set. Second, each reliable document with a true positive label contains a signal above a confidence threshold τ (i.e., a piece of set) that justifies the label. Third, there are a number of noisy documents that have been assigned a positive label where the signal with respect to that label is weaker than τ. Our pruning strategy, illustrated in Fig. 1, attempts to find and remove documents that are noisy with respect to a particular label by determining the signal strength of that label.

To do this, we rank all documents independently for each label according to the strength of the signal of this label and then determine τ as a threshold. The de-noising scheme consists of four steps for each label: (1) Splitting of documents into smaller chunks, i.e. several consecutive sentences, where the chunk size is a hyperparameter. (2) Determine whether a chunk carries a signal for the label using a prompt-based binary classification, where LLM and prompt are hyperparameters that depend on the task and the label. (3) Ranking of the documents on the basis of the absolute number of signals, i.e. the positively classified chunks. (4) Pruning of the documents with the lowest rank up to a rank or signal strength threshold τ.

4 Experimental Evaluation

We evaluate LLM-based benchmark de-noising on a multi-label classification task and evaluate the noise ratio and model effectiveness at different τ.

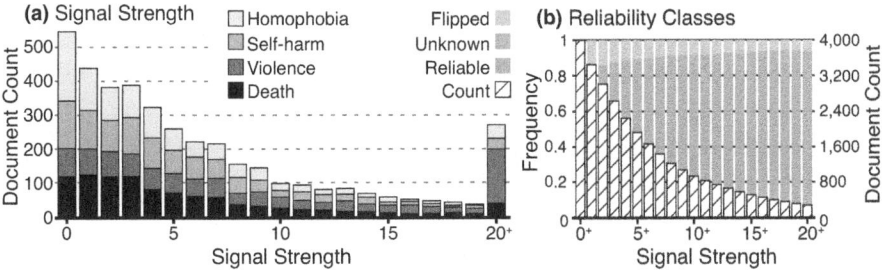

Fig. 2. (a) Signal strength distribution: Number of documents with a certain amount of positively classified 5-sentence chunks by label. (b) Number of documents and distribution of document reliability in the pruned corpus at different thresholds.

Dataset. We use evaluation data from the Webis Trigger Warning Corpus (WTWC) [16], which was used in the 2023 shared task on trigger detection [15]. The WTWC is well suited as it contains organic false positive and negative labels that emerge from human authors (sensitive human authors assign warnings for weak signals) and from weakly supervised labeling (which assigns warnings for loosely related or implied concepts). The dataset also contains additional reliability information in the "author notes" prepended to some chapters.

We sample 4,000 WTWC documents balanced across 4 warning labels *Death*, *Violence* (the two most common warnings, excluding Pornography as outlier), *Homophobia* and *Self-harm* (the two closest to median frequency with sufficient *Reliable* documents) as our evaluation dataset (cf. Table 1), which is large enough to test our method. For each label, we first sample 200 *Reliable* documents where the author note mentions either tw, cw, trigger(s), content warning within 20 tokens of a warning term (e.g. homophobia). Then, we sample 800 non-*Reliable* documents and create a subset of 200 known falsely labeled data by *Flipping* the documents' label to a different one. The reliability of the remaining 600 documents was marked as *Unknown*. We adopted all other sampling criteria from the shared task [15] (English documents; 50–10,000 words; no duplicates).

4.1 De-noising Implementation

We apply our de-noising technique (Sect. 3) using 5 consecutive sentences without overlap as chunks and *Mixtral-8x7B-v0.1* from Huggingface as LLM. We use a binary classification prompt aligned with Mistral's prompting guide:

> You are a text classification model. You determine if a given text contains death, graphic display of death, murder, or dying characters. If the given text contains intense, explicit, and graphic death, you answer: Yes. If the text contains mild or implicit death or no death at all, you answer: No.

We classify by predicting the next-token probabilities and comparing the logits of the Yes and No tokens. We rank and prune the documents by the

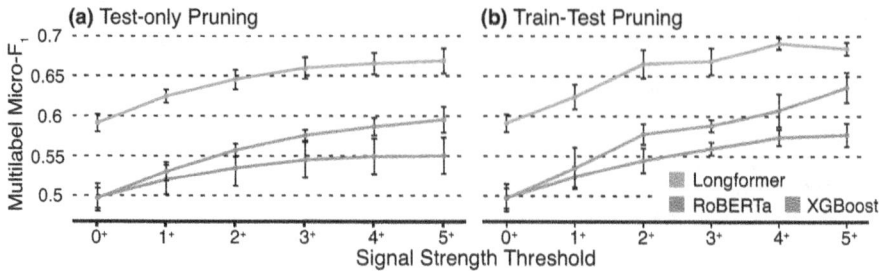

Fig. 3. Model F_1 with confidence intervals of three classification models at different pruning thresholds when (a) only test data and (b) training and test data are pruned.

absolute number of positively (Yes > No) classified chunks per document, i.e. at a τ of 5^+ all documents with less than 5 positive chunks will be pruned.

4.2 Experiments and Evaluation

To evaluate our hypotheses we conduct three experiments across three baseline classification models. First, we prune the complete dataset (with τ from 0^+ to 20^+) and observe the ratio of reliability classes. Second, we split the data 80:20 into training and test and only prune the test dataset (with τ from 0^+ to 5^+)[1] while training the models on the complete training data. Third, we prune the complete dataset (with τ from 0^+ to 5^+) before the train-test split and also train the models with pruned data. Decreasing scores in this last experiment would indicate that our method also removes (many) difficult cases, leading to both, poor models and a poor benchmark.

We train three models for multi-label classification: a fine-tuned FacebookAI/roberta-base and allenai/longformer-base-4096 [1] and a feature-based XGBoost [2] classifier (the baseline of the shared task [15]) with the top 10,000 tf · idf word 1–3-gram features selected via χ^2. The RoBERTa input was truncated to 512 tokens and the Longformer input to 4,096 tokens. We report the micro-averaged multi-label F_1 via a 5-fold Monte Carlo cross-validation and the 95% t-estimated confidence intervals. Our code repository lists training parameters and our ablation study.

5 Results and Discussion

Our first assumption is that our method removes noise from the dataset if, with increasing τ, the proportion of *Reliable* documents increases and of *Flipped* documents decreases. Figure 2(b) shows that the proportion of *Reliable* documents increases from 0.2 to 0.41 and decreases for *Flipped* documents from 0.2 to 0.05. Note that the proportion changes are strongest for smaller τ.

---

[1] At $\tau = 5^+$, half the dataset has been pruned.

Our second assumption is that de-noising improves the benchmark when the models' test scores increase with increased de-noising (for train-test and test-only pruning) and when the relative difference between models' test scores changes. Figure 3(a) shows that the F_1 of all models increases by 0.05–0.1 with $\tau = 5^+$ when pruning only the test data. The effect is strongest for XGBoost and weakest for RoBERTa (where the input documents are strongly truncated). Figure 3(a) also shows that XGBoost and RoBERTa score evenly without pruning but XGBoost improves more strongly and is significantly more effective with $\tau = 5^+$. This shows that de-noising can reveal model differences that are otherwise hidden by the noise. Figure 3(b) shows that the F_1 of all models increases when pruning all data and more strongly than when only pruning the test data.

6   Conclusion

In this paper, we investigate using rank-based pruning based on an LLMs classification signal to de-noise a document-level trigger warning classification dataset. We show that our de-noising strategy doubles the relative number of reliably labeled documents and halves the noisily labeled ones. We further show that our de-noising strategy increases the model scores and the differences between models, hence we assume that the de-noised dataset is more suited as a benchmark.

References

1. Beltagy, I., Peters, M., Cohan, A.: LongFormer: the long-document transformer. CoRR (2020)
2. Chen, T., Guestrin, C.: XGBoost: a scalable tree boosting system. In: 22nd SIGKDD (2016)
3. Frénay, B., Verleysen, M.: Classification in the presence of label noise: a survey. IEEE Trans. Neural Netw. Learn. Syst. **25**, 845–869 (2014)
4. Garg, S., Ramakrishnan, G., Thumbe, V.: Towards robustness to label noise in text classification via noise modeling. In: 30th ACM CIKM (2021)
5. Ratner, A.J., de Sa, C.M., Wu, S., Selsam, D., Ré, C.: Data programming: creating large training sets, quickly. In: Advances in Neural Information Processing Systems, vol. 29, pp. 3567–3575 (2016)
6. Liu, S., Niles-Weed, J., Razavian, N., Fernandez-Granda, C.: Early-learning regularization prevents memorization of noisy labels. In: 33rd NeurIPS (2020)
7. Liu, S., Zhu, Z., Qu, Q., You, C.: Robust training under label noise by overparameterization. arXiv abs/2202.14026 (2022)
8. Liu, Y., et al.: RoBERTa: a robustly optimized BERT pretraining approach. CoRR (2019)
9. Natarajan, N., Dhillon, I., Ravikumar, P., Tewari, A.: Learning with noisy labels. In: Neural Information Processing Systems (2013)
10. Northcutt, C.G., Wu, T., Chuang, I.L.: Learning with confident examples: rank pruning for robust classification with noisy labels. arXiv (2017)
11. Patrini, G., Rozza, A., Menon, A.K., Nock, R., Qu, L.: Making deep neural networks robust to label noise: a loss correction approach. In: 2017 IEEE CVPR (2016)

12. Ren, W., Li, Y., Su, H., Kartchner, D., Mitchell, C.S., Zhang, C.: Denoising multi-source weak supervision for neural text classification. arXiv (2020)
13. Rolnick, D., Veit, A., Belongie, S.J., Shavit, N.: Deep learning is robust to massive label noise. arXiv abs/1705.10694 (2017)
14. Sanchez, E.A., Ortego, D., Albert, P., O'Connor, N.E., McGuinness, K.: Unsupervised label noise modeling and loss correction. arXiv abs/1904.11238 (2019)
15. Wiegmann, M., Wolska, M., Potthast, M., Stein, B.: Overview of the trigger detection task at PAN 2023. In: Working Notes of CLEF. CEUR-WS, vol. 3497, pp. 2523–2536 (2023)
16. Wiegmann, M., Wolska, M., Schröder, C., Borchardt, O., Stein, B., Potthast, M.: Trigger warning assignment as a multi-label document classification problem. In: 61th ACL (2023)
17. Zhang, Z., Zhang, H., Arik, S.Ö., Lee, H., Pfister, T.: Distilling effective supervision from severe label noise. In: 2020 IEEE/CVF CVPR (2019)
18. Zhu, D., Hedderich, M.A., Zhai, F., Adelani, D.I., Klakow, D.: Is BERT robust to label noise? A study on learning with noisy labels in text classification. arXiv abs/2204.09371 (2022)

Best of CLEF 2023 Labs

Best of Touché 2023 Task 4: Testing Data Augmentation and Label Propagation for Multilingual Multi-target Stance Detection

Jorge Avila[✉], Álvaro Rodrigo[iD], and Roberto Centeno[✉][iD]

NLP & IR Group at UNED, Madrid, Spain
`javila149@alumno.uned.es`, {`alvarory,rcenteno`}`@lsi.uned.es`

Abstract. Touché 2023 task 4 evaluated stance detection in a Multilingual multi-target setting with a reduced annotated dataset. This is why we have tested different approaches focused on increasing training data by (1) including new samples from back-translating original training data and (2) adding samples from unlabeled data using label propagation. The results showed that back-translation was successful, while label propagation worsened the performance. We obtained the best results with a transformer-based model fine-tuned in two steps: the first on a related, more extensive dataset and the second on the development data. This shows the usefulness of including related data in our approach and suggests additional research based on taking advantage of other datasets and data augmentation. Besides, given that the current results were close to a 0.35 f1 score, there is still room for improvement in this task.

Keywords: Data augmentation · Label propagation · Multilingual Stance detection

1 Introduction

The task of stance detection consists of the automatic detection of the attitude, whether it is support, opposition, or neutrality, of a text towards a specific proposal [19]. This task plays a crucial role in understanding the opinion or attitude of individuals towards specific proposals and thus provides valuable information for activities such as misinformation detection [11].

The Touché Lab at CLEF 2023 proposed a series of shared tasks focused on computational argumentation and causality [5]. This paper focuses on our participation in task 4: Intra-Multilingual Multi-Target Stance Classification. The objective of this task was to classify comments on socially relevant topics that have been written on the Conference on the Future of Europe (CoFE)[1] platform. CoFE is an online platform where any user can write a proposal in

---

[1] https://futureu.europa.eu/?locale=en.

any of the 24 official languages of the EU. Other users can comment on and endorse a proposal or another comment. The task is to classify whether these comments are in favor, against, or neutral toward the proposal. The proposals, titles, and comments can be written in any of the 24 languages of the European Union.

Early tasks on stance detection, like those at SemEval-2016 Task 6 [19], only provided texts in a single language, usually English. Recently, new initiatives have proposed stance detection in other languages and included additional data. For example, SardiStance@EVALITA2020 proposes to detect stance about the Sardines movement[2] in Italian tweets, including contextual information related to users [6]. Afterward, VaxxStance@IberLEF 2021 launched a shared task in Spanish and Basque for detecting stance towards vaccines [1], including information related to the social network. These two tasks showed the importance of considering users' information when detecting the stance of a given text. However, all these tasks focused on monolingual stance detection about single topics. Thus, Touché, where comments are written in different languages, represents a real challenge given the multilingual and multi-topic nature of the data [3,4].

The data provided in the development period is mainly divided into three subsets:

- **CF_S** contains 7000 comments annotated using only two classes (favor or against).
- **CF_U** contains 12000 unlabeled comments.
- **CF_E-Dev** contains 1400 multilingual comments annotated with three classes.

One of the most significant challenges of this task is the small size of the 3-class subset, which adheres to the schema required for annotating the test set. This constraint led us to explore different alternatives to leverage the information from the other two subsets, which were unlabeled or labeled using only two classes. Our primary goal was to expand the data used for training our models. To achieve this, we primarily relied on data augmentation and label propagation techniques to create additional training data for our models.

2 Previous Work

2.1 Stance Detection

The early works on stance detection were conducted on various text genres, including political debates, online debate forums, student essays, or tweets [2,9]. These works used both traditional approaches based on support vector machines (SVM) [15] and logistic regression [6], as well as deep learning approaches based on recurrent neural networks [8] or convolutional neural networks [29]. However, since the advent of transformer-based models [27], most approaches utilize these models [14].

---

[2] https://en.wikipedia.org/wiki/Sardines_movement.

Despite considerable progress, the need for labeled training data is one of the main challenges due to the complex and tedious annotation work. Significant efforts have been made to create annotated datasets for stance detection. Regarding the languages of the annotated datasets, we can find English [18], Catalan [23], Chinese [30], Czech [12], Italian [17] or Spanish [23]. In addition to these language-specific datasets, multilingual annotated datasets are also being compiled [31]. In the context of this work, the Intra-Multilingual Multi-Target Stance Classification 2023 task proposes a new partially annotated multilingual dataset [4].

Other works have attempted to address the stance detection task from a multilingual approach. For example, we find tasks with datasets in Spanish and Catalan [24], or French and Italian [16]. Due to the difficulty in constructing these datasets in different languages, they tend to present certain limitations, such as unbalanced classes or contexts between languages, which hinders the generalization of the proposed models.

2.2 Data Augmentation

Data augmentation is a technique used to increase the quantity and diversity of training data. It involves applying transformations or modifications to existing data to generate new instances that are different but still contain the same information or labels [26]. The main goal of data augmentation is to improve the performance and generalization of models by providing them with more training samples.

Some of the most common techniques for performing data augmentation are:

- Using synonyms to replace some words and generate variants of the original text.
- Adding or deleting words at random points in the original text increases variability and forces the model to rely on other contextual cues to understand the text's meaning.
- Using machine translation to generate variants of the original text in different languages.

For example, translation-based data augmentation has been used to expand collections of tweets and improve sentiment analysis results on tweets written in languages other than English [3]. Another translation-based approach is the so-called back-translation [22]. This method is based on generating variants of the original text in different languages and then translating them back to the original language to enrich the original dataset.

2.3 Label Propagation

Label propagation is a semi-supervised machine-learning technique that can be used to propagate known labels onto unlabeled data [13]. The main objective is to utilize the information available in labeled data to assign labels to unlabeled data [28]. Label propagation is particularly useful in situations where there is a limited set of labeled data but a large amount of unlabeled data available.

This technique has been applied in previous works related to stance detection. For example, [10] used a combination of label spreading and domain adaptation to adjust the labels of 16 unsupervised datasets. Another related use case is label propagation in community detection tasks where individuals are grouped into nodes, and the propagation is carried out considering some measure of similarity between the nodes [20]. Although it does not directly address stance detection, the technique used can be adapted for stance detection tasks, where similar texts may hold the same stance.

3 Proposal

Our approach is based on training different models using additional data beyond that provided by the organizers. Our goal is to study the impact of using such additional data in different combinations. In this section, we first describe the preprocessing applied to the texts and then describe the techniques used to create additional data. Finally, we include the models that have been considered in our work.

3.1 Data Preprocessing

The textual information has been divided into two text strings. On the one hand, the comment that appears in the supplied data. On the other hand, a string has been constructed with the structure: "This is a comment about title in relation to topic." The different topics have been slightly changed in the way they were written to reflect the context of the topic in more detail (see Table 1). For example, the topic "ValuesRights" is considered better expressed as "values and rights". Each model receives these two concatenated text strings as input.

Table 1. Conversion from old to new topics

Original topics	New topics
Migration	migration
GreenDeal	sustainable development
Health	health
Economy	economy
EUInTheWorld	europe in the world
ValuesRights	values and rights
Digital	digital
Democracy	democracy
Education	education
OtherIdeas	other topics

On the other hand, to use monolingual English models with all possible data, comments that were in other languages have been translated into English using the deep-translator library[3] with the Google Translate API.

3.2 Data Augmentation

In this work, we have applied a data augmentation method based on back translation. This process introduces minor variations to the texts that do not affect the meaning while maintaining the original language. In this way, we expanded the training set while maintaining the original proportion of the different languages.

More in detail, we have worked with the CF_E-Dev dataset, which contains comments in English, French, Spanish, and German. The strategy was to translate each comment to the other three languages and then back to the original language. Therefore, we have three new versions for each comment. We have focused on this dataset because it is the only one that contains comments annotated using three classes. The translation was made using the model Helsinki-NLP/opus-mt-ine-ine[4]. As a result, the size of the training collection has been multiplied by four, increasing from 1131 to 4524 comments.

We wanted to explore this approach because, although it is better to expand the training data with entirely different messages, this is associated with a high annotation cost. By using data augmentation, we can automatically produce new data that is slightly different from the original one.

3.3 Label Propagation

We have used label spreading to propagate the labels from the development set to the unlabeled set. Thus, we increased the number of labeled comments from 1400 to over 12000, an increase close to an order of magnitude.

To perform this label spreading, we have firstly represented all the comments from the labeled and unlabeled subsets (respectively CF_E-Dev and CF_U subsets) into an embedding space. For this purpose, we have used the *paraphrase-multilingual-mpnet-base-v2*[5] model to represent the comments. This model is trained in over 50 languages, making it suitable for our multilingual data.

Then, we have applied the LabelSpreading[6] algorithm from scikit-learn. This algorithm builds a similarity matrix that includes regularization, which is more robust to noise.

3.4 Systems

Ten different systems were tested during the development period, from which the runs finally submitted were selected. The main difference between several of

---

[3] https://pypi.org/project/deep-translator/.
[4] https://huggingface.co/Helsinki-NLP/opus-mt-ine-en.
[5] https://huggingface.co/sentence-transformers/paraphrase-multilingual-mpnet-base-v2.
[6] https://scikit-learn.org/stable/modules/generated/sklearn.semi_supervised.LabelSpreading.html.

the models is the language of the input data. When all comments were in English (using both the original English comments and the translated ones), monolingual BERT-based models were used. With input data in different languages, multilingual models were used[7].

The systems tested were:

- System 1. A Roberta-base model[8] trained with the development set, translating comments from other languages to English.
- System 2. An XLM-Roberta-large model[9] trained with the original development set in different languages.
- System 3. A Roberta-base model trained with the CFS dataset (2 labels) translated into English.
- System 4. An XLM-Roberta-large model trained with the CFS dataset (2 labels).
- System 5. A Bert-base-uncased model[10] trained in two stages. In the first stage, it is trained with the dataset containing two labels in English (CFS), which is more extensive, and in the second stage, with the development dataset, containing three labels, translated into English.
- System 6. A Bert-base-multilingual-uncased model trained in two stages. The process is similar to System 5, except that, in this case, the original language datasets are used.
- System 7. A Roberta-base model trained with the CFU dataset labeled using Label Spreading and translated into English.
- System 8. An XLM-Roberta-large model trained with the CFU dataset labeled using Label Spreading.
- System 9. An XLM-Roberta-large model trained using the augmented CFE-D dataset through data augmentation.
- Ensemble System. An XGBoost model that receives the outputs of systems 1, 2, 3, 4, 5, and 6, as well as the structured information from the comments (positive and negative votes, endorsements, etc.). In more detail, the input data for this model consists of the probabilities assigned to each class by systems 1 to 6 for the development set. Additionally, the following information is added:
  - Number of positive votes
  - Number of negative votes
  - Endorsements of the proposition
  - "Last comment in thread" as a boolean variable encoded with 0 and 1
  - Language-related variables, proposition language, and topic

Systems 1 and 2 are motivated to study the influence of using data in its original language with a multilingual model versus data translated into English

---

[7] The final model selected in each case was decided based on the results obtained during previous tests.
[8] https://huggingface.co/roberta-base.
[9] https://huggingface.co/xlm-roberta-large.
[10] https://huggingface.co/bert-base-uncased.

with a monolingual model. The motivation behind Systems 3 and 4 is similar, but in this case, with a larger dataset but fewer labels.

Systems 5 and 6 again compare the influence of using translated datasets. Additionally, they intend to study whether there is knowledge transfer between the collections of two and three labels during the learning process.

With Systems 7 and 8, the aim is to evaluate the impact of training with automatically labeled data using label spreading. System 9 is proposed to study training on synthetically generated data using back-translation. The results of this system will be compared, especially with system 2, which uses the same starting dataset but without the application of data augmentation.

The motivation behind the ensemble is the unification into a single model of the information learned by the different systems, along with the rest of the structured information contained in the collection. This system should learn from the strengths of the different models and complement them with the structured information.

4 Submitted Runs

We submitted six different runs to test different approaches using the TIRA platform[11]. We selected these runs based on previous experiments using cross-validation on the CF_E dataset provided in the development period.

Furthermore, we have specifically chosen runs that enable us to evaluate different approaches. This decision was made to ensure a comprehensive understanding of the performance of each approach, beyond the results obtained previously.

The runs submitted are:

- Run 1: the ensemble model (system ensemble in Sect. 3.4).
- Run 2: system 7 described in Sect. 3.4 (a RoBERTa base model trained on CF_E adding the CF_U dataset translated into English after applying the label propagation). With this run, we wanted to study the effect of label propagation for stance detection.
- Run 3: System 8 described in Sect. 3.4 (an XLM-RoBERTa large model trained on CF_E adding the CF_U dataset after applying label propagation). In this run, we wanted to study the influence of language on label propagation in the previous run, where we translated comments into English.
- Run 4: System 9 described in Sect. 3.4 (an XLM-RoBERTa large modeltrained on the CF_E dataset augmented using back-translation). This run aimed to study the effect of data augmentation on stance detection.
- Run 5: System 1 described in Sect. 3.4 (a RoBERTa base model using the CF_E subset translated into English). We consider this run as our baseline for comparing results with those using label propagation or data augmentation.

---

[11] https://www.tira.io/.

– Run 6: System 5 described in Sect. 3.4 (a BERT base model (uncased) fine-tuned in 2 steps). With this run, we wanted to test the effect of transferring learning from a task with a more extensive dataset annotated with two classes to stance detection using three classes

The complete list of hyperparameters used in our runs is given in Appendix A.

5 Analysis of Results

We show in Table 2 the results of our runs and the two baselines proposed by the organizers. The simple baseline is a dummy system that always returns the label "in favor". The other baseline was built using XML-R [7], a multilingual transformer-based system, trained in two steps: a first fine-tuning using two classes ("in favor" or "against") over the X-Stance [25] and the CFS datasets; and a second fine-tuning over the development collection using three stance classes ("in favor", "against", or "neutral"). Thus, this baseline was similar to our system 6 described in Sect. 3.4, which we did not send for evaluation because it obtained worse results than other systems in the development period. We also include results obtained by the other participant, named queen-of-swords [21]. Results are sorted by the official measure, macro-f1 score.

Table 2. Results of the submitted runs sorted by macro f1

Run	macro f1-score	accuracy
Touche23-baseline	0.593	0.673
Run 6	0.350	0.551
Run 4	0.329	0.537
Run 1	0.323	0.524
queen-of-swords	0.324	0.616
Run 5	0.27	0.463
Run 2	0.239	0.461
Touche23-baseline-simple	0.237	0.552
Run 3	0.177	0.216

We can see in Table 2 how all the runs ranked below the most complex baseline. Unfortunately, we have not been able to evaluate our system 6 using the test set[12], which was quite similar to this baseline. However, all our runs, except run 3, ranked above the simple baseline predicting every comment "in favor".

---

[12] It is important to note that the actual labels of the test set have never been made public and therefore, no checks or additional experiments could be performed.

The results of all our runs were under 0.4, showing that there is still room for improvement in this task. However, we obtained a score of 0.6691 with run 1 in our best experiments at the development period. So, we think our models could not generalize the training data correctly.

One possible cause of the models' poorer performance on the test set is that the training set only contains 4 different languages, while the test set consists of 6 languages, of which 3 do not appear in the training data. This should not be a problem for models using data translated into English, but it could affect models using untranslated comments.

Our best results are obtained by run 6, showing the importance of using additional data for fine-tuning the model, even if this data uses a different number of labels, as well as the usefulness of fine-tuning using two steps, as the baseline has demonstrated. Besides, the good results of run 4 also show the importance of including additional training data obtained for this run using back-translation. We also have similar results with run 1, which uses an ensemble of classifiers trained on different datasets. Hence, it seems pretty promising to use approaches based on generating additional training data. All these runs outperformed run 5, which is considered our baseline.

On the other hand, the results using label propagation on the unlabeled collection were not so successful, as we can see with the results of run 3, the only run worse than the baseline, and run 2. Both runs performed worse than run 5, which only uses the CF_E subset. Therefore, we need to do further research on properly using unlabeled data for this task.

The other participant team, queen-of-swords, trained a BERT model with the dataset containing three labels and used this model to label the unlabeled dataset. Then, they used a new BERT model trained with both datasets. The results of this approach, like ours based on label spreading, show the difficulty of leveraging the unlabeled collection compared to other techniques, such as data augmentation. Therefore, a possible future work could combine these techniques to try to overcome the results obtained. It is important to note that the other participant also reported experiencing significant differences between the results obtained in development and those obtained with the test set, possibly indicating that we have suffered from overfitting.

6 Conclusions and Future Work

Stance detection is widely used to understand the opinions or attitudes expressed in texts. The Touché Lab 2023 proposed a task for stance detection in a multilingual multi-target environment with a reduced set of labeled data.

In this paper, we have studied different approaches focused on adding training data to feed our models. In more detail, we have tested two approaches: (1) data augmentation using back-translation of the development set and (2) label propagation of the unlabeled data provided by the organizers. Our best systems were those using the augmented data generated using back-translation, outperforming a similar model only using the available labeled data. However, our runs that used the unlabeled data performed poorly.

Moreover, our best system was trained in two steps with a similar dataset, showing the importance of including training-related data when facing this task. Hence, additional training data seems to be important for this task, but we need to further research how to properly generate this kind of data and take advantage of other similar labeled data.

Acknowledgments. This work has been partially funded by the Spanish Research Agency (Agencia Estatal de Investigación), DeepInfo project PID2021-127777OB-C22 (MCIU/AEI/FEDER, UE) and the HOLISTIC ANALYSIS OF ORGANISED MISINFORMATION ACTIVITY IN SOCIAL NETWORKS project (PCI2022-135026-2).

A Hyperparameters

A.1 Run 1

- subsample: 0.5
- min_child_weight: 5
- max_depth: 7
- learning rate: 0.01
- colsample_bytree: 0.5

A.2 Runs 2, 3, 4, 5 and 6

(See Table 3).

Table 3. Hyperparameters for runs 2, 3, 4 and 5

	Run 2	Run 3	Run 4	Run 5	Run 6
Batch size	14	2	2	8	8
Leaning rate	1×10^{-5}	1×10^{-5}	1×10^{-5}	6×10^{-6}	1×10^{-5}
Weight decay	0.001	0.001	0.001	0.001	0.001
Epochs	6	5	8	6	3

References

1. Agerri, R., Centeno, R., Espinosa, M.S., de Landa, J.F., Rodrigo, Á.: VaxxStance@iberLEF 2021: overview of the task on going beyond text in cross-lingual stance detection. Proces. del Leng. Nat. **67**, 173–181 (2021)
2. Anand, P., Walker, M., Abbott, R., Fox Tree, J.E., Bowmani, R., Minor, M.: Cats rule and dogs drool!: Classifying stance in online debate. In: Proceedings of the 2nd Workshop on Computational Approaches to Subjectivity and Sentiment Analysis (WASSA 2011) (2011)
3. Barriere, V., Balahur, A.: Multilingual multi-target stance recognition in online public consultations. Mathematics **11**(9), 2161 (2023)

4. Barriere, V., Jacquet, G.G., Hemamou, L.: CoFE: a new dataset of intra-multilingual multi-target stance classification from an online European participatory democracy platform. In: Proceedings of the 2nd Conference of the Asia-Pacific Chapter of the Association for Computational Linguistics and the 12th International Joint Conference on Natural Language Processing, pp. 418–422 (2022)
5. Bondarenko, A., et al.: Overview of Touché 2023: argument and causal retrieval. In: Arampatzis, A., et al. (eds.) CLEF 2023. LNCS, vol. 14163, pp. 507–530. Springer, Cham (2023). https://doi.org/10.1007/978-3-031-42448-9_31
6. Cignarella, A.T., Lai, M., Bosco, C., Patti, V., Paolo, R., et al.: SardiStance@EVALITA2020: overview of the task on stance detection in Italian tweets. In: EVALITA 2020 Seventh Evaluation Campaign of Natural Language Processing and Speech Tools for Italian, pp. 1–10. CEUR (2020)
7. Conneau, A., et al.: Unsupervised cross-lingual representation learning at scale. In: Proceedings of the 58th Annual Meeting of the Association for Computational Linguistics, pp. 8440–8451 (2020)
8. Dey, K., Shrivastava, R., Kaushik, S.: Topical stance detection for Twitter: a two-phase LSTM model using attention. In: Pasi, G., Piwowarski, B., Azzopardi, L., Hanbury, A. (eds.) ECIR 2018. LNCS, vol. 10772, pp. 529–536. Springer, Cham (2018). https://doi.org/10.1007/978-3-319-76941-7_40
9. Faulkner, A.: Automated classification of stance in student essays: an approach using stance target information and the Wikipedia link-based measure. The Florida AI Research Society (2014)
10. Hardalov, M., Arora, A., Nakov, P., Augenstein, I.: Cross-domain label-adaptive stance detection. In: Moens, M.F., Huang, X., Specia, L., Yih, S.W. (eds.) Proceedings of the 2021 Conference on Empirical Methods in Natural Language Processing, pp. 9011–9028 (2021)
11. Hardalov, M., Arora, A., Nakov, P., Augenstein, I.: A survey on stance detection for mis- and disinformation identification. In: Findings of the Association for Computational Linguistics: NAACL 2022, pp. 1259–1277. Association for Computational Linguistics, Seattle (2022)
12. Hercig, T., Krejzl, P., Hourová, B., Steinberger, J., Lenc, L.: Detecting stance in Czech news commentaries. In: Conference on Theory and Practice of Information Technologies (2017). https://api.semanticscholar.org/CorpusID:35923394
13. Iscen, A., Tolias, G., Avrithis, Y., Chum, O.: Label propagation for deep semi-supervised learning. In: Proceedings of the IEEE/CVF Conference on Computer Vision and Pattern Recognition, pp. 5070–5079 (2019)
14. Küçük, D., Can, F.: Stance detection: a survey. ACM Comput. Surv. **53**(1), 1–37 (2020)
15. Küçük, D., Can, F.: Stance detection on tweets: an SVM-based approach. arXiv preprint arXiv:1803.08910 (2018)
16. Lai, M., Cignarella, A.T., Hernández Farías, D.I., Bosco, C., Patti, V., Rosso, P.: Multilingual stance detection in social media political debates. Comput. Speech Lang. **63**, 101075 (2020)
17. Lai, M., Patti, V., Ruffo, G., Rosso, P.: Stance evolution and Twitter interactions in an Italian political debate. In: Silberztein, M., Atigui, F., Kornyshova, E., Métais, E., Meziane, F. (eds.) NLDB 2018. LNCS, vol. 10859, pp. 15–27. Springer, Cham (2018). https://doi.org/10.1007/978-3-319-91947-8_2
18. Mohammad, S., Kiritchenko, S., Sobhani, P., Zhu, X., Cherry, C.: A dataset for detecting stance in tweets. In: Calzolari, N., et al. (eds.) Proceedings of the Tenth International Conference on Language Resources and Evaluation (LREC 2016), Portorož, Slovenia (2016)

19. Mohammad, S., Kiritchenko, S., Sobhani, P., Zhu, X., Cherry, C.: SemEval-2016 task 6: detecting stance in tweets. In: Proceedings of the 10th International Workshop on Semantic Evaluation (SemEval-2016), pp. 31–41 (2016)
20. Patel, H., Verma, J.P.: Community detection using label propagation algorithm with random walk approach. In: Dhavse, R., Kumar, V., Monteleone, S. (eds.) Emerging Technology Trends in Electronics, Communication and Networking. LNEE, vol. 952, pp. 307–320. Springer, Singapore (2023). https://doi.org/10.1007/978-981-19-6737-5_25
21. Schäfer, K.: Queen of swords at Touché 2023: intra-multilingual multi-target stance classification using BERT. In: Working Notes of CLEF (2023)
22. Sugiyama, A., Yoshinaga, N.: Data augmentation using back-translation for context-aware neural machine translation. In: Proceedings of the Fourth Workshop on Discourse in Machine Translation (DiscoMT 2019) (2019)
23. Taulé, M., Martí, M.A., Pardo, F.M.R., Rosso, P., Bosco, C., Patti, V.: Overview of the task on stance and gender detection in tweets on Catalan independence. In: IberEval@SEPLN (2017)
24. Taulé, M., Pardo, F.M.R., Martí, M.A., Rosso, P.: Overview of the task on multimodal stance detection in tweets on Catalan #1oct referendum. In: IberEval@SEPLN (2018)
25. Vamvas, J., Sennrich, R.: X-stance: a multilingual multi-target dataset for stance detection. In: Proceedings of SwissText/KONVENS 2020 (2020)
26. Van Dyk, D.A., Meng, X.L.: The art of data augmentation. J. Comput. Graph. Stat. **10**(1), 1–50 (2001)
27. Vaswani, A., et al.: Attention is all you need (2023)
28. Wang, F., Zhang, C.: Label propagation through linear neighborhoods. In: Proceedings of the 23rd International Conference on Machine Learning, pp. 985–992 (2006)
29. Wei, W., Zhang, X., Liu, X., Chen, W., Wang, T.: pkudblab at SemEval-2016 task 6: a specific convolutional neural network system for effective stance detection. In: Proceedings of the 10th International Workshop on Semantic Evaluation (SemEval-2016) (2016)
30. Xu, R., Zhou, Y., Wu, D., Gui, L., Du, J., Xue, Y.: Overview of NLPCC shared task 4: stance detection in Chinese microblogs. In: Lin, C.-Y., Xue, N., Zhao, D., Huang, X., Feng, Y. (eds.) ICCPOL/NLPCC -2016. LNCS (LNAI), vol. 10102, pp. 907–916. Springer, Cham (2016). https://doi.org/10.1007/978-3-319-50496-4_85
31. Zotova, E., Agerri, R., Rigau, G.: Semi-automatic generation of multilingual datasets for stance detection in Twitter. Expert Syst. Appl. **170**, 114547 (2021)

Leveraging LLM-Generated Data for Detecting Depression Symptoms on Social Media

Ana-Maria Bucur[1,2(✉)]

[1] Interdisciplinary School of Doctoral Studies, University of Bucharest, Bucharest, Romania
ana-maria.bucur@drd.unibuc.ro
[2] PRHLT Research Center, Universitat Politècnica de València, Valencia, Spain

Abstract. In our work, we present the contribution of the BLUE team in the eRisk Lab task focused on identifying symptoms of depression in Reddit social media posts. The task consists of retrieving and ranking Reddit social media sentences that convey symptoms of depression from the BDI-II questionnaire. To augment our data and improve downstream models, we utilized synthetic data generated by GPT-3.5 and LLama-3 for each of the BDI-II symptoms. Our approach aimed to enrich the data with semantic diversity and emotional and anecdotal experiences that are specific to the more intimate way of sharing experiences on Reddit. We used semantic search and cosine similarity to rank the relevance of the sentences to the BDI-II symptoms. Our study compared the performance of two transformer-based models (MentalRoBERTa and a variant of MPNet) in embedding social media posts and the original/generated BDI-II responses for information retrieval. We found that using sentence embeddings from a model designed for semantic search outperformed the approach using embeddings from a model pre-trained on mental health data. Furthermore, the generated synthetic data were proved too specific for this task, the approach simply relying on the BDI-II responses had the best performance.

Keywords: Depression Symptoms · Beck's Depression Inventory · Large Language Models

1 Introduction

Depression is one of the most prevalent mental disorders, with 5% of adults[1] suffering from it. Even if there is effective treatment, depression remains undiagnosed in some individuals due to the lack of access to medical services or stigma around mental illnesses [11]. For depression screening, mental health professionals use different scales, such as Center of Epidemiological Scales-Depression (CES-D) [9], Patient Health Questionnaire-9 (PHQ-9) [13], Beck's Depression Inventory-II (BDI-II) [3] and Hamilton Rating Scale for Depression (HRSD)

---

[1] https://www.who.int/news-room/fact-sheets/detail/depression.

[10]. With the rise in social media use and the anonymity and support provided on these platforms [8], researchers from both natural language processing and psychology began using social media data to search for symptoms or signs of mental disorders in online users. In recent years, the field of mental illnesses detection shifted from black-box approaches providing only binary labels [5,34] to explainable, interpretable approaches [35,36] incorporating information from the depression screening scales. With the recent advancement in Large Language Models (LLMs) [4,20], there have been efforts in testing their capabilities on mental health assessment.

The eRisk lab on Early Risk Prediction on the Internet of mental disorders started in 2017, with the pilot task of early risk detection of depression from social media data. From then on, the lab organized several tasks yearly and expanded to other mental illnesses such as eating disorders, pathological gambling and self-harm. The tasks consisted of detecting these mental health problems from social media data as early as possible, or automatically filling in questionnaires used by mental health professionals to diagnose depression or eating disorders. In the current edition, the task consists in retrieving and ranking social media posts with depression symptoms from the BDI-II questionnaire.

In this work, we present our proposed method for searching symptoms of depression, as part of the eRisk Lab. Inspired by recent works on generating and augmenting data using LLMs [7,17], we follow a similar approach, and generate synthetic Reddit posts for each of the BDI-II symptoms such that the generated data has more diversity than the responses from BDI-II. We hypothesize that by generating synthetic data similar to the BDI-II responses with GPT-3.5, we will add more diversity to the data and will be able to retrieve more relevant sentences. We aim for the generated data to resemble Reddit posts, in which users share their experiences more intimately. We explore different approaches based on pre-trained transformer-based models for encoding the social media data, the BDI-II responses and the synthetic data generated by LLMs. We perform semantic search and use cosine similarity to get the most relevant social media posts to the original and generated queries. Our results infirm our hypothesis that generated data improve the results. The semantic search model utilizing the original BDI-II responses as queries performs better than the model using generated data. The data generated by GPT-3.5 and Llama-3 is too specific, and future work needs to be done to manipulate the prompt such that data is semantically similar and more diverse than the BDI-II responses, but, at the same time, has fewer specific details. However, the generated text is informative and generating mental health data with LLMs is a promising research direction.

2 Related Work

Approaches in NLP for mental disorders detection from social media data achieved state-of-the-art results by using Convolutional Neural Networks (CNN) [24,34], Recurrent Neural Networks (RNN) [25,28], Hierarchical Attention Networks (HAN) [29,30], and transformer-based architectures [5,6,16,22]. However, most methods output binary labels for classification, operate as black boxes and are not interpretable. They cannot be used in real-life scenarios due to the lack of trust from mental health professionals [19].

Recently, there have been efforts in augmenting mental disorder detection methods with information from clinical questionnaires such as CES-D, PHQ-9, BDI-II, and HRSD. [19] proposed several approaches for depression detection that were constrained by the presence of the symptoms from PHQ-9. The proposed models consisted of two components, a questionnaire model that predicted the PHQ-9 symptoms and a depression model which used the symptom features for prediction. The authors showed that the models constrained on PHQ-9 had comparable performance to unconstrained methods, could better generalize to other datasets and are interpretable. Similarly, [36] performed symptom-assisted mental disorders identification, achieving better results than baselines that use only text. Furthermore, their method was interpretable and provided symptom-based explanations for several mental health disorders, such as depression, anxiety, bipolar disorder, obsessive-compulsive disorder, eating disorders, ADHD and post-traumatic stress disorder. Psychiatric scales were also used for screening risky posts with HANs for early risk detection of depression [35]. [15] crawled data from different subreddits corresponding to 13 depression symptoms (e.g., *r/insomnia, sleep* for sleep problems, *r/chronicfatigue, r/Fatigue* for fatigue, etc.). Different models were trained on the data to detect each symptom. The predictions of these symptom detection models on Facebook data were validated against PHQ-9, General Anxiety Disorder-7 (GAD-7) and UCLA Loneliness Scale (UCLA-3) filled in by individuals. The authors showed that the automatically predicted symptoms were significantly associated with the symptoms checked by the self-report surveys, except for fatigue.

With recent advancements in LLMs [4,20], there have been efforts to evaluate them for mental health assessment [2,33]. [33] compared ChatGPT[2] with three supervised baselines and showed that, even if ChatGPT can achieve good results in a zero-shot classification setting, it lacks behind transformer-based specialized models for downstream tasks such as suicide and depression identification from social media data. [2] performed an interpretable mental health analysis through emotional reasoning using ChatGPT on 11 datasets across 5 tasks related to depression, stress and suicide ideation. Their results showed that zero-shot ChatGPT performed better than traditional neural network architectures but could not surpass the performance of specialized transformer-based models. The authors performed human evaluations and tested the impact of emotional reasoning in mental health assessment. Using emotional reasoning improved ChatGPT's performance, and the model could generate explanations for its predictions.

Besides mental health assessment, other applications of LLMs are generating and augmenting data [7,14,31]. [17] evaluated the synthetic data generated by GPT-3 [4] for conversational tasks. The authors showed that the performance of classifiers trained on synthetic data performed worse than classifiers trained on fewer samples of real user-generated data. The data generated by GPT-3 has

---

[2] https://openai.com/blog/chatgpt.

less variability than the real data. However, generating synthetic data may be a suitable approach in scenarios with very little data or resources available.

In line with these approaches of using LLMs to generate synthetic data, we use GPT-3.5 and Llama-3 to generate data similar to the BDI-II questionnaire responses, simulating how social media users disclose their feelings and experiences on Reddit. We use the original BDI-II responses and the generated data as queries for semantic search and retrieve the most relevant sentences based on their cosine similarity to the queries.

3 eRisk 2023 Task 1: Search for Symptoms of Depression

The first task from the eRisk 2023 Lab [23] consists of ranking sentences from social media posts according to their relevance to the symptoms from Beck Depression Inventory-II (BDI-II) [3]. The BDI-II is a questionnaire used by mental health professionals to screen for depression and consists of 21 questions related to symptoms of depression such as sadness, pessimism, loss of pleasure, loss of interest, tiredness and others. Each question corresponds to one of the symptoms. BDI-II is a Likert scale survey, for each question there are 4 possible responses measuring the intensity of the symptom from the absence of it, to its maximum intensity (with the exception of item 16 and 18, which have 7 possible responses). The challenge consists of ranking the sentences from Reddit by their relevance to each of the symptoms of the BDI-II. A given sentence is considered relevant to a symptom if it contains information about the user's mental state regarding the symptom, even if the user mentions that they do not suffer from the given symptom. The data for this task was compiled from the eRisk past data and was organized as TREC formatted sentences for each user. A total of approx 4 million of sentences from 3,107 users were provided for this task.

For evaluating the systems' performance and assess the sentences' relevance to the BDI-II symptoms, top-k pooling was used, with k equal to 50. The top 50 relevant sentences for each symptom from each system were combined in a pool of relevant sentences. These sentences were further assessed as being relevant or not to the symptoms by three annotators. A sentence was considered relevant to a symptom if it contained information about the state of the individual and is topically-related to the BDI-II symptoms.

4 Method

To search for symptoms of depression in Reddit data, we proposed an approach based on semantic search using as queries the corresponding responses for each item from BDI-II. Inspired by previous works that use LLMs to generate synthetic data [7,17], we also experimented with generating synthetic Reddit posts with GPT-3.5 and Llama-3 to be used as queries. We aimed for the generated data to have more diversity than the BDI-II responses, while preserving the meaning, and to be expressed more intimately, specific to Reddit.

Synthetic data provided by LLMs have been successfully used in other works [7,27,31,32] and have proved to be a reliable method for augmenting and fine-tuning downstream models. We generated synthetic data using `text-davinci-3` [4] and Llama-3 8B[3] for each item of the BDI-II questionnaire. We designed a prompt such that the answers had more diversity than the BDI-II responses and conveyed the intimate way of sharing experiences and feelings specific to Reddit [8]. In Table 1, we showcase the prompt we used to instruct the model, similar to the approach of [32]. However, our prompt was simpler, and geared towards simulating user responses, not tasks with their outcomes. We included instructions that limited the size of the text, ensured semantic diversity in the generated texts and ensured that the generated data contained emotional and anecdotal experiences that aligned with each BDI-II item. In Algorithm 1, we showcase our algorithm for generating data using the OpenAI text completion API[4]. Each completion was post-processed by removing trailing quotation marks, enumeration numbers, and splitting by *newline* to obtain individual texts. We used the same prompt to generate synthetic data with Llama-3 8B. BDI-II contains 21 items related to depression symptoms, with a total of 90 possible responses measuring the intensity of symptoms. For each of these 90 responses, we generated 30 synthetic Reddit posts, totaling 2,700 generated texts by GPT-3.5 and 2,700 generated texts by Llama-3. We show in Table 2 some GPT-3.5 generated examples for the first symptom of the BDI-II questionnaire. In Table 3 we show Llama-3 generated examples for the second symptom of the BDI-II questionnaire. The generated texts were longer than the BDI-II responses and had greater diversity. Some examples even contained self-disclosure, which is specific for Reddit data [8], such as "My cat passed away", "I just broke up with my partner". We hypothesized that, by augmenting the queries with the synthetically generated data, we would be able to retrieve more relevant sentences.

Algorithm 1. Generating synthetic data based on BDI-II items and responses

Require: instruction prompt P, questionnaire Q
   $D \leftarrow \{\}$
   for symptom s and item $q \in Q$ **do**
      $p = P$.format(N = N, item = q, symptom = s)
      $d =$ `openai.Completion.create`(model = "text-davinci-003", prompt = p)
      $d =$ `postprocess`(d)
      $D = D \cup \{d\}$
   end for
      return D

We pre-processed all the posts from each Reddit user by removing URLs and texts not in English, detected by the *polyglot* package. We computed the sentence embeddings for all texts (original posts, BDI-II responses, generated data) using

---
[3] https://huggingface.co/meta-llama/Meta-Llama-3-8B.
[4] https://platform.openai.com/docs/guides/gpt/completions-api.

Table 1. Instructions for the data generation that ensure diversity and adherence to the BDI-II responses

```
You are asked to come up with a set of "{N}" diverse reddit posts that are examples
to the BDI depression questionnaire for the "{symptom}" symptom. For this symptom,
the BDI answer of interest is "{item}".
These examples will be given to a ranking model that will compute the similarity
between the answer item text and the reddit post.

Here are the requirements:
1. The language used for the reddit posts should be diverse. For example, you should
combine descriptions of past experiences with feelings or events.
2. The reddit posts should be in English.
3. The reddit posts should be 2 to 3 sentences long.
4. The reddit posts should provide substantial content to make ranking feasible.
5. The reddit posts should be specific and not just describe general situations,
but rather specific personal experiences and self-disclosure.
6. The reddit posts should, as much as possible, not contain the
exact words of the BDI item.

List of "{N}" reddit posts:
```

Table 2. Examples generated by GPT 3.5 for the first symptom of BDI-II: Sadness

BDI-II Response	Generated text
I do not feel sad.	My cat passed away last week and although I miss her terribly, I'm not feeling any sadness about it.
	I just broke up with my partner, but I'm focusing on all the good memories we shared and not letting myself be consumed by sadness.
I feel sad much of the time.	I feel like I'm stuck in an endless cycle of sadness and emptiness.
	It feels like no matter how hard I try, I can't shake this sadness that's been following me around.
I am sad all the time.	I'm feeling so empty and down lately, like nothing can make me happy anymore.
	I feel so overwhelmed with sadness that it's hard to get out of bed in the morning.
I am so sad or unhappy that I can't stand it.	I feel like I'm stuck in a dark hole with no way out and it's suffocating me.
	I'm so overwhelmed by sadness that I can barely function anymore.

Table 3. Examples generated by LLama-3 for the second symptom of BDI-II: Pessimism

BDI-II Response	Generated text
I am not discouraged about my future.	Although life is unpredictable, I'm hopeful that I'll be able to find joy in whatever comes my way.
	No matter how difficult things get, I believe in myself and my ability to turn any situation around.
I feel more discouraged about my future than I used to be.	I'm feeling increasingly pessimistic about my prospects and it's taking a toll on my mental health.
	All my plans seem to have gone up in smoke and I'm starting to doubt whether I'll ever reach my goals.
I do not expect things to work out for me.	I know I shouldn't even bother dreaming about a better future because it'll never happen.
	I'm so used to failure that I don't even expect success anymore. It's just not an option.
I feel my future is hopeless and will only get worse.	I can't help but think that all my struggles now are leading up to an even more difficult time ahead.
	Every day I wake up with this heavy weight on my chest, knowing that things are only going to get worse.

two state-of-the-art methods, a variant of MPNet [26] and MentalRoBERTa [12]. The MPNet variant we used, `multi-qa-mpnet-base-dot-v1`[5] was explicitly designed for semantic search. MentalRoBERTa[6] was chosen because it was trained on a large corpus of mental health data, mainly from Reddit. The MentalRoBERTa model has shown good downstream performance for mental disorders detection from social media data [1,21].

We performed semantic search and used cosine similarity between embeddings to get the most relevant social media posts to the original BDI-II responses and generated queries. We retrieved the top 50 sentences with the highest cosine similarity to the queries. Given that we could submit a maximum of 1,000 results for each submission, we sorted the retrieved sentences by cosine similarity scores and kept only the most relevant 1,000 sentences.

We experimented with different queries and embedding methods The official submissions of the BLUE team are detailed below:

SemSearchOnBDI2QueriesMPNet. We performed semantic search using the original 90 BDI-II responses as queries. All texts were encoded using MPNet.

SemSearchOnGPTGeneratedQueriesMPNet. We performed semantic search using the 2,700 GPT-3.5 generated synthetic Reddit texts as queries, with MPNet embeddings.

SemSearchOnGPTGenerated & BDI2QueriesMPNet. We use all the original and GPT-3.5 generated queries and perform semantic search on texts encoded with MPNet.

SemSearchOnBDI2QueriesMentalRoberta. We use the original BDI-II responses as queries, but MentalRoBERTa is used for embedding the data.

SemSearchOnGPTGeneratedQueriesMentalRoberta. We perform semantic search using the GPT-3.5 generated data as queries; texts were encoded using MentalRoBERTa.

In addition to our official submissions, we tested two other approaches using Llama-3 generated queries:

SemSearchOnLlamaGeneratedQueriesMPNet. We performed semantic search using the 2,700 Lama-3 generated synthetic Reddit texts as queries, with MPNet embeddings.

SemSearchOnLlamaGeneratedQueriesMentalRoberta.
We perform semantic search using the Llama-3 generated data as queries; texts were encoded using MentalRoBERTa.

5 Results

The results of the eRisk Lab task on searching for depression symptoms are presented in Tables 4 and 5. We show the results of all official 5 runs submitted by our team and the best-performing run from each other team. We also

---

[5] https://huggingface.co/sentence-transformers/multi-qa-mpnet-base-dot-v1.
[6] https://huggingface.co/mental/mental-roberta-base.

Table 4. Ranking-based evaluation for Task 1 (majority voting). *Denotes that the submission was not officially submitted to the eRisk Lab.

Team	Run	AP	R-PREC	P@10	NDCG@1000
Formula-ML	SentenceTransformers_0.25	**0.319**	**0.375**	**0.861**	**0.596**
OBSER-MENH	salida-distilroberta-90-cos	0.294	0.359	0.814	0.578
uOttawa	USESim	0.160	0.248	0.600	0.382
NailP	T1_M2	0.095	0.146	0.519	0.226
RELAI	bm25—mpnetbase	0.048	0.081	0.538	0.140
UNSL	Prompting-Classifier	0.036	0.090	0.229	0.180
UMU	LexiconMultilingualSentenceTransformer	0.073	0.140	0.495	0.222
GMU	FAST-DCMN-COS-INJECT_FULL	0.001	0.003	0.014	0.005
Mason-NLP	MentalBert	0.035	0.072	0.286	0.117
BLUE	SemSearchOnBDI2QueriesMPNet	0.104	0.126	0.781	0.211
BLUE	SemSearchOnGPTGenerated&BDI2QueriesMPNet	0.065	0.086	0.629	0.160
BLUE	SemSearchOnGPTGeneratedQueriesMPNet	0.052	0.074	0.586	0.139
BLUE	SemSearchOnBDI2QueriesMentalRoberta	0.027	0.044	0.386	0.089
BLUE	SemSearchOnGPTGeneratedQueriesMentalRoberta	0.029	0.063	0.367	0.105
BLUE*	SemSearchOnLlamaGeneratedQueriesMPNet	0.007	0.023	0.142	0.005
BLUE*	SemSearchOnLlamaGeneratedQueriesMentalRoberta	0.002	0.008	0.076	0.002

Table 5. Ranking-based evaluation for Task 1 (unanimity). *Denotes that the submission was not officially submitted to the eRisk Lab.

Team	Run	AP	R-PREC	P@10	NDCG@1000
Formula-ML	SentenceTransformers_0.25	0.268	**0.360**	**0.709**	**0.615**
Formula-ML	SentenceTransformers_0.1	**0.293**	0.350	0.685	0.611
OBSER-MENH	salida-distilroberta-90-cos	0.281	0.344	0.652	0.604
uOttawa	USESim	0.139	0.232	0.438	0.380
NailP	T1_M2	0.090	0.143	0.410	0.229
UMU	LexiconMultilingualSentenceTransformer	0.059	0.125	0.333	0.209
RELAI	bm25—mpnetbase	0.039	0.069	0.343	0.124
UNSL	Prompting-Classifier	0.020	0.063	0.090	0.157
GMU	FAST-DCMN-COS-INJECT_FULL	0.001	0.003	0.014	0.006
Mason-NLP	MentalBert	0.024	0.054	0.190	0.099
BLUE	SemSearchOnBDI2Queries	0.129	0.167	0.643	0.260
BLUE	SemSearchOnAllQueries	0.067	0.105	0.452	0.177
BLUE	SemSearchOnGeneratedQueries	0.052	0.088	0.381	0.147
BLUE	SemSearchOnBDI2QueriesMentalRoberta	0.032	0.058	0.300	0.104
BLUE	SemSearchOnGeneratedQueriesMentalRoberta	0.018	0.059	0.186	0.085
BLUE*	SemSearchOnLlamaGeneratedQueriesMentalRoberta	0.003	0.010	0.052	0.001
BLUE*	SemSearchOnLlamaGeneratedQueriesMPNet	0.007	0.028	0.084	0.003

present the results of the two runs using Llama-3 generated data. The metrics used for evaluating the relevance of the sentences were Average Precision (AP), R-Precision, Precision at 10 (P@10), and Normalized Discounted Cumulative Gain at 1000 (NDCG@1000). Table 4 presents the systems' performance compared to the gold standard obtained from majority voting of the relevant sentences assessed by the annotators. Table 5 presents the systems' performance compared to the gold standard obtained from the sentences considered relevant by all three annotators. Comparing our proposed methods, the model using only the BDI-II responses as queries, SemSearchOnBDI2QueriesMPNet, performed best in both ranking-based evaluation settings, majority voting and unanimity, achieving 0.104 AP in the first scenario, and 0.129 AP in the second one. The second-best model was the one that used as queries all the texts (original and generated), SemSearchOnGPTGenerated&BDI2QueriesMPNet, with an AP of 0.065 in majority voting evaluation, and 0.067 in unanimity evaluation. The models using only generated data from GPT-3.5 and Llama-3 as queries had the lowest performance. The SemSearchOnBDI2QueriesMPNet model had a good P@10 of 0.781 for majority voting ranking-based evaluation and 0.643 for unanimity evaluation, showing that our semantic search method using MPNet embeddings on the original BDI-II queries was best at retrieving relevant sentences in top 10 documents. Even if the embeddings provided by the pre-trained model on mental health data, MentalRoBERTa, had a good performance for detection tasks [1,21], it had the lowest performance for symptoms retrieval.

Fig. 1. Comparison of most frequent words between the original BDI-II queries, generated queries with GPT-3.5 and LLama-3 8B, and the manually annotated relevant documents.

However, our proposed methods ranked fourth compared to all the systems developed by other participants in the eRisk task. Our hypothesis that the synthetically generated queries will improve performance was proved false. We aimed for variability, as the BDI-II responses were short and standard, but the texts generated by GPT-3.5 and Llama-3 might be too specific. Some of the generated texts provided too many details, which were not helpful for semantic search: "I just got back from a great vacation and it's been really hard to get back into the swing of things - not feeling particularly sad, but definitely a bit down.", "I don't know what to do with myself anymore - no matter how hard I try, I can't shake this overwhelming sense of gloom.". For future work, we would like to experiment with different prompts to generate data that are semantically similar and more diverse than the BDI-II responses, with fewer specific details. Moreover, we also consider generating synthetic mental health data that is representative of different demographic groups [18].

In Fig. 1, we present the most common words in the original BDI-II queries, generated queries with GPT-3.5 and Llama-3 8B, and the manually annotated relevant documents. Llama-3 generated queries are more diverse than the ones generated by GPT-3.5. Even though "feel" is the most common word in all the data, it is noticeable that the BDI-II queries and the relevant documents focus on more specific depression symptoms, such as "appetite", "interested", "tired", "sleep", "energy", "restless". The queries generated by GPT-3.5, on the other hand, contain anecdotal experiences (as shown in Table 2), but the most common words in all of them are generic.

6 Conclusions

In this work, we presented the contributions of the BLUE team in the eRisk Lab task on retrieving relevant social media text relevant to the symptoms of depression from the BDI-II questionnaire. We performed semantic search using the original BDI-II responses and synthetically generated texts as queries. We hypothesized that, by using GPT-3.5 and Llama-3 to generate synthetic data similar to Reddit posts in which users disclose their feelings and experiences, we could retrieve more relevant sentences for each BDI-II item. We experimented with two pretrained transformer-based methods to encode the queries and social media posts, MentalRoBERTa and a variant on MPNet designed specifically for semantic search. Our hypothesis was proved false; the model performing semantic search using as queries the original BDI-II responses outputted more relevant sentences than the one using generated data. The synthetic data generated by GPT-3.5 and Llama-3 was too specific for retrieving depression symptoms, and future work needs to be done to generate suitable data for this task.

Acknowledgements. This work was supported by the POCIDIF project in Action 1.2. "Romanian Hub for Artificial Intelligence".

References

1. Aich, A., et al.: Towards intelligent clinically-informed language analyses of people with bipolar disorder and schizophrenia. In: Findings of EMNLP, pp. 2871–2887 (2022)
2. Amin, M.M., Cambria, E., Schuller, B.W.: Will affective computing emerge from foundation models and general AI? a first evaluation on ChatGPT. IEEE Intell. Syst. **38**, 2 (2023)
3. Beck, A.T., Steer, R.A., Brown, G.: Beck depression inventory–II. Psychological assessment (1996)
4. Brown, T., et al.: Language models are few-shot learners. In: Proceedings of NeurIPS, vol. 33, pp. 1877–1901 (2020)
5. Bucur, A.M., Cosma, A., Dinu, L.P.: Early risk detection of pathological gambling, self-harm and depression using BERT. In: CLEF (Working Notes) (2021)
6. Bucur, A.M., Cosma, A., Rosso, P., Dinu, L.P.: It's just a matter of time: detecting depression with time-enriched multimodal transformers. In: Kamps, J., et al. (eds.) ECIR 2023. LNCS, vol. 13980, pp. 200–215. Springer, Cham (2023). https://doi.org/10.1007/978-3-031-28244-7_13
7. Dai, H., et al.: ChatAug: Leveraging ChatGPT for text data augmentation. arXiv preprint arXiv:2302.13007 (2023)
8. De Choudhury, M., De, S.: Mental health discourse on reddit: self-disclosure, social support, and anonymity. In: Proceedings of ICWSM, pp. 71–80 (2014)
9. Eaton, W.W., Muntaner, C., Smith, C., Tien, A., Ybarra, M.: Center for epidemiologic studies depression scale: review and revision. Psychol. Test. Treat. Plan. Outcomes Assess. (2004)
10. Hamilton, M.: A rating scale for depression. J. Neurol. Neurosurg. Psychiatry **23**(1), 56 (1960)
11. Handy, A., Mangal, R., Stead, T.S., Coffee Jr, R.L., Ganti, L.: Prevalence and impact of diagnosed and undiagnosed depression in the united states. Cureus **14**(8) (2022)
12. Ji, S., Zhang, T., Ansari, L., Fu, J., Tiwari, P., Cambria, E.: MentalBERT: publicly available pretrained language models for mental healthcare. In: Proceedings of LREC, pp. 7184–7190 (2022)
13. Kroenke, K., Spitzer, R.L., Williams, J.B.: The PHQ-9: validity of a brief depression severity measure. J. Gen. Intern. Med. **16**(9), 606–613 (2001)
14. Lee, Y.J., Lim, C.G., Choi, Y., Lm, J.H., Choi, H.J.: PERSONACHATGEN: generating personalized dialogues using GPT-3. In: Proceedings of CCGPK Workshop, pp. 29–48 (2022)
15. Liu, T., et al.: Detecting symptoms of depression on reddit. In: Proceedings of WebSci, pp. 174–183 (2023)
16. Martínez-Castaño, R., Htait, A., Azzopardi, L., Moshfeghi, Y.: Early risk detection of self-harm and depression severity using BERT-based transformers: iLab at CLEF eRisk 2020. In: CLEF (Working Notes) (2020)
17. Meyer, S., Elsweiler, D., Ludwig, B., Fernandez-Pichel, M., Losada, D.E.: Do we still need human assessors? Prompt-based GPT-3 user simulation in conversational AI. In: Proceedings of CUI, pp. 1–6 (2022)
18. Mori, S., Ignat, O., Lee, A., Mihalcea, R.: Towards algorithmic fidelity: mental health representation across demographics in synthetic vs. human-generated data. arXiv preprint arXiv:2403.16909 (2024)

19. Nguyen, T., Yates, A., Zirikly, A., Desmet, B., Cohan, A.: Improving the generalizability of depression detection by leveraging clinical questionnaires. In: Proceedings of ACL, pp. 8446–8459 (2022)
20. OpenAI: GPT-4 Technical report. arXiv (2023)
21. Owen, D., et al.: Enabling early health care intervention by detecting depression in users of web-based forums using language models: longitudinal analysis and evaluation. JMIR AI **2**(1) (2023)
22. Owen, D., Camacho-Collados, J., Anke, L.E.: Towards preemptive detection of depression and anxiety in Twitter. In: Proceedings of SMM4H Workshop, pp. 82–89 (2020)
23. Parapar, J., Martín-Rodilla, P., Losada, D.E., Crestani, F.: Overview of eRisk 2023: Early risk prediction on the internet. In: Arampatzis, A., et al. (eds.) CLEF 2023. LNCS, vol. 14163, pp. 294–315. Springer, Cham (2023). https://doi.org/10.1007/978-3-031-42448-9_22
24. Rao, G., Zhang, Y., Zhang, L., Cong, Q., Feng, Z.: MGL-CNN: a hierarchical posts representations model for identifying depressed individuals in online forums. IEEE Access **8**, 32395–32403 (2020)
25. Skaik, R., Inkpen, D.: Using Twitter social media for depression detection in the Canadian population. In: Proceedings of AICCC, pp. 109–114 (2020)
26. Song, K., Tan, X., Qin, T., Lu, J., Liu, T.Y.: MPNet: masked and permuted pre-training for language understanding. In: Proceedings of NeurIPS, vol. 33, pp. 16857–16867 (2020)
27. Taori, R., et al.: Stanford alpaca: an instruction-following llama model (2023). https://github.com/tatsu-lab/stanford$_$alpaca
28. Trotzek, M., Koitka, S., Friedrich, C.M.: Utilizing neural networks and linguistic metadata for early detection of depression indications in text sequences. IEEE Trans. Knowl. Data Eng. **32**(3), 588–601 (2018)
29. Uban, A.S., Chulvi, B., Rosso, P.: Multi-aspect transfer learning for detecting low resource mental disorders on social media. In: Proceedings of LREC, pp. 3202–3219 (2022)
30. Uban, A.S., Rosso, P.: Deep learning architectures and strategies for early detection of self-harm and depression level prediction. In: CLEF (Working Notes), vol. 2696, pp. 1–12 (2020)
31. Ubani, S., Polat, S.O., Nielsen, R.: ZeroShotDataAug: generating and augmenting training data with ChatGPT. arXiv preprint arXiv:2304.14334 (2023)
32. Wang, Y., et al.: Self-instruct: aligning language model with self generated instructions. arXiv preprint arXiv:2212.10560 (2022)
33. Yang, K., Ji, S., Zhang, T., Xie, Q., Ananiadou, S.: On the evaluations of ChatGPT and emotion-enhanced prompting for mental health analysis. arXiv preprint arXiv:2304.03347 (2023)
34. Yates, A., Cohan, A., Goharian, N.: Depression and self-harm risk assessment in online forums. In: Proceedings of EMNLP, pp. 2968–2978 (2017)
35. Zhang, Z., Chen, S., Wu, M., Zhu, K.: Psychiatric scale guided risky post screening for early detection of depression. In: Proceedings of IJCAI (2022)
36. Zhang, Z., Chen, S., Wu, M., Zhu, K.: Symptom identification for interpretable detection of multiple mental disorders on social media. In: Proceedings of EMNLP, pp. 9970–9985 (2022)

From Sentence Embeddings to Large Language Models to Detect and Understand Wordplay

Ryan Rony Dsilva[✉]

Purdue University, West Lafayette, IN, USA
dsilvar@purdue.edu

Abstract. A pun is a form of wordplay in which a word or phrase evokes the meaning of another word or phrase with a similar or identical pronunciation. In this study, we present our work for JOKER 2023, particularly the pun detection, location, and interpretation tasks. The methods used demonstrate the evolution of the field from sentence embeddings with various classifiers, sequence, and token classification using BERT-based models, to inference with prompt engineering using LLMs. Experimental results demonstrate varying effectiveness across methodologies, highlighting the strengths and limitations of each approach. Additionally, challenges such as handling nuances, diverse languages, interpreting contextually diverse word meanings, and integrating external sense dictionaries are discussed. This study provides insights into the evolution of natural language processing techniques for detecting and understanding wordplay, paving the way for future advancements in computational humor analysis.

Keywords: sentence embeddings · transformers · large language models · wordplay · puns

1 Introduction

Wordplay jokes, also known as puns, rely on the use of words that sound alike but have different meanings. The humor in these jokes stems from the conflict or surprise between the two interpretations [12]. Wordplay can occur between words that are homonyms (same spelling and pronunciation), homophones (same pronunciation, different spelling), or near-homophones (similar pronunciation, different spelling). The CLEF 2023 JOKER [8] track proposed three tasks. Task 1 involved the detection of puns in English, French, and Spanish. Task 2 entailed locating puns in English, French, and Spanish and interpreting puns in English and French. Task 3 focused on translating puns from English to French and Spanish. This work is an improved version of our original work [5] submitted to the JOKER 2023 competition. Our participation included approaches for Tasks 1 and 2. We employ sentence embeddings, BERT-based models, and large language models for our various approaches. Large language models (LLMs) have sparked significant interest in both academic and industrial circles. Unlike previous models that were limited to specific tasks, LLMs are capable of addressing a wide variety of challenges. Their exceptional performance across both general natural language tasks and domain-specific applications has led to their increasing adoption [3]. In this study, we trace the evolution of natural language processing, progressing from embedding-based methods to the surge of BERT and ultimately into the era of large language models. The methodology for using LLMs to detect and interpret humor is inspired by the work by Jentzsch and Kersting [9] wherein ChatGPT was used to detect, interpret, and generate jokes and Dsilva [6], where humor theories were given as context to a large language model using prompts to detect wordplay.

2 Methodology

2.1 Dataset and Data Preparation

The dataset from [8] was utilized, with text preprocessing applied to all input sentences. These sentences were read using UTF-8 encoding, transformed to lowercase, and stripped of all punctuation except hyphens. This decision was based on the observation that many pun words in the dataset contained hyphens. For example, after preprocessing, the instances "Lee, Chamorro", "en? - ...fermo" and, "Stan, Lee" were transformed to "lee chamorro", "en - fermo" and, "stan lee" respectively. This transformation affected proper nouns, pun words with special characters, pun words with multiple forms (like bovine | divine), and words that appeared in a different form not exactly matching their appearance in the sentences (for instance: "dégainait"). For the pun location task, such instances were excluded from the training set using a Python script, with the exclusion criteria applied only to the training data. Both training and testing data underwent the same preprocessing steps. The final training dataset for the pun location task included 4817 instances across English, Spanish, and French. For the input provided to the large language model, no preprocessing was applied, as it did not impact the outcomes in our experiments.

2.2 Pun Detection

Three main strategies were used for pun detection: sentence embeddings [11] with a binary classifier, sequence classification using XLM-RoBERTa [4], and prompting with LLMs.

The rationale for employing sentence embeddings is derived from the foundational concept of word vectors, which are utilized to encapsulate the meaning of individual words. Similarly, this study seeks to investigate whether sentence embeddings can effectively represent the entirety of a sentence, thereby enabling the learning of specific attributes of the sentence. In this particular instance, the focus is on determining whether a sentence constitutes a pun. We used a multilingual sentence embedding model to get the sentence embeddings which were then used as inputs to a classifier along with the labels: 0 for non-puns, 1 for puns. We experimented with various classifier models like SVC, Random

Forests, Logistic Regression, and also a custom Neural Network, largely inspired by the likes of ColBERT [1], which ultimately produced the best results across the three languages on the training dataset. Building on similar thoughts, the second approach involved using XLM-RoBERTa-Large, finetuned on the training dataset for sequence classification. We also conducted experiments to evaluate the impact of the training data size on our models. These results show a slight improvement in performance with an increase in dataset size. However, the results, shown in Table 1, are not convincing enough to say that increasing the dataset size will necessarily improve results.

Table 1. Pun Detection - Impact of Training Data Size (Combined)

Model with Amount of Data	Precision	Recall	Accuracy	F-Score
SentEmb-NeuralNet with 25%	0.560	0.560	0.560	0.540
SentEmb-NeuralNet with 50%	0.580	0.580	0.580	0.560
SentEmb-NeuralNet with 75%	0.570	0.570	0.570	0.570
SentEmb-NeuralNet with 100%	**0.620**	**0.620**	**0.620**	**0.610**
XLM-RoBERTa-Large with 25%	0.499	**1.000**	0.499	0.6662
XLM-RoBERTa-Large with 50%	**0.709**	0.424	**0.624**	0.529
XLM-RoBERTa-Large with 75%	0.644	0.537	0.620	0.585
XLM-RoBERTa-Large with 100%	0.526	**1.000**	0.526	**0.690**

We also experimented with LLMs using prompt engineering to detect puns. The prompt was developed to include the task description, the definition of a pun from [8], and few-shot examples for each language, one pun and one non-pun. The LLMs utilized in this study are all open-source and feature diverse architectures and parameter sizes to ensure a wide range of variability in our experiments. We decided against employing proprietary models due to the financial costs and restrictions linked to their API access. Specifically, we incorporated Mistral 7B Instruct v0.2 [10], Qwen 1.5 72B Chat [2], and the most recent Llama 3 70B Instruct [13], which will henceforth be referred to as Mistral, Qwen, and Llama 3, respectively. We summarize all the results on the training dataset in Table 2.

Table 2. Pun Detection: Training Set

	EN				FR				ES			
	P	R	A	F	P	R	A	F	P	R	A	F
SVC	0.680	0.684	0.684	0.677	0.512	0.512	0.512	0.512	**0.450**	0.458	0.458	0.452
Random Forests	0.643	0.648	0.648	0.630	0.439	0.432	0.432	0.414	0.440	0.455	0.455	0.437
Gradient Boosted Forests	0.665	0.669	0.669	0.656	**0.525**	0.525	**0.525**	0.424	0.427	0.437	0.437	0.429
Logistic Regression	0.687	0.691	0.691	0.686	0.513	0.513	0.513	0.513	0.442	0.452	0.452	0.443
Neural Network	**0.716**	**0.854**	**0.719**	**0.780**	0.499	0.740	0.499	**0.596**	0.435	**0.733**	0.476	**0.546**
BERT	0.542	0.532	0.532	0.518	**0.576**	0.570	**0.570**	0.566	**0.502**	0.533	**0.533**	0.517
XLM-RoBERTa	0.583	**1.000**	0.583	**0.737**	0.500	**1.000**	0.500	**0.666**	0.429	**1.000**	0.429	**0.600**
Mistral	0.669	0.671	0.671	0.669	0.527	0.527	0.527	0.507	0.540	0.525	0.525	0.510
Qwen	0.770	0.765	0.765	0.759	**0.588**	0.536	0.536	0.564	**0.640**	0.535	**0.535**	0.565
Llama 3	**0.776**	**0.901**	**0.790**	**0.833**	0.540	**0.904**	0.567	**0.676**	0.461	**0.939**	0.503	**0.618**

2.3 Pun Location

We utilized the token classification approach to identify pun words/phrases, as illustrated in Zou and Lu [14]. The NP tagging scheme was selected due to its ability to encompass multiple words that may collectively constitute a pun. Each word identified as part of a pun was labeled with a tag of P (1 for computation), while words not forming part of a pun were assigned a tag of N (0 for computation). The XLM-RoBERTa-Large model was employed in our experiments.

We further conducted experiments using LLMs to identify the pun word or phrase. The prompt given to the LLM instructed it to determine the word or phrase within the sentence that forms the pun, supplemented by few-shot examples encompassing a variety of scenarios observed in the training data. The results on the training dataset are summarized in Table 3. The notation A* in the tables refers to the accuracy calculated only from the attempted instances instead of the entire dataset.

Table 3. Pun Location: Training Set

	EN		FR		ES	
	A	A*	A	A*	A	A*
XLM-RoBERTa	**0.857**	**0.857**	**0.591**	**0.592**	**0.711**	**0.711**
Llama 3	0.816	0.816	0.517	0.518	0.557	0.557

2.4 Pun Interpretation

The interpretation of puns leveraged outcomes from the subtask of pun location to disambiguate the relevant meanings of the pun word within the context of the sentence and to identify synonyms for these meanings. Our sense dictionary was sourced from WordNet[1]. Specifically, our usage of WordNet through the *nltk* library led us to employ the Open Multilingual WordNet, which provided access to WordNet in English, French, and Spanish. Initially, we gathered synonyms of the target pun word from WordNet, subsequently calculating the cosine similarity between the word embeddings of these synonyms and the sentence embedding of the text. This approach is predicated on the hypothesis that the embeddings of synonyms that align most closely with the sentence will exhibit greater proximity in the vector space. Given that each synonym in WordNet corresponds to a unique concept, we selected the top two synonyms that seemed most fitting for the pun word within the sentence.

Interpretation using LLMs where a majority of the synsets used in the dataset were taken from WordNet was a challenging task since LLMs cannot directly access WordNet. The instructions for the language model involved identifying synonyms or hypernyms for the two meanings of the pun word determined in the

---

[1] https://www.nltk.org/howto/wordnet.html.

preceding task. The anticipated pun word or phrase was provided in the prompt to facilitate generation. Additionally, the prompt contained several examples corresponding to different scenarios from the training dataset.

3 Results

For Task 1, the evaluation metrics reported include precision, recall, accuracy, and f-score. For Task 2, Sub-Task 1, and Sub-Task 2, accuracy alone is reported. The metrics presented in the subsequent tables are derived from the test dataset taken from the lab overviews [7, 8].

3.1 Pun Detection

Table 4 summarizes our results. The low scores indicate that direct utilization of sentence embeddings might not adequately capture the characteristics of the sentence that define it as a pun.

Table 4. Pun Detection: Test Set

	EN				FR				ES			
	P	R	A	F	P	R	A	F	P	R	A	F
Neural Network	0.263	0.864	0.350	0.403	0.412	0.739	0.457	0.529	0.414	0.723	0.448	0.526
XLM-RoBERTa	0.254	**1.000**	0.254	0.405	0.412	**1.000**	0.412	0.584	0.426	**0.997**	0.427	0.597
Llama 3	**0.423**	0.854	**0.666**	**0.565**	**0.451**	0.897	**0.507**	**0.600**	**0.458**	0.944	**0.502**	**0.617**

Additionally, the dataset included sentences that were modified by substituting the pun word with another contextually appropriate word, thereby removing the pun element from the sentence. The techniques employed were not effective in accurately identifying these modifications. Specifically, the XLM-RoBERTa-Large model was notably deficient, leading to an excessive number of false positive predictions. As a result, it failed to identify any true negatives in both English and French.

Example 1. Surfing is a swell sport!

In this example, the XLM-RoBERTa-Large model predicted 0 which is incorrect. The model failed to recognize the nuanced meaning of the word "swell," which could mean excellent, great, or fantastic (informally) or the rising and falling motion of the waves. Informal definitions and slang might not be accurately captured.

Example 2. I used to be a banker but I lost motivation.

In this example, the XLM-RoBERTa-Large model predicted 1 which is incorrect. This example demonstrates what the authors of the dataset describe as a negative example where the pun word of "interest" was substituted to be "motivation," which has a similar sense but loses the quality that makes this sentence a pun.

In our experimental framework, large language models do not demonstrate substantial enhancements, as evidenced by results that marginally exceed random performance. The qualitative analysis highlights two primary insights: LLMs proficiently grasp commonsense knowledge and slang; however, they falter when confronted with altered versions of puns. This suggests that the structure of the pun significantly affects the LLM's ability to generate accurate predictions.

Example 3. Surfing is a swell sport!

In this example, the LLM predicted 1 which is correct. The model recognized the nuanced meaning of the word "swell," which could mean excellent, great, or fantastic (informally) or the rising and falling motion of the waves, leading us to believe that informal definitions and slang might be captured through LLMs.

Example 4. I used to be a banker but I lost motivation.

In this example, the LLM predicted 1 which is incorrect. This example demonstrates what the authors of the dataset describe as a negative example where the pun word of "interest" was substituted to be "motivation," which has a similar sense but loses the quality that makes this sentence a pun. LLMs do get these variations correct sometimes, but there are instances like these when the structure of the sentence forms a bias towards an incorrect response.

3.2 Pun Location

Table 5 summarises the results for the pun location task. One flaw noticed during experiments was that our model could not always capture instances where the pun word was a phrase of multiple words instead of a single word despite using the NP tagging scheme which we hypothesized to work well for this task. Some examples of predictions are demonstrated in Table 6.

Table 5. Pun Location: Test Set

	EN		FR		ES	
	A	A*	A	A*	A	A*
XLM-RoBERTa	0.792	0.792	0.414	0.471	**0.561**	**0.561**
Llama 3	**0.853**	**0.853**	**0.442**	**0.504**	0.560	0.560

The results of the predictions indicate diverse outcomes. The model demonstrates an ability to identify pun phrases, though its performance lacks consistency. Notably, one example is particularly intriguing; the model successfully identified the relevant concept (vision-blind) within the sentence but erroneously did not highlight the correct term that constituted the pun. The word "vision" should have been identified due to its dual significance – referring both to the faculty of sight and foresight or strategic planning.

Table 6. Pun Location: Selected Outputs (XLM-RoBERTa)

Sentence	Pun Word	Predicted
Weather forecasters have to have lots of degrees	degrees	degrees
Quand des éléphants entrent dans un bar, le patron sait qu'il peut s'attendre á des gros pour boire	des gros pour boire	gros
C'est entre mon nez et mon menton, dit Tom la bouche encœur	la bouche encœur	la bouche encœur
Some people with a lot of vision started the blind institute	vision	blind

Due to the generative nature of large language models, these models often do not meet the stringent criteria necessary for precisely identifying pun words directly from sentences, as evident in Table 7. In numerous cases, the LLM significantly alters the word, thus "choosing" a term that was not originally present in the sentence. Additionally, qualitative assessments of LLM outputs indicate that these models are more adept at identifying phrases consisting of multiple words and generally perform as well as, or slightly better than, alternative methods. This performance advantage occurs without the risk of over-fitting, as the LLM methodology does not involve any fine-tuning step.

Table 7. Pun Location: Selected Outputs (Llama 3)

Sentence	Pun Word	Predicted
Weather forecasters have to have lots of degrees	degrees	degrees
Quand des éléphants entrent dans un bar, le patron sait qu'il peut s'attendre á des gros pour boire	des gros pour boire	gros pour boire
C'est entre mon nez et mon menton, dit Tom la bouche en cœur	la bouche en cœur	en cœur
Some people with a lot of vision started the blind Institute	vision	vision
A man threatened to jump off the side of a building - alledgedly	alledgedly	allegedly
One of the tires just blew out, Tom said sparingly	sparingly	spareingly
Dollars do best when accompanied by some sense	sense	cents
"3.14159265," Tom said piously	piously	pi

3.3 Pun Interpretation

We summarize all the results for pun interpretation in Table 8. Some examples of interpretation outputs[2] are included in Table 9.

Table 8. Pun Interpretation: English

	Train		Test	
	A	A*	A	A*
SentenceEmb-WordNet	0.481	0.481	0.398	0.398

In the cases of commonly used words such as "quiver" and "leaked," our approach demonstrated strong performance. However, challenges arose with words exhibiting different forms across their various meanings. For instance, our system struggled to accurately interpret the word "amply," which in one sense is used to mean "richly," while in another, "amp" refers to "ampere." Additionally, the coverage of WordNet in languages other than English is less comprehensive, which restricts the applicability of this method. Furthermore, the system failed to produce outputs when attempting to match French synonyms that contained special characters for accents.

Table 9. Pun Interpretation: Selected Outputs (XLM-RoBERTa)

Sentence	Interpretation	Predicted
This is where I keep my arrows, said Tom, quivering	palpitate; quake; quiver/quiver	beat; pulsate; quiver/palpitate; quake; quiver
News of a coming flood was leaked	leak/leak	leak; leak out/leak
"It's a unit of electric current," said Tom amply	richly/A; amp; ampere	richly/fully

LLMs demonstrated exceptional performance on the interpretation task, a success we largely attributed to the extensive dataset used during their training. However, they did not consistently conform to the specific forms present in WordNet which could lead to lower scores during evaluation. In qualitative experiments, as evidenced in Table 10, LLMs appeared to surpass other methods used in our work. Nevertheless, a comprehensive evaluation across the entire dataset was impractical due to the late integration of this method beyond the established evaluation timelines, coupled with the significant manual effort required for such an assessment.

---

[2] In Table 9 and Table 10, the symbol/separates the two senses and; separates alternate words for the same sense.

Table 10. Pun Interpretation: Selected Outputs (Llama 3)

Sentence	Interpretation	Predicted
This is where I keep my arrows, said Tom, quivering	palpitate; quake; quiver/quiver	shaking; trembling/quiver; quivers
News of a coming flood was leaked	leak/leak	revealed; disclosed; exposed/flowed; drained; escaped
"It's a unit of electric current," said Tom amply	richly/A; amp; ampere	fully; completely; abundantly/ampere; amperage
OLD PROFESSORS never die, they just lose their class	form; grade; course/elegance	category; group; grade/style; elegance; refinement

4 Conclusion and Future Scope

Sentence embeddings tend to exhibit "noise" by failing to effectively encapsulate the ambiguous meanings integral to puns. Moreover, WordNet's coverage is limited in languages other than English. Enhancing the handling of special characters and accents in non-English languages could improve outcomes, as our study overlooked instances that could not be tokenized. For our LLM-based methods, our research did not explore various prompting strategies, such as augmenting prompts with context derived from humor theory, or employing different definitions and phrasings of "what is a pun." Predominantly, training data for LLMs is in English, leading to diminished performance in other languages. Furthermore, LLMs cannot always precisely pick the intended word or phrase due to their generative nature. Integrating sense dictionaries like WordNet with LLMs is challenging; however, providing a complete WordNet dataset for the predicted pun word may enhance interpretation accuracy. This study was limited to prompts in English, reflecting our linguistic proficiency. Finally, there remains a possibility that the LLM may have been exposed to parts of our dataset during its training, potentially influencing the results.

References

1. Annamoradnejad, I., Zoghi, G.: ColBERT: using BERT sentence embedding in parallel neural networks for computational humor. Expert Syst. Appl. **249**, 123685 (2024)
2. Bai, J., et al.: Qwen technical report. arXiv preprint arXiv:2309.16609 (2023)
3. Chang, Y., et al.: A survey on evaluation of large language models. ACM Trans. Intell. Syst. Technol. **15**(3), 1–45 (2024)
4. Conneau, A., et al.: Unsupervised cross-lingual representation learning at scale. In: Proceedings of the 58th Annual Meeting of the Association for Computational Linguistics, pp. 8440–8451 (2020)

5. Dsilva, R.R.: AKRaNLU@ CLEF JOKER 2023: using sentence embeddings and multilingual models to detect and interpret wordplay. In: Working Notes of the Conference and Labs of the Evaluation Forum (CLEF 2023), Thessaloniki, Greece, 18–21 September 2023. CEUR Workshop Proceedings, vol. 3497, pp. 1846–1853 (2023). https://ceur-ws.org/Vol-3497/paper-154.pdf
6. Dsilva, R.R.: Augmenting large language models with humor theory to understand puns. Master's thesis, Purdue University (2024). https://doi.org/10.25394/PGS.25674792.v1
7. Ermakova, L., Miller, T., Bosser, A.G., Palma, V.M.: Overview of JOKER 2023 automatic wordplay analysis task 2–pun location and interpretation (2023)
8. Ermakova, L., Miller, T., Bosser, A.G., Palma Preciado, V.M., Sidorov, G., Jatowt, A.: Overview of JOKER-CLEF-2023 track on automatic wordplay analysis. In: Arampatzis, A., et al. (eds.) CLEF 2023. LNCS, vol. 14163, pp. 397–415. Springer, Cham (2023). https://doi.org/10.1007/978-3-031-42448-9_26
9. Jentzsch, S., Kersting, K.: ChatGPT is fun, but it is not funny! Humor is still challenging large language models. In: Proceedings of the 13th Workshop on Computational Approaches to Subjectivity, Sentiment, & Social Media Analysis, pp. 325–340 (2023)
10. Jiang, A.Q., et al.: Mistral 7b (2023)
11. Reimers, N., Gurevych, I.: Sentence-BERT: sentence embeddings using Siamese BERT-networks. In: Proceedings of the 2019 Conference on Empirical Methods in Natural Language Processing and the 9th International Joint Conference on Natural Language Processing (EMNLP-IJCNLP), pp. 3982–3992 (2019)
12. Taylor, J.M., Mazlack, L.J.: Computationally recognizing wordplay in jokes. In: Proceedings of the Annual Meeting of the Cognitive Science Society, vol. 26 (2004)
13. Touvron, H., et al.: LLaMA: open and efficient foundation language models. arXiv preprint arXiv:2302.13971 (2023)
14. Zou, Y., Lu, W.: Joint detection and location of English puns. In: Proceedings of the 2019 Conference of the North American Chapter of the Association for Computational Linguistics: Human Language Technologies, Volume 1 (Long and Short Papers), pp. 2117–2123 (2019)

Replicability Measures for Longitudinal Information Retrieval Evaluation

Jüri Keller, Timo Breuer, and Philipp Schaer

Technische Hochschule Köln, Ubierring 48, 50678 Cologne, Germany
{jueri.keller,timo.breuer,philipp.schaer}@th-koeln.de
https://ir.web.th-koeln.de

Abstract. Information Retrieval (IR) systems are exposed to constant changes in most components. Documents are created, updated, or deleted, the information needs are changing, and even relevance might not be static. While it is generally expected that the IR systems retain a consistent utility for the users, test collection evaluations rely on a fixed experimental setup. Based on the LongEval shared task and test collection, this work explores how the effectiveness measured in evolving experiments can be assessed. Specifically, the persistency of effectiveness is investigated as a replicability task. It is observed how the effectiveness progressively deteriorates over time compared to the initial measurement. Employing adapted replicability measures provides further insight into the persistence of effectiveness. The ranking of systems varies across retrieval measures and time. In conclusion, it was found that the most effective systems are not necessarily the ones with the most persistent performance.

Keywords: Retrieval Effectiveness · Longitudinal Evaluation · Continuous Evaluation · Replicability

1 Introduction

The environment of a retrieval system changes constantly. Not only but especially web retrieval systems are exposed to this due to the dynamic nature of the web. Documents, i.e., websites, get created, updated, or deleted [4,12]. But besides the evolving collection, the other components underlay change as well, from the information needs [13] to the relevance of search results [9,27]. These changes raise questions about the generalizability, temporal validity, and the persistency of Information Retrieval (IR) system effectiveness evaluations.

The LongEval shared task [1][1] seeks to investigate the temporal persistence of retrieval systems in a longitudinal evaluation. It, therefore, provides a first-of-its-kind web retrieval collection with sub-collections from different points in time [14]. These sub-collections resemble the Evaluation Environment (EE) a retrieval system is exposed to and allow to investigate how temporal changes

---

[1] https://clef-longeval.github.io.

influence retrieval systems [25]. The overall goal of the LongEval lab is to examine the *temporal persistence* of retrieval systems. While the influence of temporal changes on the retrieved results are undeniable, it is unclear how the changes in effectiveness should be valued. For example, an over time increasing effectiveness would yield reliably good results. In this case, the users may profit, but the effectiveness would still change and quickly become unknown. Therefore, we argued that from an evaluation point of view, it can be desirable to investigate temporal reliability as persistence. In this work, we investigate the temporal persistence as a replicability task. Oriented at the ACM definition of replicability[2], the goal is to achieve the same measurements in a different experimental setup, in this case, at a proceeded point in time. We investigate the temporal persistence of five advanced retrieval systems as a replicability problem. The systems are not specifically adapted to changes in the LongEval dataset to validate the temporal reliability of system-oriented IR evaluations following the Cranfield paradigm. To facilitate reproducibility we make the code publicly available on GitHub.[3]

2 Related Work

The LongEval dataset [2] and shared task [1] provides the first test bed for investigating the temporal persistence of IR systems. In the ongoing shared task, IR systems are evaluated across three points in time and the relative change in effectiveness based on nDCG is measured by the Result Delta ($\mathcal{R}_e\Delta$). Based on the submitted systems, no connection between effectiveness and temporal robustness was found but substantial correlation between the ranking of systems across the different points in time. González-Sáez [24] described different strategies for comparing IR systems in evolved environments. Beyond tracking one system across time also different systems are compared at different points in time. To maintain comparability different strategies are explored that use a pivot system, project scores to a common scale, or group topics into grains.

Directly related to the comparison strategy proposed in this work, González-Sáez et al. [25] achieve comparability by relating the results of different systems at different points in time to the same pivot system and compare only the measured deltas. In this work, also a pivot system is used but the same systems are compared in an environment with reduced dynamics.

Besides the comparability of effectiveness, the temporal influences on test collections was investigated earlier by Soboroff [26]. He used the bpref measure to achieve a robuster ranking of systems on an evolving version of the GOV2 test collection. Further, he proposes indicators that describe how the test collection changed which can help to maintain it. Tonon et al. [28] describe test collection maintenance in an "evaluation as a service" methodology. To achieve reliable evaluations it is quantified how fair the current state of a test collection assesses a new system and estimates the cost of updating the test collection.

---

[2] https://www.acm.org/publications/policies/artifact-review-and-badging-current.
[3] https://github.com/irgroup/CLEF2023-LongEval-IRC.

More works directly describe the changes in datasets, focusing on different components and granularity's [9,13,15,17,27]. These works are valuable sources to relate the changes in effectiveness back to the changes in the EE.

3 Temporal Replicability

To analyze how the effectiveness evolves over time, we cast the longitudinal evaluation into a replicability task, i.e., we evaluate the same set of systems on different data. Naturally, a direct comparison of the measured effectiveness scores of the different EEs is difficult since the recall base is not the same anymore. This makes it difficult to directly compare scores, and it remains unclear if the observed effects should be attributed to the system or the changing EE. An advanced comparison strategy is necessary to overcome this problem [24]. In this work, we further explore the pivot strategy [5,25] in which the results of one system in one EE are related to a pivot system that is evaluated in the same EE. The delta between the experimental and the pivot system is then compared to a delta between the same systems measured in an evolved EE. To align the terminology, the pivot system is a baseline run, BM25 for simplicity in this example, and the advanced run is the experimental system investigated. The intuition behind this evaluation strategy is that since the pivot system is exposed to the same EE as the experimental system, hence encountering the same difficulties, it represents a neutral reference point that makes the results more comparable.

In the LongEval shared task, the $\mathcal{R}_e\Delta$ is used to describe how the effectiveness of the retrieval systems evolves over time. In this setting, the $\mathcal{R}_e\Delta$ is defined as reproduced and will serve as a baseline measure [1]:

$$\mathcal{R}_e\Delta = \frac{\overline{M^{EE}(S)} - \overline{M^{EE'}(S)}}{\overline{M^{EE}(S)}}. \tag{1}$$

The $\mathcal{R}_e\Delta$ directly compares the mean retrieval effectiveness of a system S quantified by a measure M between the sub-collection EE and EE'. Improved effectiveness is denoted by a negative $\mathcal{R}_e\Delta$, and values closer to 0 denote smaller changes which indicate more persistent systems.

In addition to the $\mathcal{R}_e\Delta$, we adapt the Delta Relative Improvement (ΔRI) and the Effect Ratio (ER), initially proposed by Breuer et al. [5] as replicability measures, to investigate the temporal persistence of retrieval effectiveness. The replicability measures are implemented with the help of repro_eval [6], which is a dedicated reproducibility and replicability evaluation toolkit.

The ΔRI describes how the effectiveness relatively changed from one EE to an evolved EE'. It is based on the Relative Improvements (RI) of an experimental system S over the pivot system P. The RI is adapted to the LongEval definitions as follows:

$$\text{RI} = \frac{\overline{M^{EE}(S)} - \overline{M^{EE}(P)}}{\overline{M^{EE}(P)}}, \quad \text{RI}' = \frac{\overline{M^{EE'}(S)} - \overline{M^{EE'}(P)}}{\overline{M^{EE'}(P)}}. \tag{2}$$

M^{EE} denotes the effectiveness score of a measure M, e.g., nDCG, determined on the sub-collection EE or EE' respectively. The ΔRI is then defined as:

$$\Delta\text{RI} = \text{RI} - \text{RI}'. \tag{3}$$

Comparing different sub-collections is straightforward. The ideal ΔRI of 0 is achieved if the RI is the same between both sub-collections, indicating a system that performs robustly over time. The more ΔRI deviates from 0, the less robust is the system, whereas negative scores indicate a more effective experimental system S in the evaluation environment EE', and higher scores correspond to a less effective experimental systems than in the evaluation environment EE.

While the ΔRI describes the change in effectiveness, the ER describes the persistence of the effectiveness. It is originally defined by the ratio between relative improvements of an advanced run over a baseline run. The relative improvements are based on the per-topic improvements, which are adapted for changing EEs as follows:

$$\Delta M_j^{EE} = M_j^{EE}(S) - M_j^{EE}(P) \tag{4}$$

where ΔM_j^{EE} denotes the difference in terms of a measure M between the pivot system P and the experimental system S for the j-th topic of the evaluation environment EE. Correspondingly, $\Delta' M_j^{EE'}$ denotes the topic-wise improvement in the evaluation environment EE'. The ER is then defined as:

$$\text{ER}(\Delta' M^{EE'}, \Delta M^{EE}) = \frac{\overline{\Delta' M^{EE'}}}{\overline{\Delta M^{EE}}} = \frac{\frac{1}{n_{EE'}}\sum_{j=1}^{n_{EE'}} \Delta' M_j^{EE'}}{\frac{1}{n_{EE}}\sum_{j=1}^{n_{EE}} \Delta M_j^{EE}}. \tag{5}$$

More specifically, the mean improvement per topic between the pivot and experimental system on one sub-collection EE in comparison to the effect on the other sub-collection EE' is measured. Thereby, the ER is sensitive to the effect size. If the effect size is completely replicated in the second sub-collection, the ER is 1, i.e., the retrieval system is robust. If the ER is between 0 and 1, the effect is smaller, indicating a less robust system with performance drops. If the ER is larger than 1, the effect is larger, indicating performance gains caused by the change of the EE.

4 Experimental Evaluation

The proposed measures are tested in an experimental evaluation based on the LongEval test collection. The test collection is limited to the queries that are present in all sub-collections to reduce the dynamics and improve interpretability. Five retrieval systems and a BM25 baseline are compared, and the results for different effectiveness, persistency, and replicability measures are reported.

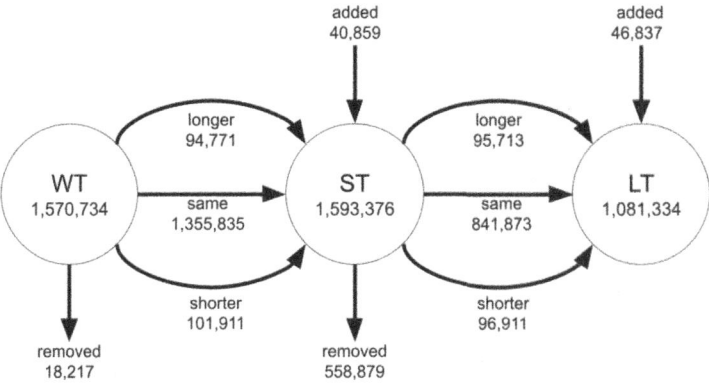

Fig. 1. The evolution of the LongEval test collection documents across the three sub-collections. Over time, documents are added, removed, or updated. All documents were harmonized by their URLs.

4.1 LongEval Test Collection

To our knowledge, the LongEval test collection [14] is the first dataset specifically designed to investigate temporal changes in IR. It consists of consecutive sub-collections that represent snapshots of a web search scenario evolving over time. The documents, topics, and qrels originate from the French, privacy-focused search engine Qwant.[4] Logged user queries are selected as topics for the test collection, and the qrels are created from logged user interactions based on the Cascade Click Model [8,11]. Therefore, the documents and queries are mostly in French, but there are also English machine translations available, which are mainly used in this work. The collections are organized into three sub-collections. The within time (WT) sub-collection was created in June 2022. The short-term (ST) sub-collection was created in July 2022, immediately after the WT collection. The third sub-collection, long term (LT), contains more distant data as it was created with a two-month gap from ST in September 2022. The changes in the document component are classified on a high level based on the string length in Fig. 1. We note that between ST and LT considerably more documents are removed from the collections than between WT and ST. The topic sets also change across sub-collections, leaving a core set of 124 queries present in all sub-collections. The queries are typical keyword queries composed of at least one word and up to 11 words with few outliers. On average, a query consists of 2.5 words. The qrels classify the documents' relevance on a three-graded scale, including *not relevant*, *relevant*, and *highly relevant* labels. In general, the dataset has few assessed documents per topic. While the mean number of qrels is 14 per topic, the absolute number fluctuates between 2 and 59. Most of the documents are marked as not relevant, and the distribution of relevant and highly relevant qrels is skewed as well. Highly relevant documents are rare, with a maximum

---

[4] https://www.qwant.com/.

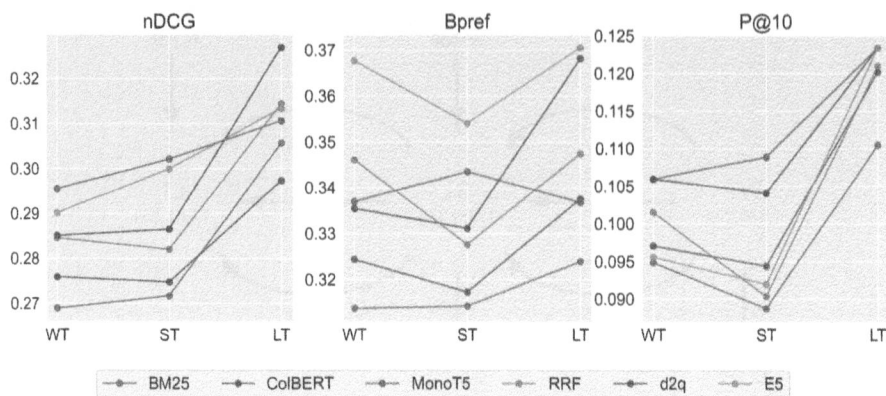

Fig. 2. The P@10, bpref, and nDCG results based on the core queries.

of only four and a mean of only one highly relevant document per topic. In the evaluations, these single documents heavily influence the final outcome as their position in the ranking especially impacts the score of rank-based measures like nDCG. For this work, we entirely rely on the English automatic translations of the test collection.

4.2 Experimental Systems

We compared different ranking functions and multi-stage retrieval systems on the WT train slice of the LongEval dataset. The systems were selected as they represent state-of-the-art, off-the-shelf methods that are used in many recent IR experiments. Therefore, it is especially interesting how these systems behave over time without being specifically adapted to a changing environment. The BM25 [23] ranking function is used as the baseline and first-stage ranker for the advanced systems colBERT [20] and monoT5 [22]. Further, Reciprocal Rank Fusion (RRF) [10] of the runs from BM25 with Bo1 [3] reranking, DFR χ^2 and PL2, E5_base [29] as a dense retrieval system on the full dataset and d2q with ten expanded queries per document and BM25 as the retriever are tested. For a detailed description of the experimental systems, we refer the reader to the working notes [18] and the GitHub repository.[5]

4.3 Results

For the evaluation of the result, the main goal is not a high but rather persistent performance. Therefore, the Average Retrieval Performance (ARP) across EEs is compared to the $\mathcal{R}_e \Delta$, and also the replicability measures ΔRI, ER, and the p-values of unpaired t-tests. The results measured by P@10, nDCG [16], and bpref [7] are reported in Table 1, and the ARP is visualized in Fig. 2.

---

[5] https://github.com/irgroup/CLEF2023-LongEval-IRC.

Table 1. Results of the persistency of effectiveness, measured on the core queries of the LongEval test collection. The replicability measures can not measure any persistancy for BM25 since this system is also used as the pivot. The ideal values of the replicability measures are noted at WT, the most persistent results are highlighted in bold, and results significantly different from BM25 at the same sub-collection are denoted by *.

		P@10					bpref					nDCG				
		ARP	$\mathcal{R}_e\Delta$	ΔRI	ER	p-val	ARP	$\mathcal{R}_e\Delta$	ΔRI	ER	p-val	ARP	$\mathcal{R}_e\Delta$	ΔRI	ER	p-val
BM25	WT	0.095	0	–	–	–	0.314	0	–	–	–	0.269	0	–	–	–
	ST	0.089	0.064	–	–	–	0.314	−0.002	–	–	–	0.272	−0.010	–	–	–
	LT	0.110	−0.165	–	–	–	0.324	−0.033	–	–	–	0.306	−0.137	–	–	–
colBERT	WT	0.097	0	0	1	1	0.324	0	0	1	1	0.276	0	0	1	1
	ST	0.094	0.028	−0.040	2.540	0.858	0.317	0.022	0.024	0.286	0.826	0.275	**0.004**	0.015	0.441	**0.967**
	LT	0.120	−0.238	−0.064	4.355	0.178	0.338	−0.041	−0.008	1.278	0.668	0.297	−0.078	0.053	−1.198	0.412
monoT5	WT	**0.106**	0	0	1	1	0.337	0	0	1	1	**0.295**	0	0	1	1
	ST	0.109	−0.028	−0.110	1.815	0.857	0.344	−0.019	−0.019	1.261	0.850	**0.302**	−0.023	−0.013	1.146	0.817
	LT	**0.123**	**−0.165**	**0.000**	**1.161**	**0.332**	0.337	**0.000**	0.034	0.553	**0.997**	0.311	**−0.051**	0.083	0.187	0.580
RRF	WT	0.101	0	0	1	1	0.346*	0	0	1	1	0.285*	0	0	1	1
	ST	0.090	0.110	0.052	0.242	0.453	0.328	0.054	0.032	0.574	0.784	0.282	0.009	**0.003**	**0.925**	**0.945**
	LT	0.121	−0.192	−0.025	1.573	0.237	0.347*	−0.004	**0.002**	**1.007**	**0.756**	0.314	−0.105	0.013	**0.786**	0.227
d2q	WT	**0.106***	0	0	1	1	0.335	0	0	1	1	0.285	0	0	1	1
	ST	0.104*	0.018	−0.056	1.379	**0.911**	0.331	0.013	**0.015**	**0.779**	**0.894**	0.287	−0.005	0.006	0.916	0.960
	LT	**0.123**	**−0.165**	**0.000**	**1.161**	0.326	0.368*	−0.098	−0.067	2.034	0.300	**0.327***	−0.147	**−0.010**	1.317	0.150
E5	WT	0.096	0	0	1	1	**0.368***	0	0	1	1	0.290	0	0	1	1
	ST	0.092	0.038	**−0.029**	**4.355**	0.815	**0.354**	0.037	0.045	0.738	0.692	0.300	−0.034	−0.025	1.333	0.720
	LT	**0.123**	−0.291	−0.109	17.419	0.111	**0.371**	−0.008	0.028	0.863	0.931	0.313	−0.080	0.054	0.362	0.382

The effectiveness is similar for the systems but varies across EEs. Overall, the results of the tested systems improves in the long run with few exceptions, as measured by bpref. Mainly in the second EE (ST), weaker results are achieved. Also, the ranking of systems varies across time and measure. In the first two EEs, monoT5 performs well, only outperformed by E5 as measured by bpref. In the last EE (LT), the d2q, RRF, and E5 systems perform better than monoT5, except on P@10.

The $\mathcal{R}_e\Delta$ reflects the general upward trend in effectiveness indicated by decreasing negative values. While the $\mathcal{R}_e\Delta$ at ST is negative for all systems except RRF and colBERT measured by nDCG, regarding bpref it is also positive for E5 and d2q. The more the $\mathcal{R}_e\Delta$ diverges from 0, the larger is the relative change and the less persistent the system performs. Regarding the different measures the $\mathcal{R}_e\Delta$ is instantiated with, no strong agreement for the most persistent system can be found in ST. d2q, BM25, and ColBER achieve the most persistent results on P@10, bpref and nDCG. For the LT EE monoT5 achieves the most persistent results on all measures, accompanied by BM25 and d2q in P@10.

The ΔRI and ER complement the $\mathcal{R}_e\Delta$. For instance, monoT5 achieved similar bpref scores on WT and LT, resulting in a $\mathcal{R}_e\Delta$ score of 0, which indicates perfect robustness in terms of $\mathcal{R}_e\Delta$. However, when comparing ΔRI and also ER, more granular analysis is possible. In this case, the scores are close to but different from the perfect scores of 1 and 0, respectively, which would indicate perfect robustness. Regarding bpref, d2q achieves the best persistency according to ΔRI and ER in ST and RRD in LT. For the other measures, less agreement can be found. The full potential of the ER and ΔRI can be seen if plotted against each other as in Fig. 3. The closer the systems are located to the point (1, 0), the

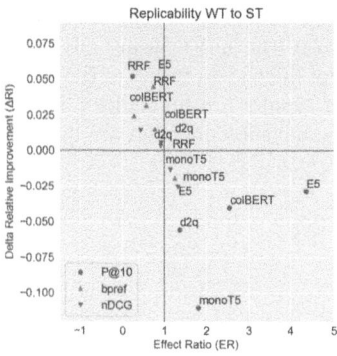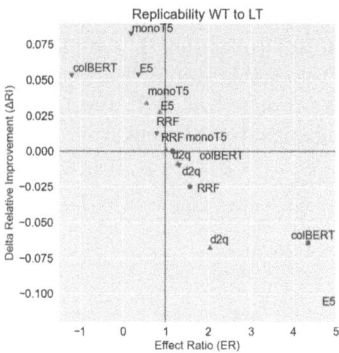

Fig. 3. The ER plotted against the ΔRI for the replication WT to ST (left) and WT to LT (right). The ER for E5 is excluded as an outlier.

more persistent they are, with the preferable regions bottom right and top left. For the comparison of WT to ST, the monoT5 system performs well on bpref and nDCG. However, the effect and the absolute scores are slightly larger. E5 and monoT5 show large differences measured by P@10, with a larger effect for E5 (ER) and a stronger improvement for monoT5 (ΔRI). The RRF system, like most others, shows smaller absolute scores according to the ΔRI and a slightly decreased ER. The plot regarding WT to LT shows more outliers with larger effect sizes for P@10 for the E5 system (ER = 17.419) and bpref for the d2q system. The systems are shifted to the top right of the plot, a trend similar to the increased $\mathcal{R}_e\Delta$ for WT to LT.

5 Discussion and Limitations

As initially mentioned, the notion of temporal persistence remains challenging to grasp. From the user's perspective, it might likely be desirable to always get the best results possible, even if the utility varies. Therefore, improving a system to perform more persistent is not beneficial, and direct implications for system design can not be derived. Instead, the potential in persistence evaluations lies in learning about the evaluation and test bed, quantifying the temporal validity of results, and the influence of the point in time when a test collection is created.

Comparing retrieval systems across time is difficult due to the changes in the experimental setups. It is unclear how to attribute the measured differences. Depending on the degree of change, a direct comparison of the ARP might not be sufficient or even meaningful since the recall base changes. As described before, in direct comparison, for example, through the $\mathcal{R}_e\Delta$, the effect of the evolved environment is mainly extracted [25]. The replicability measures provide a method to abstract this effect to some extent and make the results comparable through the pivot system. The experimental results showed that, in general, the $\mathcal{R}_e\Delta$ scores do not always agree on the most robust system with ER and ΔRI. Based on these findings, we conclude that the replicability measures provide another

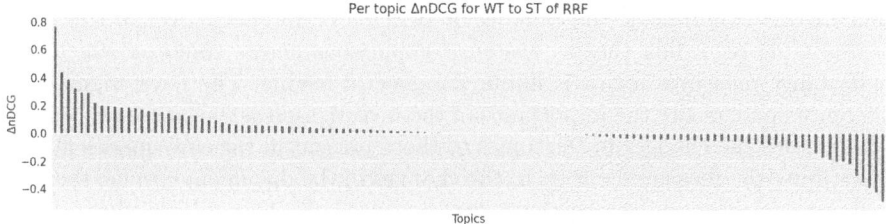

Fig. 4. RRF $\Delta nDCG$ results per topic for WT to ST. The topics are ordered according to the delta.

robustness perspective. We further see that it is not enough to consider the differences of a single retrieval measure like nDCG. Depending on the evaluation measure, different systems perform best in terms of robustness. For instance, $\mathcal{R}_e\Delta$ on ST of nDCG is lower for colBERT and RRF than that of monoT5, while $\mathcal{R}_e\Delta$ of P@10 is lower or equal for monoT5. Similarly, the replicability measures should be instantiated with different retrieval measures to get a more comprehensive understanding of robustness. While the RRF system achieves the best ER instantiated with nDCG on both EEs, monoT5 is the most robust system in terms of ER instantiated with P@10. Likewise, ER and ΔRI identify different systems as the most robust for the same measures and tasks, which shows that it is insightful to evaluate both replicability measures.

In addition, we also included the p-values of unpaired tests based on the topic score distributions from different EE that were determined with the same experimental system as proposed in [5]. The general idea of these evaluations proposes to assess the quality of replicability (in our case, robustness) by the p-values. It follows the assumption that lower p-values give a higher probability of failed replications or systems that are not robust. As can be seen, the highest p-values are achieved for the monoT5, colBERT, or d2q, which generally agrees with our earlier observations.

The $\mathcal{R}_e\Delta$ directly compares the results averaged across topics, but this ARP may hide differences between the topic score distributions [5]. For example, the RRF system achieved a high nDCG (0.285) at WT and is relatively stable at ST considering the $\mathcal{R}_e\Delta$ of 0.009. However, the per-topic results fluctuate between −0.4 and 0.8, as shown in Fig. 4. For some topics, the retrieval performance improves, while the changes in the EE harm retrieval performance for other topics. We note that these circumstances require a more in-depth evaluation.

The experimental setup in this work limits the topic set to of the LongEval test collection to the core queries that are present in all sub-collections, thereby reducing the number of changing factors. In comparison, the effectiveness measured using the full test collection with all queries appears to be higher and demonstrates a stronger increase [1,18]. Generally, in this setting, the results for the different systems tend to be more similar. This is also reflected in the fewer significant differences per sub-collection between the experimental and the BM25 baseline system. Consequently, since only a few improvements are significant in this experiment, the ranking of systems is unreliable. While this may be negligible regarding the per-system comparisons across time, on which the replicability measures focus, it limits the general results. The fewer significant differences underscore the importance of the investigated retrieval scenario. Narrowing down the changes in the topics to those present in the core queries allows to attribute the measured effects to the changes in the document corpus, thereby improving interpretability. However, the measured effect also diminishes.

Further questions regard the relation between sub-collections. The disagreement between the $\mathcal{R}_e\Delta$ and the replicability measures might indicate the differences between sub-collections. While the sub-collections are related in time, it remains unclear what constitutes this context, especially regarding the effectiveness. This fosters the need to investigate what differentiates a longitudinal evaluation from a cross test collection evaluation.

This study is limited as it only considers the queries present in all sub-collections of LongEval, and no attempts were made to generalize across further test collections or retrieval scenarios. We note that the interpretation of results remains difficult, among others, because of the unintuitive notion of effectiveness persistence. Also, only BM25 was considered as pivot system for the replicability measures.

6 Conclusion

In this work, we investigated the utility of replicability measures to describe how persistent retrieval systems perform over time. We applied five retrieval systems to the LongEval test collection and quantified how the effectiveness changes. The results showed that the retrieval effectiveness for most systems and measures increased over time on the LongEval dataset. The measured effectiveness deteriorates over time, which aligns with the natural assumption that results spanning longer timeframes are more different. Further, we report preliminary results applying replicability measures to quantify temporal persistence, an extension on common practices of these measures and their interpretation [21]. It was shown that the results based on different measures and likewise for different topics do not necessarily agree with each other. Therefore, we see great potential in using replicability measures to gain further insights into robustness and also saw similarities to the measured result deltas. All in all, the strong influence of the experimental setup on the system's results could be shown and was analyzed. Since temporal persistence is a new challenge, interpreting the results is difficult.

While these results are limited to the LongEval scenario, future work will extend the evaluation to further evaluation scenarios with different changes and dynamics [19]. Aligning the documents of different sub-collections would enable to investigate the persistence on an even more specific level, for example, by casting the problem as a reproducibility task. Further open questions regard the selection of the pivot system to make the scores comparable and the selection of queries that allow meaningful temporal comparisons. Since the notion of temporal change remains difficult future work should regard generalizing persistence to temporal change. Lastly, an overall goal would be to employ the gained insights about temporal change to assess the temporal validity of evaluations.

Acknowledgments. We would like to express our gratitude to the LongEval Shared Task organizers for their invaluable efforts in constructing the LongEval dataset used in this study. Their dedication and hard work have provided an essential foundation for our research. We also gratefully acknowledge the support of the German Research Foundation (DFG) through project grant No. 407518790.

Disclosure of Interests. The authors have no competing interests to declare that are relevant to the content of this article.

References

1. Alkhalifa, R., et al.: Overview of the CLEF-2023 LongEval lab on longitudinal evaluation of model performance. In: Arampatzis, A., et al. (eds.) CLEF 2023. LNCS, vol. 14163, pp. 440–458. Springer, Cham (2023). https://doi.org/10.1007/978-3-031-42448-9_28
2. Alkhalifa, R., et al.: Extended overview of the CLEF-2023 LongEval lab on longitudinal evaluation of model performance. In: CLEF (Working Notes). CEUR Workshop Proceedings, vol. 3497, pp. 2181–2203. CEUR-WS.org (2023)
3. Amati, G.: Probability models for information retrieval based on divergence from randomness. Ph.D. thesis, University of Glasgow, UK (2003)
4. Bar-Ilan, J.: Criteria for evaluating information retrieval systems in highly dynamic environments. In: WebDyn@WWW. CEUR Workshop Proceedings, vol. 702, pp. 70–77. CEUR-WS.org (2002)
5. Breuer, T., et al.: How to measure the reproducibility of system-oriented IR experiments. In: SIGIR, pp. 349–358. ACM (2020)
6. Breuer, T., Ferro, N., Maistro, M., Schaer, P.: `repro_eval`: a Python interface to reproducibility measures of system-oriented IR experiments. In: Hiemstra, D., Moens, M.-F., Mothe, J., Perego, R., Potthast, M., Sebastiani, F. (eds.) ECIR 2021. LNCS, vol. 12657, pp. 481–486. Springer, Cham (2021). https://doi.org/10.1007/978-3-030-72240-1_51
7. Buckley, C., Voorhees, E.M.: Retrieval evaluation with incomplete information. In: SIGIR, pp. 25–32. ACM (2004)
8. Chapelle, O., Zhang, Y.: A dynamic Bayesian network click model for web search ranking. In: WWW, pp. 1–10. ACM (2009)
9. Clarke, C.L.A., et al.: Novelty and diversity in information retrieval evaluation. In: SIGIR, pp. 659–666. ACM (2008)
10. Cormack, G.V., Clarke, C.L.A., Büttcher, S.: Reciprocal rank fusion outperforms condorcet and individual rank learning methods. In: SIGIR, pp. 758–759. ACM (2009)
11. Craswell, N., Zoeter, O., Taylor, M.J., Ramsey, B.: An experimental comparison of click position-bias models. In: WSDM, pp. 87–94. ACM (2008)
12. Dumais, S.T.: Temporal dynamics and information retrieval. In: CIKM, pp. 7–8. ACM (2010)
13. Dumais, S.T.: Putting searchers into search. In: SIGIR, pp. 1–2. ACM (2014)
14. Galuscáková, P., et al.: LongEval-retrieval: French-English dynamic test collection for continuous web search evaluation. In: SIGIR, pp. 3086–3094. ACM (2023)

15. Hopfgartner, F., et al.: Continuous evaluation of large-scale information access systems: a case for living labs. In: Ferro, N., Peters, C. (eds.) Information Retrieval Evaluation in a Changing World. TIRS, vol. 41, pp. 511–543. Springer, Cham (2019). https://doi.org/10.1007/978-3-030-22948-1_21
16. Järvelin, K., Kekäläinen, J.: Cumulated gain-based evaluation of IR techniques. ACM Trans. Inf. Syst. **20**(4), 422–446 (2002)
17. Jensen, E.C., Beitzel, S.M., Chowdhury, A., Frieder, O.: Repeatable evaluation of search services in dynamic environments. ACM Trans. Inf. Syst. **26**(1), 1 (2007)
18. Keller, J., Breuer, T., Schaer, P.: Evaluating temporal persistence using replicability measures. In: CLEF (Working Notes). CEUR Workshop Proceedings, vol. 3497, pp. 2441–2457. CEUR-WS.org (2023)
19. Keller, J., Breuer, T., Schaer, P.: Evaluation of temporal change in IR test collections. In: ICTIR. ACM (2024)
20. Khattab, O., Zaharia, M.: ColBERT: efficient and effective passage search via contextualized late interaction over BERT. In: SIGIR, pp. 39–48. ACM (2020)
21. Maistro, M., Breuer, T., Schaer, P., Ferro, N.: An in-depth investigation on the behavior of measures to quantify reproducibility. Inf. Process. Manag. **60**(3), 103332 (2023)
22. Pradeep, R., Nogueira, R.F., Lin, J.: The expando-mono-duo design pattern for text ranking with pretrained sequence-to-sequence models. CoRR abs/2101.05667 (2021)
23. Robertson, S.E., Walker, S., Jones, S., Hancock-Beaulieu, M., Gatford, M.: Okapi at TREC-3. In: TREC. NIST Special Publication, vol. 500-225, pp. 109–126. National Institute of Standards and Technology (NIST) (1994)
24. Saez, G.G.: Continuous evaluation framework for information retrieval systems. Theses, Université Grenoble Alpes [2020-....] (2023). https://theses.hal.science/tel-04547265
25. González-Sáez, G.N., Mulhem, P., Goeuriot, L.: Towards the evaluation of information retrieval systems on evolving datasets with pivot systems. In: Candan, K.S., et al. (eds.) CLEF 2021. LNCS, vol. 12880, pp. 91–102. Springer, Cham (2021). https://doi.org/10.1007/978-3-030-85251-1_8
26. Soboroff, I.: Dynamic test collections: measuring search effectiveness on the live web. In: SIGIR, pp. 276–283. ACM (2006)
27. Tikhonov, A., Bogatyy, I., Burangulov, P., Ostroumova, L., Koshelev, V., Gusev, G.: Studying page life patterns in dynamical web. In: SIGIR, pp. 905–908. ACM (2013)
28. Tonon, A., Demartini, G., Cudré-Mauroux, P.: Pooling-based continuous evaluation of information retrieval systems. Inf. Retr. J. **18**(5), 445–472 (2015)
29. Wang, L., et al.: Text embeddings by weakly-supervised contrastive pre-training. CoRR abs/2212.03533 (2022)

SimpleText Best of Labs in CLEF-2023: Scientific Text Simplification Using Multi-prompt Minimum Bayes Risk Decoding

Andrianos Michail[(✉)], Pascal Severin Andermatt[(✉)], and Tobias Fankhauser

Department of Computational Linguistics/Informatics, University of Zurich,
Andreastrasse 15, 8050 Zürich, Switzerland
andrianos.michail@cl.uzh.ch, pandermatt@ifi.uzh.ch

Abstract. We investigate the use of large language models (LLMs) for scientific text simplification in the context of SimpleText in CLEF-2023 Shared Task 3. Our methodology integrates fine-tuning of the Alpaca LoRA 7B model, collecting candidate simplifications from different prompt designs and multistep prompts using data from Task 2, which includes complex terms and their definitions. In this way, we generate a variety of simplification candidates, including candidates that provide definitions of complex terms in the text. The multi-prompt candidates are then re-ranked using Minimum Bayes Risk Decoding with LENS as the utility function, resulting in a number of interesting source-prompt distributions and better SARI. An additional perturbation ablation study is performed, which shows that the efficient reference-less metric Simplicity Level Estimate (SLE) doesn't rate ungrammatical simplifications lower, revealing its inadequacy as a selection criterion. Finally, we observe an ablation between the simplifications of the domain-adapted Alpaca LoRA and the newer LLama3 Instruct, indicating the adequacy of older models to compete with newer stronger models through in-domain instruction tuning and Minimum Bayes Risk Decoding.

Keywords: Scientific Text Simplification · Generative Language Models · Minimum Bayes Risk Decoding · Multi Prompt Ensembling · Prompt Engineering · Large Language Models · SimpleText@CLEF-2023 · Best of CLEF 2023 Labs

1 Introduction

In the rapidly evolving landscape of NLP, the simplification of complex text remains a pervasive challenge. While LLMs have been shown to be adequate for text simplification, there appears to be a large variation in performance across different domains and prompting strategies [10]. When we consider the domain

---

A. Michail and P. S. Andermatt—Equal contribution.

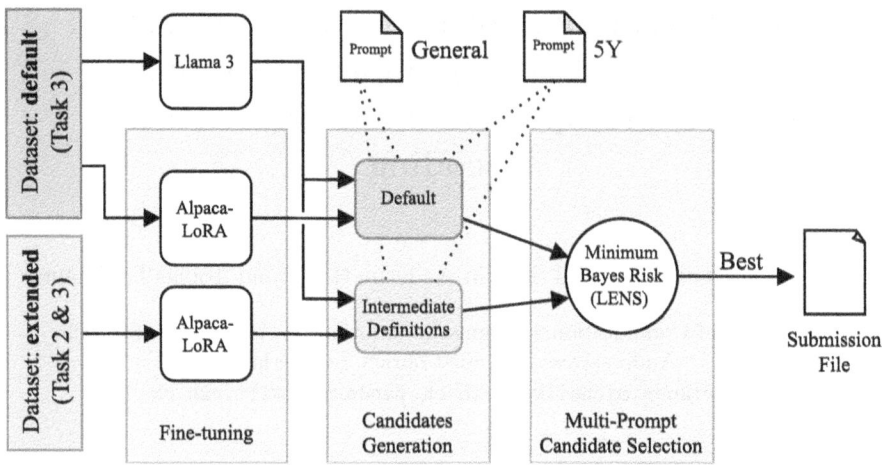

Fig. 1. Complete schematic of the Simplification pipeline.

of scientific texts, where this instability is compounded by the inherently difficult nature of such texts, we find that simple prompting techniques are insufficient to achieve consistent results.

To improve this, we propose a novel approach involving multi-prompting followed by re-ranking. Specifically, we prompt the models with different prompts and then perform Minimum Bayes Risk re-ranking using the Learnable Evaluation Metric for Text Simplification (LENS) [14] as the utility function. We investigate two models, Alpaca LoRA 7B [21] further fine-tuned on pairs of scientific abstract sentences and their simplified versions from the Shared Task SimpleText 2023 [4,5], and off-the-shelf Llama 3 Instruct [1]. For evaluation, we use the test set from the shared task for both models. The resulting simplifications are the four candidates re-ranked separately for each model. The entire schematic of the process can be seen in Fig. 1.

To generate a diverse set of strong candidates, we first fine-tune an Alpaca LoRA model using the data from Task 2 to create explanations of difficult terms and concepts as an intermediate generation step. Then, these predictions are concatenated with the prompt of the simplification model to encourage explanations for the difficult terms. We also compare the results of further fine-tuned Alpaca LoRA with an off-the-shelf Llama 3 Instruct.

This work is a continuation of the shared task submission of UZH_Pandas [2] on the SimpleText 2023 Shared Task [5] and extends the methodology to also examine a newer version of the Llama family, off-the-shelf Llama 3 Instruct [1], as an alternative. Additionally, we assess the suitability of the Simplicity Level Estimate (SLE) metric [3] for candidate re-ranking in simplification tasks. We provide code to replicate this methodology in the repository[1]. Overall, the main contributions and findings of this study are:

---

[1] https://github.com/pandermatt/simpletext-clef2023-2024.

- We show the effectiveness of using an intermediate step of explanation of difficult terms when performing scientific text simplification.
- We compare the results of a fine-tuned version of an older member of the Llama family against performing this task with the latest off-the-shelf Llama 3 Instruct.
- We perform a perturbation ablation study between two methods of scoring candidates and show that ungrammatical texts could be selected if we use a less comprehensive scoring metric.

2 Background

Over the past few years, many new LLMs have been released, providing a diverse range of models from which to choose, with evaluation papers comparing their performance on many different sets of tasks. In a study of summarization, it is claimed that specialized models are now redundant, as LLMs can now perform the task adequately [16]. In this paper, we show that fine-tuning "weaker" models of equal size can remain competitive with their modern counterparts, which exhibit impressive off-the-shelf zero-shot capabilities.

As our main model, we use Alpaca 7B [18]. This is an instruction-following language model, fine-tuned from the original Llama base model [19], pretrained on 1 trillion tokens and instruction-tuned using Low Rank Adaptation (LoRA) [9], which introduces trainable decomposition matrices, which are added at each transformer layer for more efficient training. We further fine-tune this model with the training data from both Task 2 (Explanation of Difficult Terms)and Task 3 (Scientific Text Simplification) [5] for better domain adaptation.

For comparison, we also consider Llama 3 8B Instruct. Compared to the original LLaMA, Llama 3 Instruct is pre-trained on over 15 trillion tokens and combines supervised fine-tuning with reinforcement learning from human feedback with an additional 10 million human-commented examples. The authors showcase that the model achieves strong performance on multiple NLP benchmarks [1].

Within the field of Machine Translation, recent studies showed that re-ranking [7] multiple translation hypothesis through utilizing Minimum Bayes Risk (MBR) decoding [11,12,15] utilizing a quality estimation metric such as COMET as the utility function. Further works [6,20] instead merged such candidates through a text-to-text model, achieving similar performance gains as re-ranking.

Inspired by the previous work, we hypothesize that simplification of scientific texts could benefit from MBR selection of a set of candidates that were generated with different prompts. We examine the following two metrics as utility functions:

Learnable Evaluation Metric for Text Simplification (LENS) is an evaluation metric trained on the SimpEval [14] corpus, a robust dataset featuring human ratings of text simplifications from multiple sources, including GPT-3.5. Through this method, LENS captures nuanced aspects of text simplification.

Input *snt_id: G06.2_2805209921_6*: we have proven that transfer learning is not only applicable in this field, but it requires smaller well-prepared training datasets, trains significantly faster and reaches similar accuracy compared to the original method, even improving it on some aspects.

Prompt General, off-the-shelf Alpaca: Transfer learning is not only applicable in this field, but it requires smaller well-prepared training datasets, trains significantly faster and reaches similar accuracy compared to the original method, even improving it on some aspects.

Prompt General, Domain Adapted Alpaca: Transfer learning requires smaller well-prepared training datasets, trains significantly faster and reaches similar accuracy compared to the original method, even improving it on some aspects.

Prompt 5Y, Domain Adapted Alpaca: Transfer learning is better than the original method.

Fig. 2. Example simplifications generated by the off-the-shelf & Domain Adapted Alpaca with different prompt configurations. The responses show varying degrees of simplifications and details.

Their evaluation shows it aligns more closely with human judgment than traditional metrics, making LENS a promising tool for evaluating and advancing text simplification technologies [14].

Simplicity Level Estimate (SLE) is an efficiently computed, learned reference-less evaluation metric based on Flesch-Kincaid Grade Level (FKGL) [8]. It operates on a regression model architecture fine-tuned from a pretrained RoBERTa base [13], specifically trained on the Newsela dataset [23], which consists of news articles rewritten at various reading levels. SLE has shown high correlation with human judgments of simplicity, making it a promising evaluation metric for selecting the best candidate from our generations [3].

3 Methodology

In this section, we outline our research methodology, first explaining the dataset we use and then describing how we design our simplification prompts and details on fine-tuning the Alpaca model. We then present an ablation on the sensitivity of LENS and SLE to perturbations in texts.

3.1 Dataset

To fine-tune Alpaca, we used the dataset from the SimpleText Shared Task [4]. The primary simplification dataset, from Task 3, consists of a parallel corpus of simplified sentences from the fields of medicine and computer science. The simplification process was carried out either by a Master's student in Technical Writing and Translation or by a duo of a computer scientist and a professional translator. Despite its high quality, the Task 3 dataset was challenging to fine-tune on due to its small size of only 648 text pairs.

Input *snt_id: G06.2_2805209921_6*: We have proven that transfer learning is not only applicable in this field, but it requires smaller well-prepared training datasets, trains significantly faster and reaches similar accuracy compared to the original method, even improving it on some aspects.

Prompt 5Y, fine-tuned Alpaca through ID: Transfer learning is a technique used in machine learning which involves taking a pre-trained model and using it as a starting point for a new model. It requires smaller well-prepared training datasets, trains significantly faster and reaches similar accuracy compared to the original method, even improving it on some aspects.

Prompt 5Y, off-the-shelf Llama3 Instruct through ID We learned that transfer learning is a special trick for teaching computers, which involves using pre-trained models as a starting point for training on a new task, leveraging the knowledge and features learned from the original task to improve performance on the new one, helping them learn faster and better with less information.

Prompt 5Y, off-the-shelf Llama3 Instruct We learned that a special trick for teaching computers helps them learn faster and better with less information.

Fig. 3. Example simplifications generated by the off-the-shelf Llama3 Instruct & the Domain Adapted Alpaca with the 5Y prompt configurations ID and non ID. This example shows the diversity of simplification candidates produced by our multi-prompt approach.

To address this challenge and generate explanations of complex terms, we extend the fine-tuning data by including training instances from Task 2. This dataset, also from medicine and computer science, includes scientific abstracts annotated by a master's student, resulting in 453 annotated samples [4].

3.2 Candidate Hypothesis Generation Through Multiple Prompts

We generate a diverse set of hypothesis candidates using various prompts, aiming to better exploit the properties of Minimum Bayes Risk to obtain a good simplification prediction.

Within our initial observations, the off-the-shelf Alpaca 7B model demonstrated the potential of directly producing text simplification outputs given appropriate instructions. An example is illustrated in Fig. 2. However, by further fine-tuning the model for the task of text simplification, we anticipate that the simplifications produced will be of higher quality and more contextually appropriate. We illustrate stronger simplification candidates generated by the domain-adapted Alpaca LoRA and off-the-shelf Llama3 Instruct in Fig. 3.

We conducted experiments using two distinct prompt templates, selected for their efficacy in preliminary tests. The prompts templates are available in Table 1. For a detailed list of the prompt templates used in our experiments and their full corresponding instruction prompts, please refer to Appendix A.

Table 1. Prompt templates used with and without intermediate definitions. The full prompts are visible in Appendix A.

Target	Prompt	Intuition
P1: General	Simplify the following scientific sentence to make it more understandable for a general audience:	This prompt aims to rephrase complex scientific content into a format that is easier for a general audience to grasp, enhancing overall accessibility and comprehension
P2: 5Y	Simplify the following scientific sentence. Explain it as if you were talking to a 5-year-old, using simple words and concepts:	Break down complex words into their most fundamental elements, using very simple language and concepts that even a child could understand

Simplification Through Intermediate Definitions. We hypothesize that prefixing the simplification model with intermediate definitions of complex terms during inference provides valuable candidates. These definitions are generated by the same LLM in a separate session. This process is visualized through Fig. 4. The inspiration for this approach comes from Chain-of-Thought Prompting, which has been shown to improve complex reasoning tasks [22]. We refer to the simplifications generated with this approach as being generated through *Intermediate Definitions (ID)*.

3.3 Fine-Tuning Hyperparameters

Each fine-tuning iteration of Alpaca LoRA was conducted on an off-the-shelf Alpaca 7B model, resulting in the fine-tuning of a single prompt on each model.

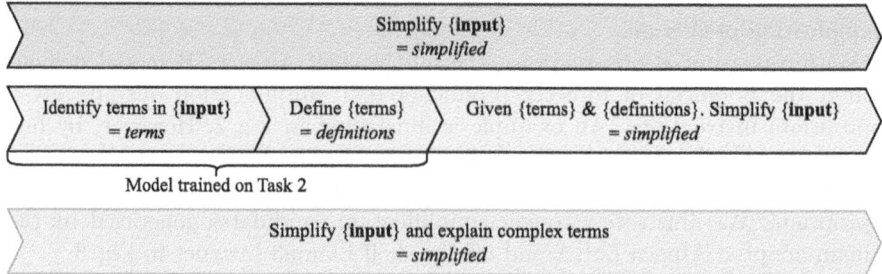

Fig. 4. Illustration of the Default, Intermediate Definitions pipeline. The Intermediate Definition process involves identifying and explaining difficult terms first and then performing the text simplification task given the source and the definitions.

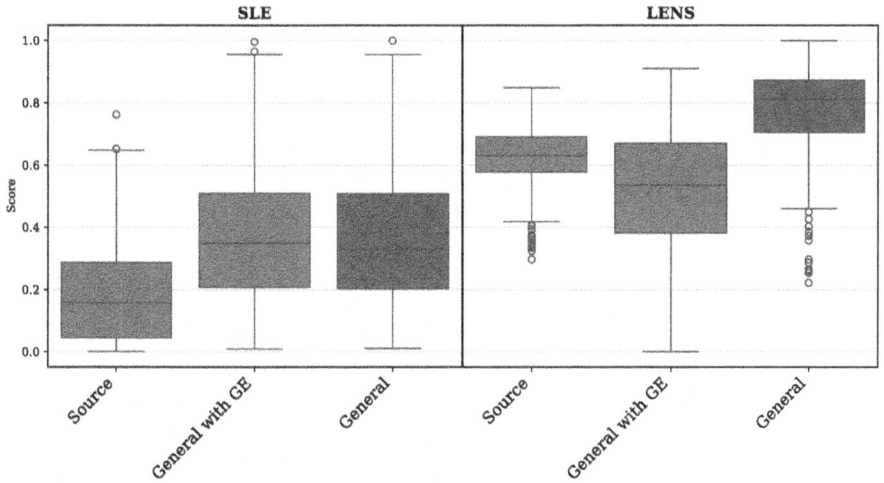

Fig. 5. Box plots of SLE and LENS quality estimation scores of different texts.

An initial exploration of 3–10 epochs and batch sizes of 32 and 64 was conducted. A 16 GB Tesla T4 GPU was utilized, with each training epoch taking approximately 8 min.

3.4 Utility Function Selection

As both LENS [14] and SLE [3] show high correlation with human judgement, we run a simple ablation to understand how they rank different simplifications and perturbations of text. We perform a comparative evaluation ablation to observe the differences in behaviors between evaluations produced by LENS and SLE.

We compare the distributions of simplifications generated by off-the-shelf Llama 3 Instruct [1] using the SLE and LENS metrics to understand their responsiveness to different types of text perturbations. The illustrated text variations include:

- (Source) Source Text
- (General with GE) Introduction of grammatical errors[2] to already simplified texts.
- (General) Simplification Candidates

We illustrate these distributions in Fig. 5, where we observe that these two learned metrics have a common behavior of giving a lower score when the simplification contains elaborate text, while these explanations might actually make the text more understandable to a human reader.

Another important finding is that while SLE shows a similar distribution between original simplifications and grammar error-induced versions of them,

---

[2] Through distortion-prompt A.4.

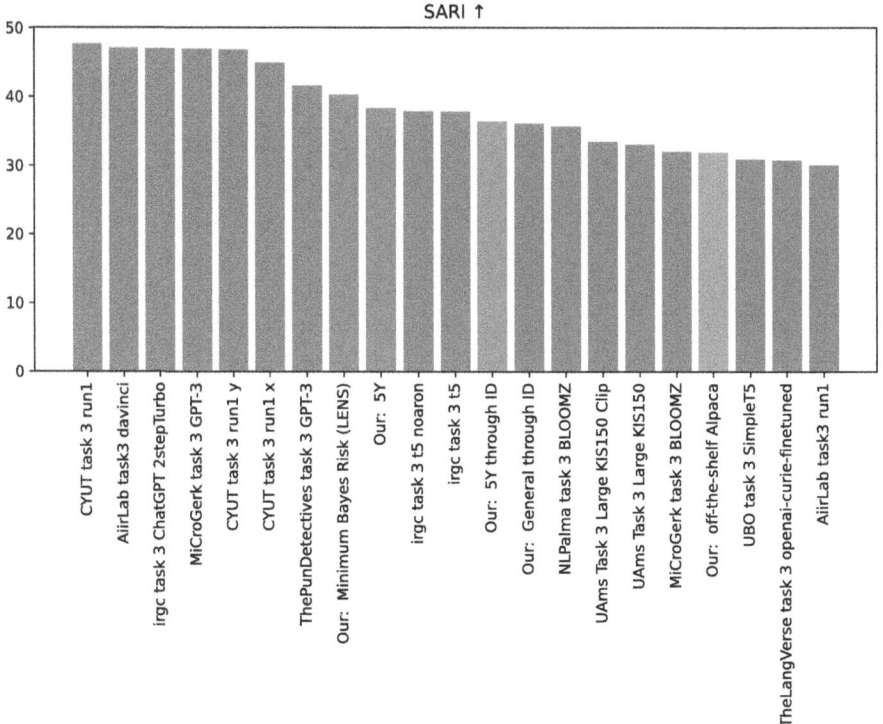

Fig. 6. SARI Evaluation of the submissions in the CLEF 2023 SimpleText Shared Task.

LENS responds to this manipulation by drastically reducing their distribution. This suggests that SLE has inherited some of the limitations of FKGL [17]. Despite its high inference speed, we consider SLE to be inadequate for selecting the best candidate, as the metric seems to disregard the grammaticality of the text. Due to these limitations of SLE, LENS has been chosen as the utility function for our Minimum Bayes Risk re-ranking.

4 Results

Within the Results section, we first present our Alpaca LoRA submissions to the SimpleText2023 Shared Task. We then present an ablation analysis of the performance of Alpaca LoRA and LLama3. Lastly, we analyse the distribution of methods that generate the selected simplification candidate through Minimum Bayes Risk in each model.

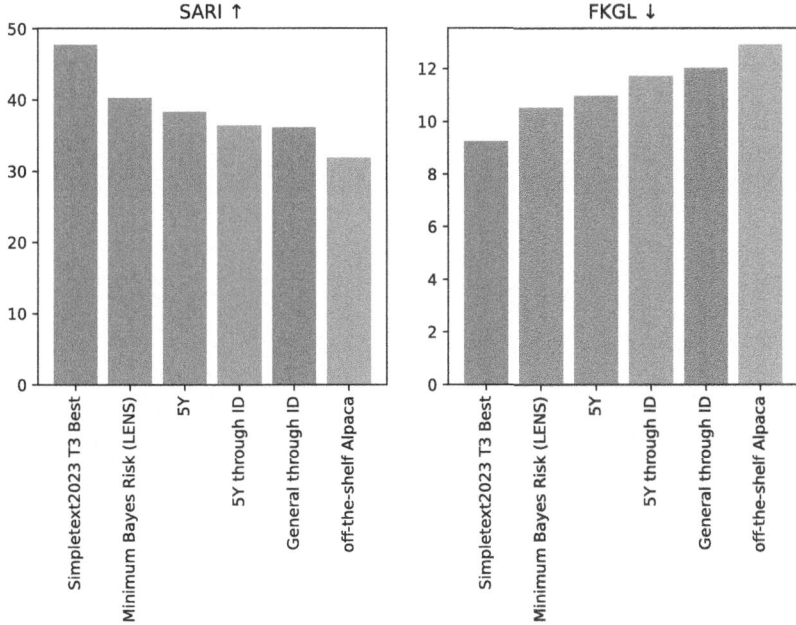

Fig. 7. Domain-adapted Alpaca LoRA models performance on SimpleText2023.

4.1 Shared Task Results

As part of the official submission to the CLEFF2023 SimpleText Shared Task, the predictions were evaluated using automated metrics. The evaluations in terms of SARI are presented in Fig. 6 while a more comprehensive suite of automated metrics is available in Appendix B.

In terms of SARI, the popular non-neural simplification metric [24], the MBR submission demonstrated one of the highest performances on the test set. Furthermore, it can be observed that domain-adapted models yielded higher scores than their off-the-shelf equivalents, demonstrating the benefit of fine-tuning Alpaca to the domain.

Another important observation is that the run that used Minimum Bayes Risk Decoding for re-ranking received a higher SARI, speaking for the value of MBR in already domain-adapted Language Models.

4.2 Performance Ablation Studies

To understand how the independent prompts and how MBR affects the simplification performance, we illustrate a closeup of our 2023 Alpaca LoRA models in Fig. 7.

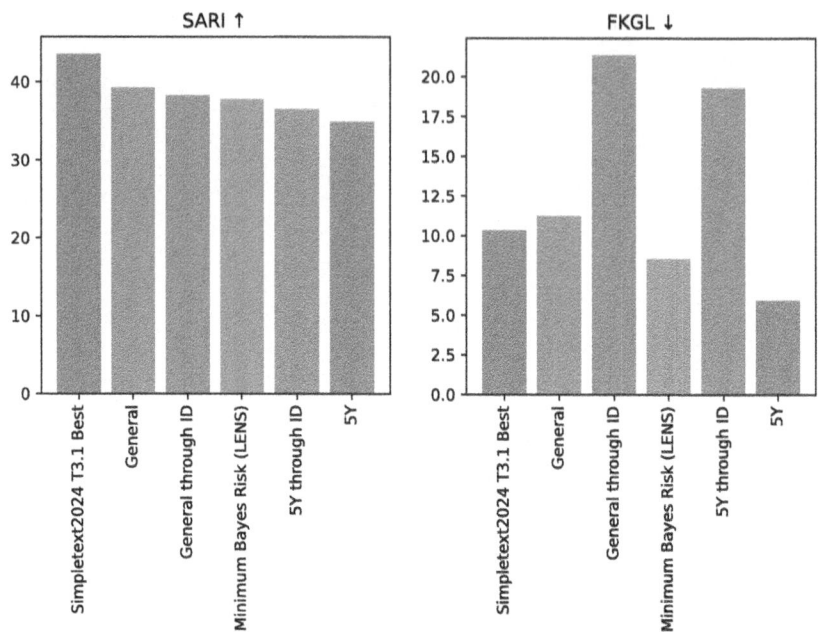

Fig. 8. Off-the-shelf Llama3 performance on SimpleText2024.

Key results show that Minimum Bayes Risk performs better than any independent candidate, demonstrating the strength of this approach with the domain-adapted model. Another interesting observation is that SARI and FKGL seem to perfectly agree on the ordering of the approaches.

Additionally, we evaluate the ability of off-the-shelf LLama3 to generate the simplification candidates. We perform this evaluation on the SimpleText2024 evaluation campaign, a superset of the SimpleText2023 test set. We visualize the comparative results in Fig. 8.

Through the Figure, we see that simplifications generated through the *General* prompt score a higher SARI value, closely followed by the prompt *General through Intermediate Definitions*. Contrary to Alpaca LoRA, MBR does not improve the performance and lands third in the evaluation[3].

An interesting observation is the discrepancy between the SARI and FKGL scores for our candidate methods. FKGL gives lower (better) scores to texts with shorter sentences and words with fewer syllables. Therefore, it's understandable that a text simplification that includes intermediate definitions of difficult concepts within the sentence receives a worse FKGL score.

---

[3] We have reasons to believe there was an error in the calculation of SARI of the MBR selected candidates. To be updated once resolved.

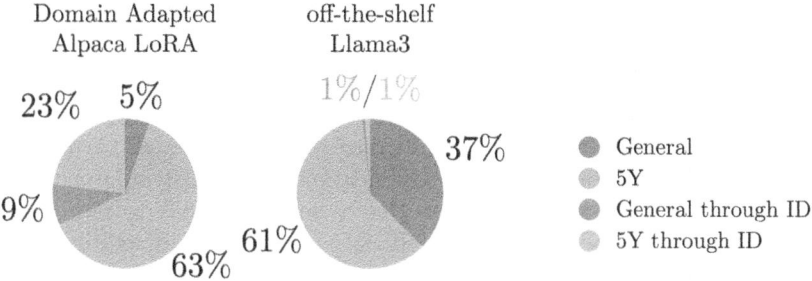

Fig. 9. Selection rate of the simplification pipelines through MBR.

This discrepancy further highlights the limitations of FKGL as a simplification metric, as also discussed in [17]. Through manual observation, we see that simplifications from Alpaca LoRA intermediate definition pipelines include the explanations of intermediate definitions in a separate sentence, thus avoiding the penalty through FKGL score. In contrast, LLama3, an instruction-tuned generative language model from META AI, performs the explanation of difficult terms within the text in a similar way to this sentence instead, resulting in high FKGL scores.

4.3 Selection Rate of Source Prompts Through MBR

To understand the selection behavior of MBR, we look at the source pipeline that generated the candidates selected as most promising through MBR. We visualize the percentage of samples selected from each source pipeline in Fig. 9.

Within the fine-tuned Alpaca LoRA, we see that MBR, for 63%, chose a candidate generated through the *5Y* prompt. The second most selected(23%) pipeline was the *5Y* that included intermediate definitions, showcasing that in some cases, LENS found these definitions to be beneficial to the final simplification quality. The two variations of the *General* prompt were overall selected (14%) less frequently, but with the added effects of the intermediate definitions to be deemed preferable more often than not.

In contrast, the LLama3 simplifications had a different distribution of MBR selections. Only a total of 2% of the selected samples were generated by a pipeline that used an intermediate definition step. The *5Y* prompt simplification is the most selected prompt with 61%, while the *General* prompt is lagging behind with 37% of the samples.

Whilst there exist differences between the two evaluation sets, MBR domain-adapted Alpaca seems to perform as well as off-the-shelf Llama3. Whilst this speaks for the impressive off-the-shelf capabilities of newer models without the need of complex engineering, it also showcases the potential improvements of weaker models when finetuned in a small domain specific dataset.

5 Conclusion

Within text simplification, many answers can be deemed of high quality. This is often not captured in the set of human-written references. We deploy different prompts to feed candidates to MBR that utilize a metric that aims to understand whether these candidates are good simplifications. The MBR approach allows us to select a simplification prediction that is most probable to represent the qualities of all given candidates. This prediction aims to capture multiple aspects of good text simplification. Our results demonstrate that this approach improves the quality of our simplifications in comparison to a single prediction of a given prompt. This highlights the benefit of designing diverse prompts as a candidate set. Another result of this study shows that fine-tuning a weaker/older LLM on a small training set may result in performance levels comparable to those of more recent LLM models that have been pre-trained on significantly larger datasets.

6 Limitations

In terms of deployability, the approach we propose might result in errors within simplifications that could be avoided through simple engineering. While Minimum Bayes Risk decoding is also a method of avoiding bad generations and improving consistency, we only explore a few prompts for generating candidates due to computation limitations.

We used the Intermediate Explanation simplification method so that texts with more complex terms requiring explanation would be more likely to be simplified. Nevertheless, it is possible that the influence of this candidate is minimal, and that in the end this candidate is not selected due to significant differences in surface similarity, or simply the difficulty of the LLM to follow these instructions, which may lead to irregular generations within the candidate pool.

A core principle of LENS is the assumption that a single metric can assess simplifications of different simplification approaches. Our methodology inherently requires this assumption. Therefore, the validity of our methodology depends on the validity of LENS.

Acknowledgement. We express our deepest gratitude and sincere appreciation to Simon Clematide and the Department of Computational Linguistics for their unwavering support, computational resources and constructive guidance during the creation of this work. Andrianos Michail acknowledges funding by the SNSF (213585) under the "impresso 2" project. We also like to express our appreciation to Tannon Kew for an exchange of ideas and reviewing an early version of this work.

A Prompt Engineering

This section details the different prompt templates used in our experiments, categorized by model and task specificity.

A.1 Prompt Templates for Alpaca LoRA

General. General prompt template provided to Alpaca LoRA [21].

```
Below is an instruction that describes a task,
paired with an input that provides further
context. Write a response that appropriately
completes the request.
###
Instruction:
{instruction}
###
Input:
{input}
###
Response:
```

5Y. Modified prompt template specifically for simplification tasks.

```
Below is an instruction that describes a
simplification task, paired with an input
that provides further context.
Write a simple response that appropriately
completes the request. Write your response
as you would talk to a 5-year-old.
###
Instruction:
{instruction}
###
Input:
{input}
###
Response:
```

A.2 Instructions Alpaca

This section shows the prompts for the two datasets used in this paper. The instructions are interpolated in the previously provided template.

Default. For the default evaluation process, we used a simple instruction prompt.

```
Simplify the following sentence
```

Intermediate Definitions (ID). For the complex terms evaluation, we used chained the model using the following two instruction prompts.

1. To identify the difficult *terms*:

   > Decide which terms (up to 5) require
   > explanation and contextualization to
   > help a reader understand a complex
   > scientific text

2. To obtain definitions for the previously identified *terms*:

   > Provide a short (one/two sentence)
   > explanations/definitions for the detected
   > difficult terms: {term} in
   > the context of the following sentence:

A.3 Prompt Templates for Llama3 Instruct

General. General prompt template.

```
{
 "role": "system",
 "content": "As a text simplification assistant, your
    task is to simplify the scientific sentence to
    make it more understandable for a general audience
    . Return only the simplified sentence, without any
    additional information."
},
{
 "role": "user",
 "content": "Simplify the following scientific sentence
    to make it more understandable for a general
    audience: {source_snt}.\nSimplified Sentence:"
}
```

5Y. Modified prompt template specifically for simplification tasks.

```
{
 "role": "system",
 "content": "Simplify the scientific sentence. Explain
    it as if you were talking to a 5-year-old, using
    simple words and concepts. Return only the
    simplified sentence, without any additional text
    or information."
},
```

```
{
  "role": "user",
  "content": "Simplify the following scientific sentence
      . Explain it as if you were talking to a 5-year-
      old, using simple words and concepts: {source_snt}
      .\nSimplified Sentence:"
}
```

Intermediate Definitions (ID). Used for identifying and defining complex terms within a sentence.

```
{
  "role": "system",
  "content": "Decide which terms (up to 5) require
      explanation and contextualization to help a reader
      to understand a complex scientific text. Return
      only the identified terms, without any additional
      text or information."
},
{
  "role": "user",
  "content": "Decide which terms (up to 5) require
      explanation and contextualization to help a reader
      understand a complex scientific text: {source_snt
      }.\nIdentified terms:"
}
```

```
{
  "role": "system",
  "content": "Provide a short, one or two sentence
      explanation for each of the difficult terms
      identified. Ensure the definitions are concise and
      contextualized within the scope of the sentence.
      Return only the definition for each of the terms,
      without any additional text or information."
},
{
  "role": "user",
  "content": "Provide a short (one/two sentence)
      explanation/definition for the detected difficult
      terms: '{terms}' in the context of this sentence:
      {source_snt}. Definitions:"
}
```

Chain-of-Thought inspired Prompts (CoT). These prompts involve a detailed simplification process, including the explanation of complex terms within a single prompt.

```
{
 "role": "system",
 "content": "Simplify the scientific sentence by
     integrating explanations of any complex terms
     directly within the text. The goal is to produce a
      single, coherent text that not only simplifies
     the content but also explains difficult terms in a
      way that is easily understandable for a general
     audience. Return only this integrated text,
     without any additional text or information."
},
{
    "role": "user",
    "content": "Simplify the following scientific
        sentence: '{source_snt}'. If it contains
        complex terms, explain them directly within the
         simplified sentence:\nSimplified Sentence:"
}
```

```
{
 "role": "system",
 "content": "Simplify the scientific sentence as if you
      were explaining it to a 5-year-old, using simple
     words and concepts. Seamlessly integrate
     explanations of any complex terms directly within
     the simplified text. The goal is to produce a
     single, coherent text that is understandable to a
     child. Return only this integrated text, without
     any additional text or information."
},
{
 "role": "user",
 "content": "Simplify the following scientific sentence
     : '{source_snt}'. Explain it as if you were
     talking to a 5-year-old and directly explain any
     complex terms within the simplified sentence:\
     nSimplified Sentence:"
}
```

A.4 Distortion Prompts for Assessing Metric Robustness

Distortion. The prompt used for introducing grammatical errors and disfluencies to evaluate the robustness of simplification metrics:

```
{
 "role": "system",
 "content": "As a text manipulation assistant, your
     task is to modify simplified scientific sentences
     by introducing grammatical errors and disfluencies
     . The goal is to subtly alter the syntax and
     insert errors without completely distorting the
     overall meaning of the text. Return only the
     altered sentence, without any additional
     information."
},
{
 "role": "user",
 "content": "Modify the following simplified sentence
     from a scientific abstract to include grammatical
     errors and disfluency: {simplified_llama3}.
     Altered Sentence:"
}
```

B Official Assessment

This section presents a detailed analysis of the official evaluation of the SimpleText CLEF shared task. All figures in this section show a graphical representation of the ranking of all submissions (Figs. 10, 11, 12, 13, 14, 15, 16 and 17).

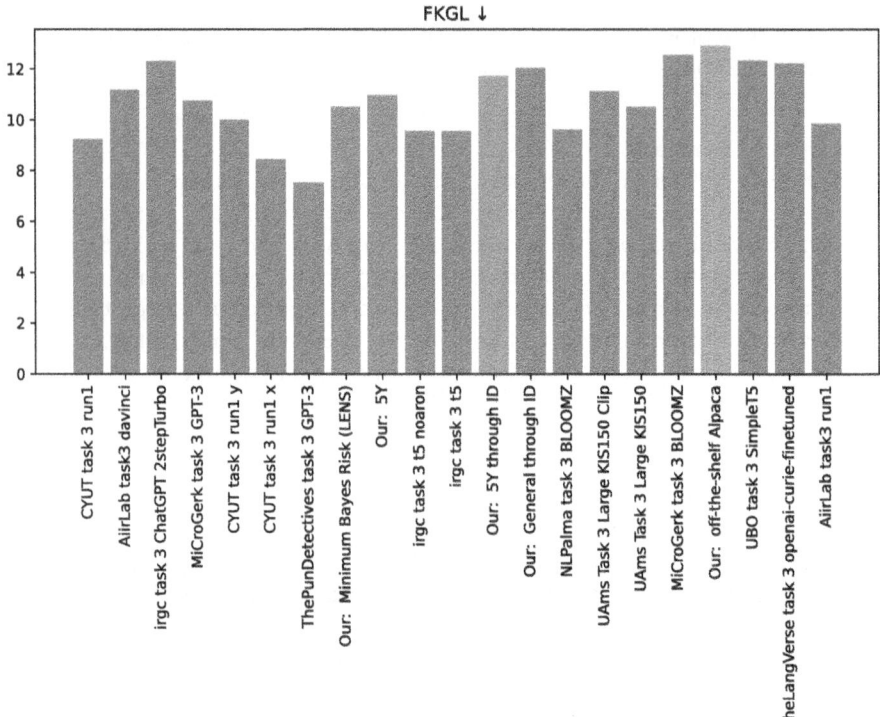

Fig. 10. FKGL (Flesch-Kincaid Grade Level) scores for the text simplification models. The FKGL metric measures the grade level required to understand the text, with lower scores indicating simpler and more accessible language. Runs are sorted in descending SARI performance.

Scientific Text Simplification Using Multi-prompt MBR Decoding   245

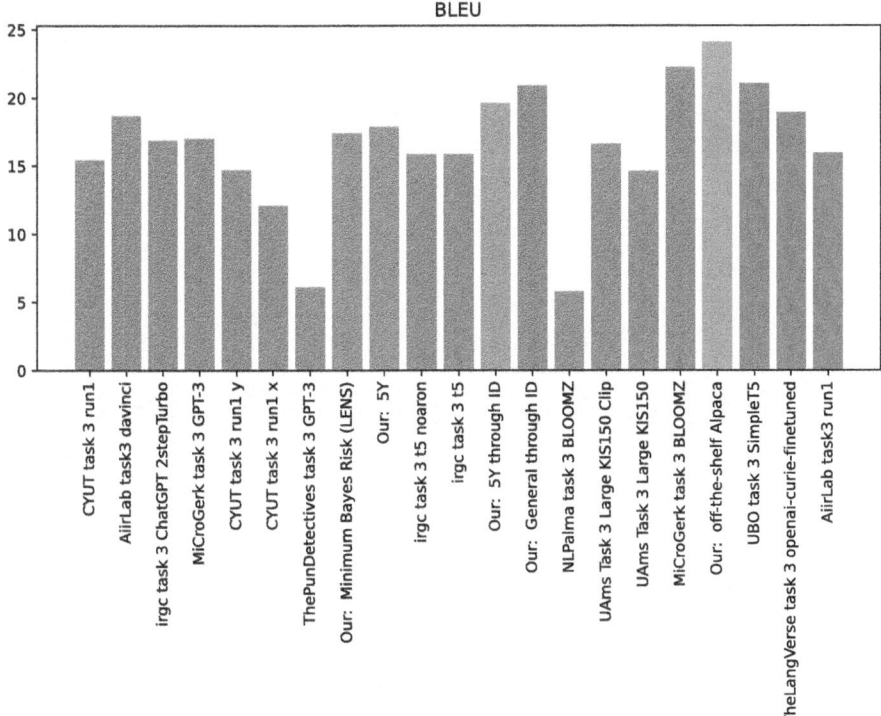

Fig. 11. BLEU (Bilingual Evaluation Understudy) scores for the text simplification models. BLEU measures the overlap between the generated simplified text and reference simplifications, with higher scores indicating better similarity. Runs are sorted in descending SARI performance.

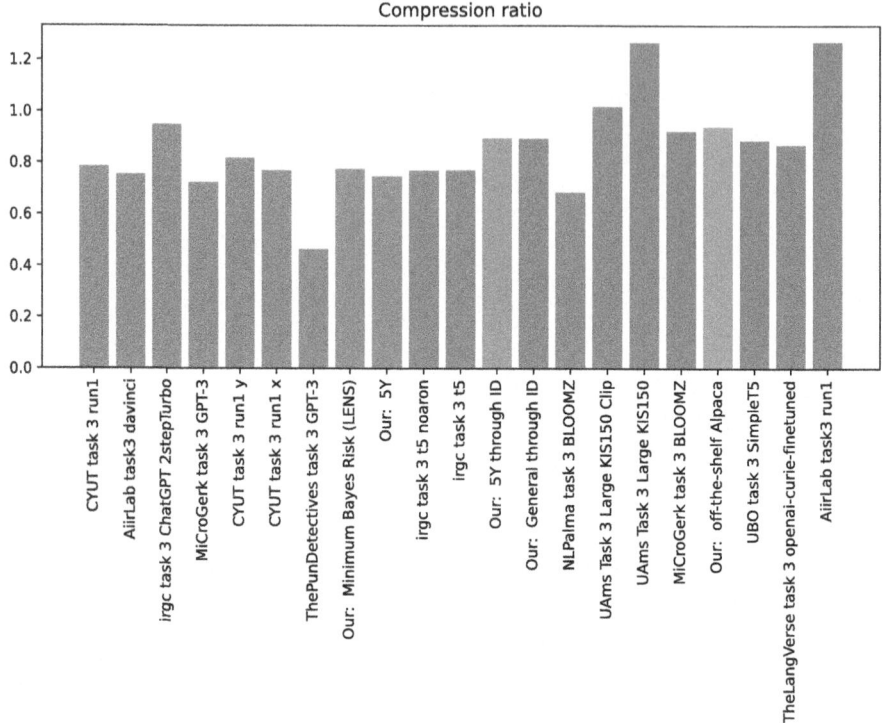

Fig. 12. Compression ratios for the text simplification models. Compression ratio measures the reduction in sentence length achieved by the text simplification models, with higher values indicating more significant simplification. Runs are sorted in descending SARI performance.

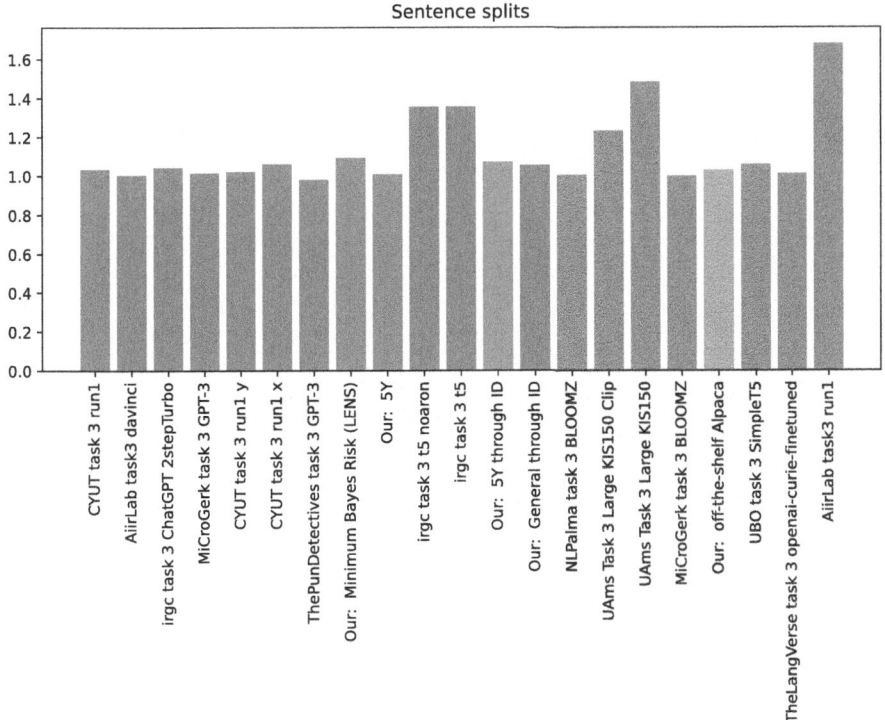

Fig. 13. Number of sentence splits for the text simplification models. Sentence splits measure the extent to which the original sentences were divided during the simplification process, with lower values indicating better preservation of sentence structure. Runs are sorted in descending SARI performance.

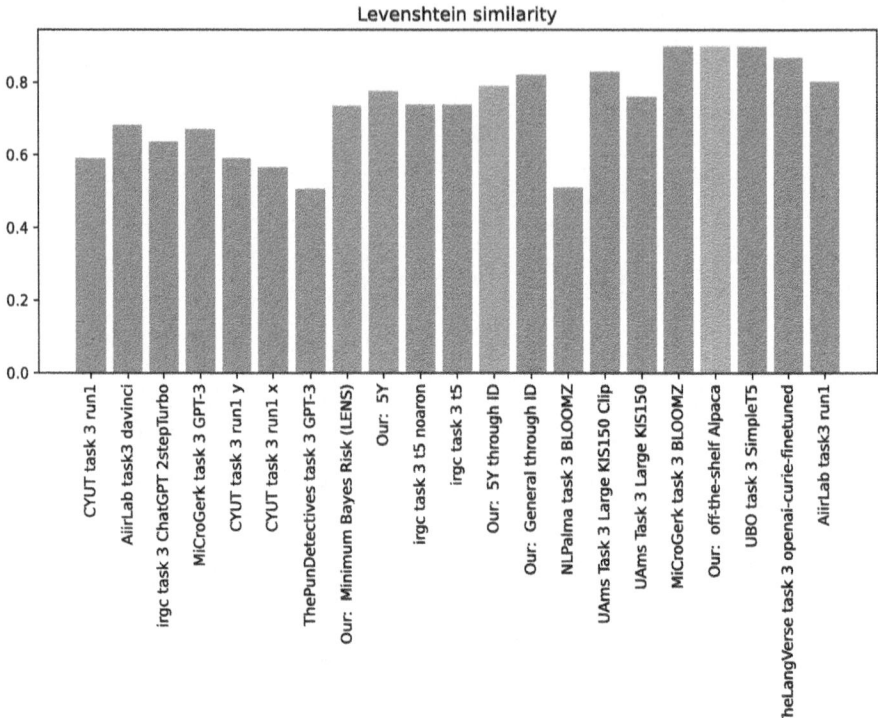

Fig. 14. Levenshtein similarity scores for the text simplification models. Levenshtein similarity measures the similarity between the generated simplified text and the original text, with higher values indicating better preservation of the original content. Runs are sorted in descending SARI performance.

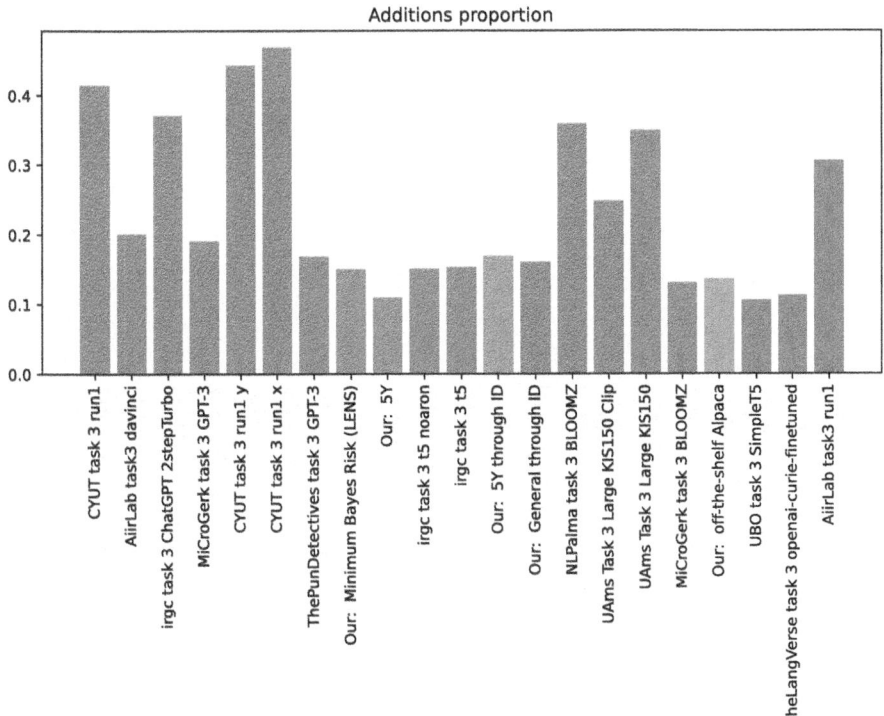

Fig. 15. Proportion of additions for the text simplification models. This metric measures the extent to which additional information was introduced during the simplification process, with lower values indicating better adherence to simplicity. Runs are sorted in descending SARI performance.

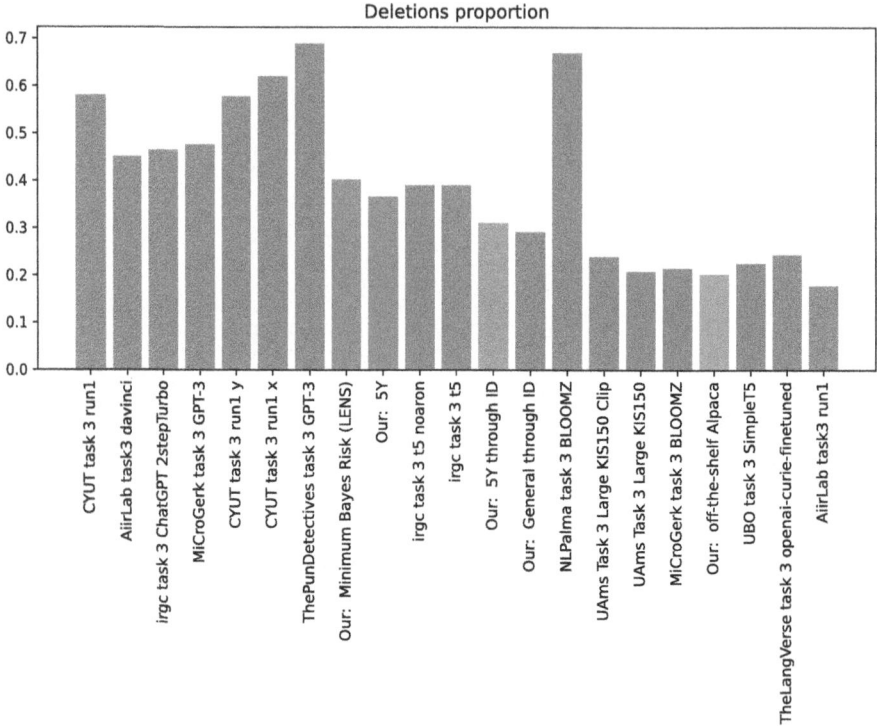

Fig. 16. Proportion of deletions for the text simplification models. This metric measures the extent to which unnecessary or redundant information was removed during the simplification process, with lower values indicating better conciseness. Runs are sorted in descending SARI performance.

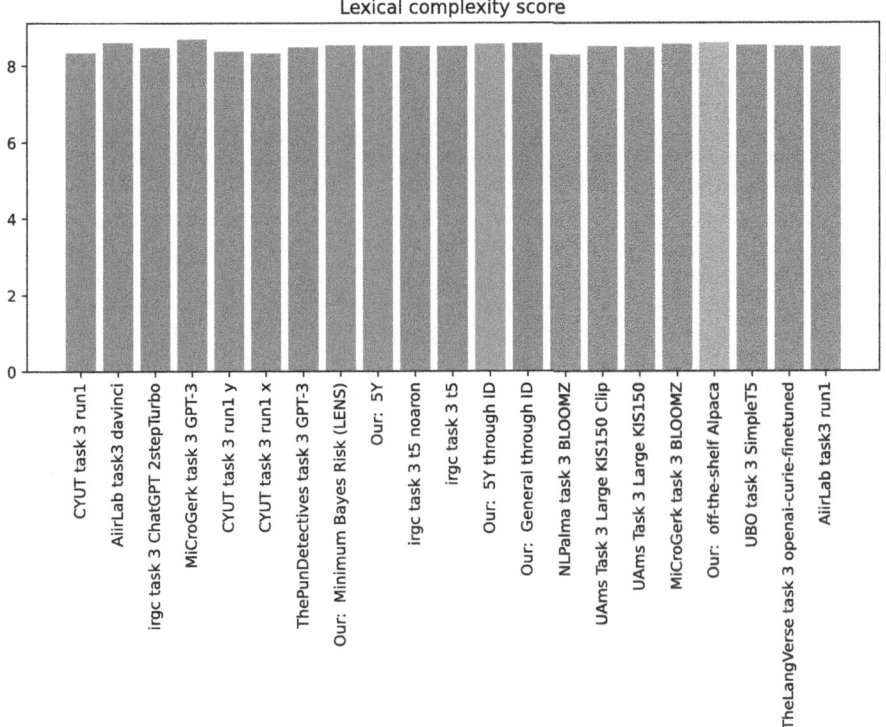

Fig. 17. Lexical complexity scores for the text simplification models. This metric measures the complexity of the vocabulary used in the generated simplified text, with lower scores indicating simpler and more accessible language.

References

1. AI@Meta: Llama 3 model card (2024). https://github.com/meta-llama/llama3/blob/main/MODEL_CARD.md
2. Andermatt, P.S., Fankhauser, T.: Uzh_pandas at simpletext@clef-2023: alpaca lora 7b and lens model selection for scientific literature simplification. In: Conference and Labs of the Evaluation Forum (2023). https://api.semanticscholar.org/CorpusID:264441298
3. Cripwell, L., Legrand, J., Gardent, C.: Simplicity level estimate (SLE): a learned reference-less metric for sentence simplification. In: Proceedings of the 2023 Conference on Empirical Methods in Natural Language Processing, pp. 12053–12059 (2023). https://doi.org/10.18653/v1/2023.emnlp-main.739
4. Ermakova, L., SanJuan, E., Huet, S., Augereau, O., Azarbonyad, H., Kamps, J.: Overview of simpletext - CLEF-2023 track on automatic simplification of scientific texts. In: Arampatzis, A., et al. (eds.) Experimental IR Meets Multilinguality, Multimodality, and Interaction. Proceedings of the Fourteenth International Conference of the CLEF Association. CLEF Association (2023). https://doi.org/10.1007/978-3-031-42448-9_30

5. Ermakova, L., SanJuan, E., Huet, S., Azarbonyad, H., Augereau, O., Kamps, J.: Overview of the CLEF 2023 simpletext lab: automatic simplification of scientific texts. In: Arampatzis, A., et al. (eds.) CLEF 2023. LNCS, vol. 14163, pp. 482–506. Springer, Cham (2023). https://doi.org/10.1007/978-3-031-42448-9_30
6. Farinhas, A., de Souza, J., Martins, A.: An empirical study of translation hypothesis ensembling with large language models. In: Bouamor, H., Pino, J., Bali, K. (eds.) Proceedings of the 2023 Conference on Empirical Methods in Natural Language Processing, Singapore, pp. 11956–11970. Association for Computational Linguistics (2023). https://doi.org/10.18653/v1/2023.emnlp-main.733. https://aclanthology.org/2023.emnlp-main.733
7. Fernandes, P., et al.: Quality-aware decoding for neural machine translation. In: Carpuat, M., de Marneffe, M.C., Meza Ruiz, I.V. (eds.) Proceedings of the 2022 Conference of the North American Chapter of the Association for Computational Linguistics: Human Language Technologies, Seattle, USA, pp. 1396–1412. Association for Computational Linguistics (2022). https://doi.org/10.18653/v1/2022.naacl-main.100. https://aclanthology.org/2022.naacl-main.100
8. Flesch, R.: Marks of readable style; a study in adult education. Teachers College Contributions to Education (1943)
9. Hu, E.J., et al.: LoRA: low-rank adaptation of large language models. In: International Conference on Learning Representations (2022). https://openreview.net/forum?id=nZeVKeeFYf9
10. Kew, T., et al.: BLESS: benchmarking large language models on sentence simplification. In: Bouamor, H., Pino, J., Bali, K. (eds.) Proceedings of the 2023 Conference on Empirical Methods in Natural Language Processing, Singapore, pp. 13291–13309. Association for Computational Linguistics (2023). https://doi.org/10.18653/v1/2023.emnlp-main.821. https://aclanthology.org/2023.emnlp-main.821
11. Kumar, S., Byrne, W.: Minimum bayes-risk word alignments of bilingual texts. In: Proceedings of the 2002 Conference on Empirical Methods in Natural Language Processing (EMNLP 2002), pp. 140–147. Association for Computational Linguistics (2002). https://doi.org/10.3115/1118693.1118712. https://aclanthology.org/W02-1019
12. Kumar, S., Byrne, W.: Minimum bayes-risk decoding for statistical machine translation. In: Proceedings of the Human Language Technology Conference of the North American Chapter of the Association for Computational Linguistics: HLT-NAACL 2004, Boston, Massachusetts, USA, pp. 169–176. Association for Computational Linguistics (2004). https://aclanthology.org/N04-1022
13. Liu, Y., et al.: RoBERTa: a robustly optimized BERT pretraining approach. arXiv preprint arXiv:1907.11692 (2019)
14. Maddela, M., Dou, Y., Heineman, D., Xu, W.: LENS: a learnable evaluation metric for text simplification. In: Rogers, A., Boyd-Graber, J., Okazaki, N. (eds.) Proceedings of the 61st Annual Meeting of the Association for Computational Linguistics (Volume 1: Long Papers), pp. 16383–16408. Association for Computational Linguistics, Toronto, Canada (2023). https://doi.org/10.18653/v1/2023.acl-long.905. https://aclanthology.org/2023.acl-long.905
15. Müller, M., Sennrich, R.: Understanding the properties of minimum bayes risk decoding in neural machine translation. In: Zong, C., Xia, F., Li, W., Navigli, R. (eds.) Proceedings of the 59th Annual Meeting of the Association for Computational Linguistics and the 11th International Joint Conference on Natural Language Processing (Volume 1: Long Papers), pp. 259–272. Association for Computational

Linguistics, Online (2021). https://doi.org/10.18653/v1/2021.acl-long.22. https://aclanthology.org/2021.acl-long.22
16. Pu, X., Gao, M., Wan, X.: Summarization is (almost) dead. arXiv e-prints, pp. arXiv–2309 (2023). https://doi.org/10.48550/arXiv.2309.09558
17. Tanprasert, T., Kauchak, D.: Flesch-Kincaid is not a text simplification evaluation metric. In: Bosselut, A., et al. (eds.) Proceedings of the 1st Workshop on Natural Language Generation, Evaluation, and Metrics (GEM 2021), pp. 1–14. Association for Computational Linguistics, Online (2021). https://doi.org/10.18653/v1/2021.gem-1.1. https://aclanthology.org/2021.gem-1.1
18. Taori, R., et al.: Stanford alpaca: An instruction-following LLaMA model (2023). https://github.com/tatsu-lab/stanford_alpaca
19. Touvron, H., et al.: LLaMA: open and efficient foundation language models (2023). https://doi.org/10.48550/arXiv.2302.13971 [cs]
20. Vernikos, G., Popescu-Belis, A.: Don't rank, combine! combining machine translation hypotheses using quality estimation. arXiv e-prints pp. arXiv–2401 (2024). https://doi.org/10.48550/arXiv.2401.06688
21. Wang, E.J.: Alpaca-lora (2023). https://github.com/tloen/alpaca-lora. Accessed 27 May 2023
22. Wei, J., et al.: Chain-of-thought prompting elicits reasoning in large language models (2023). https://doi.org/10.48550/arXiv.2201.11903, [cs]
23. Xu, W., Callison-Burch, C., Napoles, C.: Problems in current text simplification research: new data can help. Trans. Assoc. Comput. Linguist. **3**, 283–297 (2015). https://doi.org/10.1162/tacl_a_00139. https://aclanthology.org/Q15-1021
24. Xu, W., Napoles, C., Pavlick, E., Chen, Q., Callison-Burch, C.: Optimizing statistical machine translation for text simplification. Trans. Assoc. Comput. Linguist. **4**, 401–415 (2016)

Large Language Model Cascades and Persona-Based In-Context Learning for Multilingual Sexism Detection

Lin Tian, Nannan Huang, and Xiuzhen Zhang[✉]

RMIT University, Melbourne, VIC 3000, Australia
{lin.tian2,nannan.huang}@student.rmit.edu.au,
xiuzhen.zhang@rmit.edu.au

Abstract. This paper presents an approach for detecting and categorising sexism in social media posts using large language models (LLMs) and ensemble methods. The sEXism Identification in Social neTworks (EXIST) shared task, part of CLEF 2023, consists of three sub-tasks: Sexism Identification, Source Intention, and Sexism Categorisation. We formulate sexism detection in English and Spanish as text classification problems. A distinctive feature of the EXIST datasets is that, in addition to that each social media post is assigned a hard label from majority vote from all annotations, the soft labels – the labels by all annotators of different gender and age profiles – are also included. We propose cascade strategies to leverage LLMs for learning from hard labels. Our hard label-based system Mario is ranked first for the hard label evaluation on both sexism identification (Task 1) and source intention classification (Task 2) of the EXIST 2023 shared task. Our experiments show that more advanced base LLMs (e.g. Llama-3) can further improve the performance of Mario. To learn from soft labels for sexism identification, we further propose fine-grained in-context learning strategies based on personas of different age and gender profiles. Our experiments show that our few-shot persona-based in-context learning strategy leveraging Llama-3 can achieve reasonable performance for soft label prediction for sexism identification, and outperforms previous approaches of directly fine-tuning ensemble of BERT-based models.

Keywords: Sexism Detection · Automatic Sexism Categorisation · LLMs · Persona · In-context learning

1 Introduction

The pervasive nature of sexism on social media platforms has become an increasingly pressing issue, necessitating the development of effective methods for automatic detection and classification of sexist content [20]. The sEXism Identification in Social neTworks (EXIST) shared tasks, organised as part of CLEF 2023 [11], aim to address this problem by providing a framework for identifying, understanding, and categorising various forms of sexist expressions and behaviours in social media posts. The EXIST 2023 shared task consists of three sub-tasks: Sexism Identification, Source Intention, and Sexism Categorisation. Task 1 focuses on the binary classification of tweets into sexist and non-sexist categories. Task 2 aims to discern the underlying intent behind sexist tweets, classifying them as direct perpetration of sexism, reporting of experienced or observed sexism, or judgemental commentary on sexist situations or behaviours. Task 3 involves categorising sexist tweets into five distinct forms: ideological and inequality, stereotyping and dominance, objectification, sexual violence, and misogyny and non-sexual violence.

A distinctive feature of the EXIST datasets is the hard-label and soft-label settings for the ground truth labels for sexism. In the hard-label setting, each social media post is assigned one label from majority vote from all human annotators. In addition, social media posts are also annotated with soft labels – the labels by all annotators of different gender and age profiles – are also included. The soft-label setting is to represent that judgement of sexism varies across different demographic backgrounds.

We approach sexism detection in social media as a text classification problem, where the goal is to predict predefined labels for given texts – tweets – from their textual contents. To learn from hard labels, we use an ensemble of LLMs to capture diverse linguistic features and improve classification accuracy. Additionally, we introduce a cascading strategy that sequentially applies the models trained for related tasks, allowing for a more fine-grained analysis of sexist content. Our system Mario is ranked first among all 74 runs for Task 1 and Task 2 on hard label evaluations and achieved a F_1 score of 0.8109 and a F_1 score for task 2 of 0.575. This shows that large language based cascade models are able to handle sexism identification and categorisation tasks confidently.

To learn from the soft labels provided in the dataset, which represent the variability in human annotations, we adopt a persona-based in-context learning framework with LLMs. This approach enables our system to better capture the nuances and subjectivity inherent in the perception of sexism by different annotators, characterised by their age and gender. We evaluate our proposed approach on both English and Spanish datasets.

2 Related Work

Sexism identification can be seen as abusive language detection and it shares close relationship with abusive language detection, including racism, hate speech, personal attacks, and others. There have been studies done to identify sexism in the text on social media platforms [7,9,18,20,21]. Most approaches use deep learning based methods to tackle this task. [9] included simple machine learning baselines (Support Vector Machines and FastText classifier) to classify tweets into three categories (hostile, benevolent and others). [21] adopted text augmentation techniques and text generation data from ConceptNet and Wikidata to boost the model performance. Some related datasets have also been released to promote further research in this line [10,13,14].

Table 1. Illustration of the three sub-tasks and data for EXIST

Attribute	Description
id_EXIST	Unique identifier for the tweet
lang	Language of the tweet text ("en" or "es")
tweet	Text content of the tweet
number_annotators	Number of annotators who labeled the tweet
annotators	Unique identifier for each annotator
gender_annotators	Gender of each annotator ("F" for female, "M" for male)
age_annotators	Age group of each annotator ("18–22", "23–45", or "46+")
labels_task1	Set of labels indicating presence of sexist expressions or behaviours (YES or NO)
labels_task2	Set of labels indicating the intention of the tweet author ("direct", "reported", "judgemental", "-", or "unknown")
labels_task3	Set of arrays indicating the type(s) of sexism found in the tweet ("ideologicalinequality", "stereotyping-dominance", "objectification", "sexualviolence", "misogyny-non-sexual-violcence", "-", or "unknown")

Learning with disagreements is an approach that leverages the variability in human annotations to improve the performance of machine learning models [22]. [16] introduced this concept in the context of part-of-speech tagging, showing that incorporating annotator disagreements during training can lead to improved model performance. [22] extended this idea to hate speech detection, demonstrating that modelling annotator disagreements helps capture the nuances and subjectivity involved in identifying hate speech. In the context of sexism detection, [20] explored the use of learning from disagreements to address the challenges posed by the subjective nature of sexist content. These approaches aim to capture the inherent subjectivity and ambiguity in the data by treating disagreements as valuable information rather than noise. By incorporating annotator disagreements into the learning process, models can better handle the subjectivity present in various tasks.

In-context learning allows LLMs to perform new tasks by conditioning on a few demonstrations without modifying model parameters [5]. Recent studies have explored in-context learning from various perspectives, such as during pre-training [23], recursive learning [3], and meta-learning [12]. However, the potential of in-context learning, which involves using LLMs as persona-based models on soft label prediction task has not been fully explored yet.

3 Tasks

The EXIST 2023 shared task comprises three sub-tasks that aim to identify, understand, and categorise sexism in English and Spanish tweets. The dataset, as illustrated in Table 1, contains tweets labelled with hard and soft labels, indicating the presence of sexism content (Task 1), the intention behind the post (Task 2), and the categories of sexism expressed (Task 3).

Task 1: Sexism Identification. The first task is a binary classification problem that requires systems to determine whether a given tweet contains sexism expressions or behaviours. Tweets are labeled as either sexism or not sexism. Sexism tweets are those that contain sexism content, describe a sexism situation, or criticise sexism behaviour.

It is worth noting that we removed the "NOT_FOUND" instances when training all the models for this task. The distribution of the YES and NO is well-balanced.

Task 2: Source Intention. This task is a multi-class hierarchical classification problem that aims to categorise sexism tweets according to the intention of the author. The hierarchy of classes has two levels: the first level distinguishes between sexism and non-sexism tweets, while the second level further categorises sexism tweets into three mutually exclusive subcategories:

- Direct: The tweet is inherently sexism or incites sexism behaviour.
- Reported: The tweet reports or shares a sexism situation experienced by a woman or women in the first or third person.
- Judgemental: The tweet describes and condemns sexism situations or behaviours.

For Task 2, we follow the same approach by removing "NOT_FOUND" hard labelled instances from the training set when training the cascades model. For both English and Spanish data, it has an unbalanced distribution across the "Direct", "Judgement", and "Reported" labels. The "No" instances are shared across all jobs because they all use the same source data.

Task 3: Sexism Categorisation. This is a multi-class hierarchical multi-label classification task that requires systems to categorise sexism tweets into one or more of the following categories:

- Ideological and Inequality: The tweet discredits the feminist movement, rejects inequality between men and women, or presents men as victims of gender-based oppression.
- Stereotyping and Dominance: The tweet expresses false ideas about women's roles or claims that men are superior to women.
- Objectification: The tweet presents women as objects or focuses on their physical attributes.
- Sexual Violence: The tweet contains sexual suggestions, requests for sexual favours, or sexual harassment.
- Misogyny and Non-sexual Violence: The tweet expresses hatred or non-sexual violence towards women.

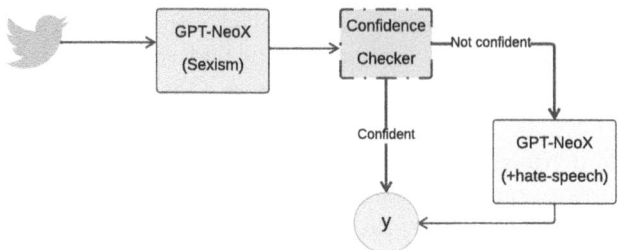

Fig. 1. LLM Cascades Model Architecture for Hard Label Prediction.

4 Methodology

We describe next our models for learning from hard labels and soft labels respectively.

4.1 A Cascade Model for Learning with Hard Labels

Our proposed approach, as illustrated in Fig. 1, incorporates two GPT-based large language models (LLMs) in a cascade architecture. The first model is finetuned using in-domain training data specifically curated for three distinct tasks. This fine-tuning process allows the model to acquire task-specific knowledge and adapt to the domain. The second model in the cascade is a hate-speech boosted model, which undergoes a two-stage fine-tuning process. Initially, the model is sequentially fine-tuned on a diverse range of hate speech datasets [1,6,8,19], as well as an open-source hate-speech tweets dataset from the Hugging Face library[1] in the target language (English or Spanish). This stage enables the model to develop a robust understanding of hate speech patterns and characteristics specific to the target language. Subsequently, the model undergoes further finetuning using in-domain task-specific training data, aligning its knowledge with the specific requirements of the tasks at hand.

To effectively differentiate between easy and hard samples, we introduce a confidence checker module that acts as a confidence-score based filter. The confidence checker assigns a confidence score to each sample processed by the LLMs. We employ a threshold-based approach, where samples with confidence scores above a predetermined threshold are considered easy and are processed by the first model in the cascade. Conversely, samples with confidence scores below the threshold are deemed hard and are passed on to the hate-speech boosted model for further processing. The confidence threshold is treated as a hyper-parameter in our experimental setup. Its value is determined through a rigorous optimization process, where different threshold values are evaluated on a development set. The threshold that yields the best performance on the development set is selected as the final value for the confidence checker.

---

[1] https://huggingface.co/datasets/tweets_hate_speech_detection.

One of the key practical benefits of our cascade architecture is its ability to save computation costs and improve inference speed compared to traditional ensemble models. By strategically routing samples through the cascade based on their confidence scores, we can allocate computational resources more efficiently. Easy samples are processed by the first model, which is computationally lighter, while hard samples are delegated to the more resource-intensive hate-speech boosted model only when necessary.

4.2 Label Smoothing

One of the common problems with LLMs is their tendency to become overconfident in prediction tasks. Overconfidence can lead to poor generalisation and suboptimal performance, especially when dealing with unseen or out-of-distribution data. To mitigate this issue, we apply label smoothing when training the LLMs.

Label smoothing is a simple yet effective approach. Instead of using hard one-hot encoded labels, where the correct class has a probability of 1 and all other classes have a probability of 0, label smoothing assigns a small non-zero probability to the incorrect classes. This is typically done by redistributing a portion of the probability mass from the correct class to the other classes.

Specifically, instead of using one-hot encoded vectors ([0, 1] in this case), we introduce noise distribution $u(y|x)$. Our new ground truth label for data (x_i, y_i) would be

$$p'(y \mid x_i) = (1-\varepsilon)p(y \mid x_i) + \varepsilon u(y \mid x_i)$$
$$= \begin{cases} 1 - \varepsilon + \varepsilon u(y \mid x_i) & \text{if } y = y_i \\ \varepsilon u(y \mid x_i) & \text{otherwise} \end{cases} \quad (1)$$

where ε is a weight factor, $\varepsilon \in [0,1]$ and note that $\sum_{y=1}^{K} p'(y \mid x_i) = 1$.

In this case, $u(y|x)$ is a uniform distribution, which does not depend on the data.

$$u(y|x) = \frac{1}{K} \quad (2)$$

where K is the total number of classes/labels.

By applying this technique, the model becomes less confident with extremely confident labels. By introducing a small amount of uncertainty in the labels, the model is encouraged to learn more robust and generalisable representations. This helps in reducing overfitting and improving the model's ability to handle ambiguous or noisy data.

4.3 Persona-Based In-Context Learning with Soft Labels

In our last year's submission, we did not consider the soft labels with Mario_1, Mario_2, and Mario_3 for all the tasks. To examine the in-context learning ability of the recently released LLM, Llama-3 8B[2], we propose a persona-based approach to learn with both hard and soft labels. In-context learning has emerged as a

---
[2] https://huggingface.co/meta-llama/Meta-Llama-3-8B.

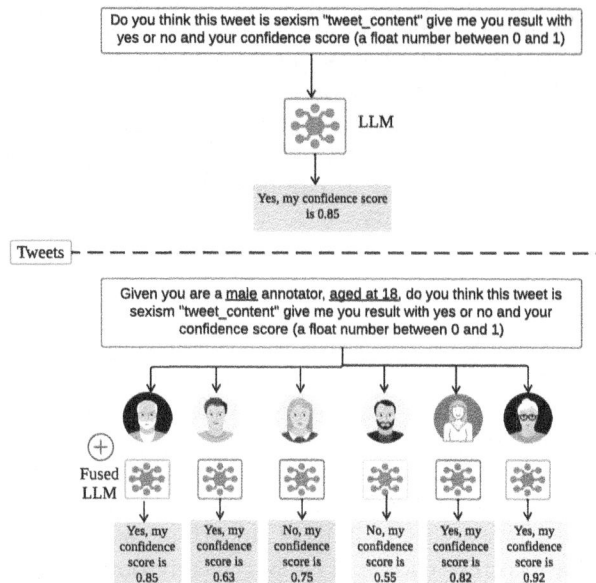

Fig. 2. Persona-based In-context Learning Architecture.

powerful technique for enabling large language models to perform tasks without explicit fine-tuning. By providing a few examples of the desired behaviour within the input prompt, the model can adapt and generate responses that follow the given pattern. Our approach, as depicted in Fig. 2, extends this concept by incorporating persona-based information, allowing the model to generate responses aligned with specific demographic attributes.

The motivation behind using persona-based in-context learning is to capture the nuances and variations in language patterns, preferences, and behaviours associated with different age groups and genders. By providing the model with a diverse set of persona-based prompts, we aim to enhance its understanding and generation of responses that are more representative of the target demographics. The aggregated label of each group will be used as soft label predictions in this case (we take the average in this case).

As illustrated on the bottom part of Fig. 2, we craft multiple prompts for each age group, considering individual ages within the range. For instance, when dealing with the 18–22 age group, we generate distinct prompts for ages 18, 19, 20, 21, and 22. Each of these age-specific prompts is then paired with the corresponding gender, resulting in a matrix of prompts that captures the nuances and variations within the demographic spectrum.

5 Experiments

5.1 Settings

In our experiments, we used open-source language models as our backbone for all tasks, instead of using the paid APIs by GPT-3, ChatGPT[3], or GPT-4[4]. For English, we used GPT-NeoX[5], while for Spanish, we used BERTIN-GPT-J-6B[6], which shares the same architecture as GPT-NeoX but is trained on Spanish data[7]. In our cascades framework, we applied GPT-NeoX for English and GPT-NeoX-Spanish for Spanish. To improve the robustness of our models and mitigate performance variance, we incorporated five additional public datasets into our training data, following the approach of [4]. Prior to training and inference, we preprocessed all tweet texts by removing @username mentions to focus on the content of the tweets.

We further test out most recently released Llama-3 model with 8B parameters[8]. For experiments involving in-context learning, we used the same Llama-3 model. Our original framework used two separate language models, one for English and another for Spanish, to handle the respective languages. However, Llama-3 represents the latest advancements in large language models and has demonstrated impressive performance across a wide range of natural language tasks, including its ability to handle both English and Spanish.

The evaluation of the tasks will be reported using the ICM (Information Contrast Measure) metric [2], which computes the similarity between system outputs and ground truth categories. The evaluation will be conducted in three settings: hard-hard (hard system output and hard ground truth), hard-soft (hard system output and soft ground truth), and soft-soft (soft system output and soft ground truth), with an extension of ICM (ICM-soft) being used to handle soft system outputs and soft ground truth assignments in the hierarchical multi-label classification problems.

5.2 Results

The results for our LLM cascade models on the development set are shown in Table 2, where hard predictions are evaluated against hard labels using ICM-hard and F_1. Across all three tasks and both languages, the Llama-3 variant (Mario_L) consistently outperforms the original Mario framework. For Task 1, Mario_L achieves an ICM-hard of 0.6768 and an F_1 score of 0.8203, surpassing Mario's scores of 0.6525 and 0.8063, respectively. Similar improvements are observed for Task 2 and Task 3. This finding highlights the potential for simplifying our

---

[3] https://openai.com/blog/chatgpt/.
[4] https://cdn.openai.com/papers/gpt-4.pdf.
[5] https://huggingface.co/docs/transformers/en/model_doc/gpt_neox.
[6] https://huggingface.co/bertin-project/bertin-gpt-j-6B.
[7] https://huggingface.co/datasets/bertin-project/mc4-es-sampled.
[8] https://huggingface.co/meta-llama/Meta-Llama-3-8B.

Table 2. Results in the dev set. Mario_L is replacing base model with Llama-3-8B.

Task 1				Task 2				Task 3			
Lang	Model	ICM-H	F_1	Lang	Model	ICM-H	F_1	Lang	Model	ICM-H	F_1
All	Mario	0.6525	0.8063	All	Mario	0.4849	0.5672	All	Mario	0.1689	0.5286
All	Mario_L	0.6768	0.8203	All	Mario_L	0.5199	0.5976	All	Mario_L	0.2125	0.5541
English	Mario	0.5952	0.7695	English	Mario	0.3650	0.5193	English	Mario	0.0564	0.4941
English	Mario_L	0.6163	0.7880	English	Mario_L	0.3980	0.5488	English	Mario_L	0.1013	0.5218
Spanish	Mario	0.6903	0.8341	Spanish	Mario	0.5665	0.6011	Spanish	Mario	0.2545	0.5541
Spanish	Mario_L	0.7191	0.8472	Spanish	Mario_L	0.6032	0.6320	Spanish	Mario_L	0.2961	0.5782

Table 3. Results on development data for prompted learning with Llama-3 8B

Task	Lang	Zero-Shot			One-Shot			5-Shot		
		ICM-hard	ICM-soft	F_1	ICM-hard	ICM-soft	F_1	ICM-hard	ICM-soft	F_1
Task 1	All	0.5572	0.4127	0.7438	0.5921	0.4539	0.7649	0.6269	0.5083	0.7859
Task 1	English	0.5012	0.0523	0.7120	0.5415	0.1067	0.7358	0.5817	0.1542	0.7596
Task 1	Spanish	0.6132	0.6483	0.7756	0.6426	0.6957	0.7939	0.6720	0.7496	0.8122
Task 2	All	0.4182	−1.6793	0.5327	0.4422	−1.0235	0.5491	0.4663	−0.5137	0.5655
Task 2	English	0.3235	−4.3627	0.4941	0.3482	−3.0148	0.5117	0.3729	−2.0093	0.5293
Task 2	Spanish	0.5128	−1.7684	0.5712	0.5362	−1.0317	0.5864	0.5596	−0.5042	0.6016
Task 3	All	0.1618	−3.2957	0.5010	0.1718	−2.5124	0.5131	0.1817	−2.0083	0.5251
Task 3	English	0.0824	−3.3015	0.4742	0.0886	−2.5067	0.4855	0.0948	−2.0114	0.4968
Task 3	Spanish	0.2412	−2.3042	0.5278	0.2549	−2.0053	0.5406	0.2686	−1.5129	0.5534

framework by using a single, versatile language model like Llama-3, rather than relying on separate models for each language.

To investigate the performance of large language models in handling soft labels, we use a persona-based in-context learning approach as an alternative to the existing language model cascade framework. Table 3 presents the performance on the development data for the Llama-3 8B model using zero-shot, one-shot, and 5-shot in-context learning settings across three tasks and two languages (English and Spanish). The 5-shot persona-based in-context learning results demonstrate the potential of the Llama-3 8B model to handle soft labels more effectively than the previous year's approach of directly tuning soft labels using an ensemble of BERT-based models [15]. Although the ICM-soft scores are negative for Task 2 and Task 3, they show improvement in the 5-shot setting compared to the zero-shot and one-shot settings, indicating that providing more examples helps the model generate more accurate soft labels. For instance, in Task 2, the 5-shot ICM-soft score is −0.5137, which is better than the scores of −1.6793 and −1.0235 for the zero-shot and one-shot settings, respectively. These findings highlight the potential of large language models to handle soft labels through our persona-based in-context learning, eliminating the need for modifying the model architecture or fine-tuning on soft labels.

Table 4. Results for the blind test data in EXIST 2023. The three submitted runs of our Mario model used base LLMs GPT-NeoX for English and BERTIN-GPT-J-6B for Spanish respectively. The official ICM scores for each task and evaluation setting are shown, along with the corresponding rankings in parentheses.

Task	Lang	Mario_1			Mario_2			Mario_3		
		ICM-hard	ICM-soft	F_1	ICM-hard	ICM-soft	F_1	ICM-hard	ICM-soft	F_1
Task 1	All	0.6540 (2)	0.4507(14)	0.8058(2)	0.6120	0.3634(15)	0.8029(3)	0.6575 (1)	0.4719(16)	0.8109(1)
Task 1	English	0.5880 (2)	0.1009	0.7626	0.5459	0.0038	0.7650	0.5996 (1)	0.1280	0.7734
Task 1	Spanish	0.6995 (1)	0.6826	0.8383	0.6552	0.6071	0.8300	0.6959 (2)	0.6998	0.8387
Task 2	All	–	–	–	0.4887 (1)	−5.8157(14)	0.5715(1)	–	–	–
Task 2	English	–	–	–	0.3677 (1)	−7.1029	0.5224(1)	–	–	–
Task 2	Spanish	–	–	–	0.5711 (1)	−5.1329	0.6059(1)	–	–	–
Task 3	All	0.0896	−9.1398(16)	0.5011(9)	0.1228	−9.6735(17)	0.5145(8)	0.1700	−10.2297(18)	0.5323(6)
Task 3	English	−0.0269	−10.8847	0.4595	0.0133	−11.4612	0.4772	0.0568	−11.9003	0.4971
Task 3	Spanish	0.1779	−7.7970	0.5305	0.2040	−8.2903	0.5405	0.2562	−8.9369	0.5578

5.3 Discussions

In the EXIST 2023 shared task [11], we submitted three runs of our cascade model: Mario_1, Mario_2, and Mario_3. The performance of these models on the blind test data is shown in Table 4. The cascade framework can only produce hard labels, while the submitted soft predictions are based on model confidence scores. Interestingly, the ranking of ICM-soft for all variants of Mario is not consistent with their performance on the hard labels, as evident from Table 4. This finding aligns with the observations made by [17], suggesting that there is no correlation between human disagreement and the confidence of large language models.

The three Mario variants differ in their training settings. Mario_1 is trained directly with binary labels for Task 1, while Mario_2 is trained with more detailed labels from Task 2. Mario_3 combines the predictions from Mario_1 and Mario_2, prioritising the prediction with the higher confidence score in case of conflicts. For Task 3, each Mario variant uses a different confidence threshold. Comparing the performance among these three variants, we observe that the mixture of experts approach employed by Mario_3 performs better on Task 1 and Task 2. This suggests that the mixture of experts strategy may be more effective for text-based sexism detection.

When compared to other teams, Mario_3 outperforms them on Task 1 and Task 2, indicating that the LLM-based cascade model can effectively handle hard label predictions for both binary and multi-class classification tasks. Furthermore, we explored a persona-based in-context learning approach, and the results presented in Table 2 suggest that this approach may offer a way to utilise LLMs for learning from disagreements.

6 Conclusion

In this paper, we proposed a novel approach for detecting and categorising sexism in social media posts using LLMs and ensemble methods. Our system, designed for the EXIST 2023 shared task, employs open-source LLMs as the backbone and introduces a cascading strategy that sequentially applies the models trained for each sub-task. Our experiments demonstrated the effectiveness of using LLMs, confidence-based cascade models, and persona-based in-context learning with Llama-3 for handling soft labels.

For future work, we plan to explore the integration of multi-modal features, such as images and user metadata, to further enhance the accuracy and generalisability of sexism detection and categorisation models.

Acknowledgement. The authors acknowledge the CLEF EXIST shared task organisers for the hard work of curating the data. This research was supported in part by the Australian Research Council Discovery Project DP200101441.

References

1. Álvarez-Carmona, M.Á., et al.: Overview of MEX-A3T at IberEval 2018: authorship and aggressiveness analysis in Mexican Spanish tweets. In: Notebook papers of 3rd SEPLN Workshop on Evaluation of Human Language Technologies for Iberian Languages (IberEval), Seville, Spain, vol. 6 (2018)
2. Amigo, E., Delgado, A.: Evaluating extreme hierarchical multi-label classification. In: Proceedings of the 60th Annual Meeting of the Association for Computational Linguistics, pp. 5809–5819 (2022)
3. Coda-Forno, J., Binz, M., Akata, Z., Botvinick, M., Wang, J., Schulz, E.: Meta-in-context learning in large language models. Adv. Neural. Inf. Process. Syst. **36**, 65189–65201 (2023)
4. Do, C.B., Ng, A.Y.: Transfer learning for text classification. Adv. Neural. Inf. Process. Syst. **18** (2005)
5. Dong, Q., et al.: A survey on in-context learning. arXiv preprint arXiv:2301.00234 (2022)
6. Fersini, E., Rosso, P., Anzovino, M.: Overview of the task on automatic misogyny identification at IberEval 2018. In: IberEval@ SEPLN, vol. 2150, pp. 214–228 (2018)
7. Frenda, S., Ghanem, B., Montes-y Gómez, M., Rosso, P.: Online hate speech against women: automatic identification of misogyny and sexism on twitter. J. Intell. Fuzzy Syst. **36**(5), 4743–4752 (2019)
8. Gautam, A., Mathur, P., Gosangi, R., Mahata, D., Sawhney, R., Shah, R.R.: #MeTooMA: multi-aspect annotations of tweets related to the MeToo movement. In: Proceedings of the International AAAI Conference on Web and Social Media, vol. 14, pp. 209–216 (2020)
9. Jha, A., Mamidi, R.: When does a compliment become sexist? Analysis and classification of ambivalent sexism using twitter data. In: Proceedings of the Second Workshop on NLP and Computational Social Science, pp. 7–16 (2017)
10. Karlekar, S., Bansal, M.: SafeCity: understanding diverse forms of sexual harassment personal stories. arXiv preprint arXiv:1809.04739 (2018)
11. Plaza, L., et al.: Overview of EXIST 2023 - learning with disagreement for sexism identification and characterization (extended overview). In: Aliannejadi, M., Faggioli, G., Ferro, N., Vlachos, M. (eds.) Working Notes of CLEF 2023 - Conference and Labs of the Evaluation Forum (2023)

12. Min, S., et al.: Rethinking the role of demonstrations: What makes in-context learning work? In: Proceedings of the 2022 Conference on Empirical Methods in Natural Language Processing, pp. 11048–11064 (2022)
13. Parikh, P., et al.: Multi-label categorization of accounts of sexism using a neural framework. arXiv preprint arXiv:1910.04602 (2019)
14. Parikh, P., Abburi, H., Chhaya, N., Gupta, M., Varma, V.: Categorizing sexism and misogyny through neural approaches. ACM Trans. Web (TWEB) **15**(4), 1–31 (2021)
15. de Paula, A., Rizzi, G., Fersini, E., Spina, D., et al.: AI-UPV at exist 2023–sexism characterization using large language models under the learning with disagreements regime. In: Working Notes of CLEF 2023 - Conference and Labs of the Evaluation Forum, vol. 3497, pp. 985–999. CEUR-WS (2023)
16. Plank, B., Hovy, D., Søgaard, A.: Linguistically debatable or just plain wrong? In: Proceedings of the 52nd Annual Meeting of the Association for Computational Linguistics, pp. 507–511 (2014)
17. Qu, Y., Roitero, K., Barbera, D.L., Spina, D., Mizzaro, S., Demartini, G.: Combining human and machine confidence in truthfulness assessment. ACM J. Data Inf. Qual. **15**(1), 1–17 (2022)
18. Rodríguez-Sánchez, F., Carrillo-de Albornoz, J., Plaza, L.: Automatic classification of sexism in social networks: an empirical study on twitter data. IEEE Access **8**, 219563–219576 (2020)
19. Sachdeva, P., Barreto, R., Bacon, G., Sahn, A., Von Vacano, C., Kennedy, C.: The measuring hate speech corpus: leveraging Rasch measurement theory for data perspectivism. In: Proceedings of the 1st Workshop on Perspectivist Approaches to NLP@ LREC2022, pp. 83–94 (2022)
20. Samory, M., Sen, I., Kohne, J., Flöck, F., Wagner, C.: "call me sexist, but...": revisiting sexism detection using psychological scales and adversarial samples. In: Proceedings of the International AAAI Conference on Web and Social Media, vol. 15, pp. 573–584 (2021)
21. Sharifirad, S., Jafarpour, B., Matwin, S.: Boosting text classification performance on sexist tweets by text augmentation and text generation using a combination of knowledge graphs. In: Proceedings of the 2nd Workshop on Abusive Language Online (ALW2), pp. 107–114 (2018)
22. Uma, A.N., Fornaciari, T., Hovy, D., Paun, S., Plank, B., Poesio, M.: Learning from disagreement: a survey. J. Artif. Intell. Res. **72**, 1385–1470 (2021)
23. Xie, S.M., Raghunathan, A., Liang, P., Ma, T.: An explanation of in-context learning as implicit Bayesian inference. In: International Conference on Learning Representations (2022)

Correction to: Under-Sampling Strategies for Better Transformer-Based Classifications Models

Marcin Sawiński⬤, Krzysztof Węcel⬤, and Ewelina Księżniak⬤

Correction to:
Chapter 8 in: L. Goeuriot et al. (Eds.): *Experimental IR Meets Multilinguality, Multimodality, and Interaction*, **LNCS 14958, https://doi.org/10.1007/978-3-031-71736-9_8**

In the original version of this chapter the letter "ę" is missing in the author names Krzysztof Węcel and Ewelina Księżniak. This has been corrected.

The updated version of this chapter can be found at
https://doi.org/10.1007/978-3-031-71736-9_8

© The Author(s), under exclusive license to Springer Nature Switzerland AG 2024
L. Goeuriot et al. (Eds.): CLEF 2024, LNCS 14958, p. C1, 2024.
https://doi.org/10.1007/978-3-031-71736-9_19

Author Index

A

Afzal, Zubair I-74
Aidos, Helena II-118
Alam, Firoj II-28
Alkhalifa, Rabab II-208
Alshomary, Milad II-308
Amigó, Enrique II-93
Andermatt, Pascal Severin I-227
Andrei, Alexandra-Georgiana II-140
Arcos, Iván I-61
Avila, Jorge I-181
Ayele, Abinew Ali II-231
Azarbonyad, Hosein I-74, II-283

B

Babakov, Nikolay II-231
Barrón-Cedeño, Alberto II-28
Ben Abacha, Asma II-140
Bergamaschi, Roberto II-118
Bevendorff, Janek II-231
Birolo, Giovanni II-118
Bloch, Louise II-140
Bondarenko, Alexander I-100
Bonnet, Pierre II-183
Borkakoty, Hsuvas II-208
Bosoni, Pietro II-118
Bosser, Anne-Gwenn II-165
Botella, Christophe II-183
Bracke, Benjamin II-140
Breuer, Timo I-215
Brüngel, Raphael II-140
Bucur, Ana-Maria I-193
Burdisso, Sergio I-127
Buscaldi, Davide I-152

C

Capari, Artemis I-74
Carrillo-de-Albornoz, Jorge II-93
Casals, Xavier Bonet II-231
Caselli, Tommaso II-28
Cassani, Luca I-88
Cavalla, Paola II-118
Centeno, Roberto I-181
Chakraborty, Tanmoy II-28
Chiò, Adriano II-118
Chulvi, Berta I-152, II-93, II-231
Çöltekin, Çağrı II-308
Cremonesi, Paolo II-260
Crestani, Fabio II-73

D

D'Souza, Jennifer II-283
Da San Martino, Giovanni II-28
Dagliati, Arianna II-118
Damm, Hendrik II-140
Davydova, Vera II-3
de Carvalho, Mamede II-118
De Longueville, Bertrand II-308
Dementieva, Daryna II-231
Denton, Tom II-183
Deveaud, Romain II-208
Di Camillo, Barbara II-118
Di Nunzio, Giorgio Maria II-118, II-283
Dominguez, Jose Manuel García II-118
Drăgulinescu, Ana-Maria II-140
Dsilva, Ryan Rony I-205
Dunham, Judson I-74
Dürlich, Luise II-53

E

Eggel, Ivan II-183
El-Ebshihy, Alaa I-113, II-208
Elnagar, Ashaf II-231
Elsayed, Tamer II-28
Erjavec, Tomaž II-308
Ermakova, Liana II-165, II-283

Esperança-Rodier, Emmanuelle II-140
Espinosa-Anke, Luis II-208
Espitalier, Vincent II-183
Estopinan, Joaquim II-183

F
Faggioli, Guglielmo II-118
Fankhauser, Tobias I-227
Fariselli, Piero II-118
Farré-Maduell, Eulàlia II-3
Ferrari Dacrema, Maurizio II-260
Ferro, Nicola I-3, II-118, II-260
Fink, Tobias II-208
Freitag, Dayne II-231
Friedrich, Christoph M. II-140
Fröbe, Maik II-231, II-308
Fu, Yujuan II-140

G
Galuščáková, Petra II-208
García Seco de Herrera, Alba II-140
Glotin, Hervé II-183
Goëau, Hervé II-183
Goeuriot, Lorraine II-208
Gogoulou, Evangelia II-53
Gohsen, Marcel I-166
Gonzalez-Saez, Gabriela II-208
Gonzalo, Julio II-93
Gromicho, Marta II-118
Guazzo, Alessandro II-118
Guillou, Liane II-53

H
Hagen, Matthias I-166
Halvorsen, Pål II-140
Hanbury, Allan I-113
Handke, Nicolas II-308
Haouari, Fatima II-28
Hasanain, Maram II-28
Heinrich, Maximilian II-140, II-308
Hicks, Steven A. II-140
Hrúz, Marek II-183
Huang, Nannan I-254
Huet, Stéphane II-283

I
Idrissi-Yaghir, Ahmad II-140
Iommi, David II-208
Ionescu, Bogdan II-140

J
Jatowt, Adam II-165
Joly, Alexis II-183

K
Kahl, Stefan II-183
Kamps, Jaap I-74, II-283
Karimi, Mozhgan I-159
Karlgren, Jussi II-53
Karpenka, Dzmitry II-140
Katsimpras, Georgios II-3
Keller, Jüri I-215
Kiesel, Johannes I-166, II-140, II-308
Klinck, Holger II-183
Kopp, Matyáš II-308
Korenčić, Damir II-231
Kovalev, Vassili II-140
Krallinger, Martin II-3
Krithara, Anastasia II-3
Księżniak, Ewelina I-139

L
Larcher, Théo II-183
Leblanc, Cesar II-183
Lecouteux, Benjamin II-140
Li, Chengkai II-28
Liakata, Maria II-208
Lima-López, Salvador II-3
Livraga, Giovanni I-88
Ljubešić, Nikola II-308
Longato, Enrico II-118
Losada, David E. II-73
Loukachevitch, Natalia II-3

M
Macaire, Cécile II-140
Madabushi, Harish Tayyar II-208
Madeira, Sara C. II-118
Maeso, Alba II-93
Manera, Umberto II-118
Marchesin, Stefano II-118
Marcos, Diego II-183
Martín-Rodilla, Patricia II-73
Mayerl, Maximilian II-231
Meden, Katja II-308
Medina-Alias, Pablo II-208
Menotti, Laura II-118
Merker, Jan Heinrich I-100
Merker, Lena I-100

Author Index

Michail, Andrianos I-227
Miller, Tristan II-165
Mirzakhmedova, Nailia I-166
Mirzhakhmedova, Nailia II-308
Morante, Roser II-93
Morkevičius, Vaidas II-308
Moskovskiy, Daniil II-231
Motlicek, Petr I-127
Mukherjee, Animesh II-231
Mulhem, Philippe II-208
Müller, Henning II-140, II-183

N

Nakov, Preslav II-28
Nentidis, Anastasios II-3
Ningtyas, Annisa Maulida I-113
Nivre, Joakim II-53

P

Pakull, Tabea M. G. II-140
Paliouras, Georgios II-3
Palma Preciado, Victor Manuel II-165
Panchenko, Alexander II-231
Parapar, Javier II-73
Pasin, Andrea II-260
Picek, Lukáš II-183
Piroi, Florina I-113, II-208
Piskorski, Jakub II-28
Planqué, Robert II-183
Plaza, Laura II-93
Popel, Martin II-208
Potthast, Martin I-172, II-140, II-231, II-308
Prokopchuk, Yuri II-140
Przybyła, Piotr II-28

R

Radzhabov, Ahmedkhan II-140
Rangel, Francisco II-231
Reitis-Münstermann, Theresa II-308
Riegler, Michael A. II-140
Rizwan, Naquee II-231
Rodrigo, Álvaro I-181
Rosso, Paolo I-61, I-152, II-93, II-231
Rückert, Johannes II-140
Ruggeri, Federico II-28
Ruiz, Víctor II-93

S

Sahlgren, Magnus II-53
Sánchez-Cortés, Dairazalia I-127
SanJuan, Eric II-283
Sawiński, Marcin I-139
Schaer, Philipp I-215
Schäfer, Henning II-140
Scharfbillig, Mario II-308
Schmidt, Cynthia Sabrina II-140
Schneider, Florian II-231
Schwab, Didier II-140
Servajean, Maximilien II-183
Sidorov, Grigori II-165
Silvello, Gianmaria II-118
Smirnova, Alisa II-231
Song, Xingyi II-28
Spina, Damiano II-93
Stakovskii, Elisei II-231
Stamatatos, Efstathios II-231
Stefanovitch, Nicolas II-308
Stein, Benno I-166, I-172, II-140, II-231, II-308
Storås, Andrea II-140
Struß, Julia Maria II-28
Šulc, Milan II-183
Sun, Zhaoyi II-140
Suwaileh, Reem II-28

T

Talman, Aarne II-53
Taulé, Mariona II-231
Tavazzi, Eleonora II-118
Tavazzi, Erica II-118
Thambawita, Vajira II-140
Tian, Lin I-254
Trescato, Isotta II-118
Tsatsaronis, Georgios I-74
Tutubalina, Elena II-3

U

Ustalov, Dmitry II-231

V

Vellinga, Willem-Pier II-183
Vettoretti, Martina II-118
Vezzani, Federica II-283
Villatoro-Tello, Esaú I-127
Viviani, Marco I-88

W

Wachsmuth, Henning II-308
Wang, Ting I-152
Wang, Xintong II-231
Węcel, Krzysztof I-139
Wiegmann, Matti I-172, II-231

X

Xia, Fei II-140

Y

Yetisgen, Meliha II-140
Yim, Wen-Wai II-140
Yimam, Seid Muhie II-231

Z

Zahra, Shorouq II-53
Zangerle, Eva II-231
Zhang, Xiuzhen I-254
Zubiaga, Arkaitz II-208

SPRINGER NATURE

GPSR Compliance

The European Union's (EU) General Product Safety Regulation (GPSR) is a set of rules that requires consumer products to be safe and our obligations to ensure this.

If you have any concerns about our products, you can contact us on ProductSafety@springernature.com

In case Publisher is established outside the EU, the EU authorized representative is:

Springer Nature Customer Service Center GmbH
Europaplatz 3
69115 Heidelberg, Germany

The manufacturer's authorised representative in the EU is Springer Nature Customer Service Centre GmbH, Europaplatz 3, 69115 Heidelberg, Germany. If you have any concerns regarding our products, please contact ProductSafety@springernature.com

Printed and bound by CPI Group (UK) Ltd, Croydon, CR0 4YY

25/03/2026

02078187-0015